Creationism's Trojan Horse

Creationism's Trojan Horse

THE

WEDGE

OF

INTELLIGENT

DESIGN

Barbara Forrest & Paul R. Gross

OXFORD
UNIVERSITY PRESS
2004

Oxford University Press, Inc., publishes works that further
Oxford University's objective of excellence
in research, scholarship, and education.

Oxford New York
Auckland Cape Town Dar es Salaam Hong Kong Karachi
Kuala Lumpur Madrid Melbourne Mexico City Nairobi
New Delhi Shanghai Taipei Toronto

With offices in
Argentina Austria Brazil Chile Czech Republic France Greece
Guatemala Hungary Italy Japan Poland Portugal Singapore
South Korea Switzerland Thailand Turkey Ukraine Vietnam

Published by Oxford University Press Inc.,
198 Madison Avenue, New York, New York 10016

www.oup.com

Library of Congress Cataloging-in-Publication Data
Forrest, Barbara, 1952–
Creationism's Trojan horse : the wedge of intelligent design /
by Barbara Forrest, Paul R. Gross.
p. cm.
Includes bibliographical references.
ISBN-13 978-0-19-515742-0
ISBN 0-19-515742-7
1. Creationism—United States. 2. Evolution (Biology)—Religious aspects—Christianity.
3. Creationism—Study and teaching—United States. 4. Evolution (Biology)—Study and
teaching—United States. 5. Center for Science & Culture. I. Gross, Paul R. II. Title.
BS659 .F67 2003
231.7'652—dc21 2002192677

9 8 7

Printed in the United States of America
on acid-free paper

To the memory of

Robert J. Schadewald

February 19, 1943–March 12, 2000

Acknowledgments

In writing this book, we have incurred numerous debts, not least to our families, whose forbearance enabled us to devote the necessary time to this project. We thank them for recognizing the importance of making the case.

With respect to the book itself, our first debt is to Tim Rhodes, who brought the "Wedge Document" to public view on the Internet early in 1999. Among the first to recognize the significance of this document was James Still, then president of the Secular Web. His request for an article about the Wedge's activities for the Secular Web is the direct reason for the book's existence. We are grateful for Still's recognition of the importance of discussing the Wedge more fully, in book form, and for his graciousness in foregoing the Secular Web article. He provided valuable feedback on the earliest draft of the book. We thank Ursula Goodenough for her recognition of the contribution this work might make to the effort against creationism and for helping to make contacts that led to publication. Dr. William Robison, chair of the Department of History and Political Science at Southeastern Louisiana University, contributed support and encouragement throughout the entire process of research and writing. Michael Rodgers of Oxford University Press-UK was very helpful in the early stages of planning. And we appreciate greatly the patience, tact, and encouragement of Kirk Jensen, executive editor of Oxford University Press-USA, during the process of writing and revision.

Every book can benefit from the knowledge of experts; fortunately, a number of such people have devoted their expertise to the threat of creationism as it has emerged in the intelligent design (ID) movement. We received indispensable help from the following scholars, all of whom shared the fruits of their own labor, offered professional appraisals of the claims of intelligent design proponents, and presented their views on our arguments: Richard Dawkins, Wesley Elsberry, Kenneth Miller, David Ussery, George Gilchrist, John M. Lynch, Thomas J. Wheeler (University of Louisville), Kevin Padian, David Bottjer, Nigel Hughes, Alan Gishlick, Richard Wein, Victor Stenger, Gert Korthoff, Matt Young, Mark Perakh, Jeffrey Shallit, William C. Wimsatt, and Jason Rosenhouse.

We also thank the National Center for Science Education, whose staff never failed to provide information we requested. Of special significance is the input from citizens attempting to preserve the integrity of public science education in state and local school systems: Carl Johnson and Burlington-Edison Committee for Science Education, Jack Krebs and Kansas Citizens for Science, Dave Thomas of New Mexicans for Science and Reason, Marilyn Savitt-Kring of the Coalition for Excellence in Science and Math Education (New Mexico), and Steve Brugge of Eisenhower Middle School (New Mexico), all of whom we thank for their help with parts of the text that discuss the battles against ID in their states.

In relating the details of the Wedge's execution of its program and in our analyses of the relevant issues, we have made every effort to be accurate. We bear the responsibility for any remaining errors, and we thank our reviewers for their assistance in improving the quality of the published work. We are much indebted to the following people for their contributions and feedback on the manuscript: David Applegate, John Cole, Kurt Corbello, Russell Durbin, Marjorie Esman, Dennis Hirsch, Molleen Matsumura, Jeffrey K. McKee, Robert Pennock, and Roger D. K. Thomas. Special appreciation goes to Paul Haschak, reference librarian at Sims Memorial Library at Southeastern Louisiana University, for his kind assistance with our research.

Finally, we wish to thank many fellow citizens not named here who, in addition to the responsibilities of work and family, have shouldered the task in their communities of defending public science education with a fraction of the resources available to their opponents. In defending science education today, they are also defending some of the core values of Western society.

Contents

Creationism's Trojan Horse

Introduction

It used to be obvious that the world was designed by some sort of intelligence. What else could account for fire and rain and lightning and earthquakes? Above all, the wonderful abilities of living things seemed to point to a creator who had a special interest in life. Today we understand most of these things in terms of physical forces acting under impersonal laws. We don't yet know the most fundamental laws, and we can't work out the consequences of all the laws we know. The human mind remains extraordinarily difficult to understand, but so is the weather. We can't predict whether it will rain one month from today, but we do know the rules that govern the rain, even though we can't always calculate the consequences. I see nothing about the human mind any more than about the weather that stands out as beyond the hope of understanding as a consequence of impersonal laws acting over billions of years.

Steven Weinberg,
1979 Nobel Laureate in Physics

Dr. Fox's Lecture

Nearly thirty years ago one of the funniest articles ever published in a respectable medical journal appeared. Of course, it was not meant to be funny. Its purposes were serious and sober enough. The conclusions, moreover, were trustworthy and had important implications for education at all levels. In fact, the conclusions had implications for *all* conveyance of knowledge by experts to intelligent, but nonexpert, audiences. In the *Journal of Medical Education*, D. H. Naftulin, M.D., and colleagues published a research study entitled "The Doctor Fox Lecture: A Paradigm of Educational Seduction."[1] There is no better way to explain the intention and the results of this work than to quote from its abstract:

> [T]he authors programmed an actor to teach charismatically and *nonsubstantively on a topic about which he knew nothing.* The authors hypothesized that given a sufficiently impressive lecture paradigm, even experienced educators participating in a new learning experience can be seduced into feeling satisfied that they have learned despite irrelevant, conflicting, and meaningless content conveyed by the lecturer. The hypothesis was supported when 55 subjects responded favorably at the significant level to an eight-item questionnaire concerning their attitudes toward the lecture.(emphasis added)

For purposes of this experiment, the investigators hired a mature, respectable, scholarly looking fellow, a professional actor. He memorized a prefabricated nonsense lecture entitled "Mathematical Game Theory as Applied to Physician Education." The better popular science magazines had recently covered (real) game theory and its possible applications, so the title was appropriate. The silver-haired actor was trained to answer affably all audience questions following his lecture—by means, as the authors explain, of "double talk, neologisms, non sequiturs, and contradictory statements. All this was to be interspersed with parenthetical humor and meaningless references to unrelated topics."[2] In two of the three trials of this experiment, the audience consisted of "psychiatrists, psychologists, and social-worker educators," while that of the third trial "consisted of 33 educators and administrators enrolled in a graduate level university educational philosophy course." This counterfeit scholar of "Mathematical Game Theory" was called Dr. Myron L. Fox, and a fraudulent but respectful and laudatory introduction was supplied.

Very interesting data followed from the survey and questionnaire administered after each session in which Fox's (and other) presentations were made. These were simply the detailed statistics of approval or disapproval. The phony Dr. Fox's presentations of discoveries in mathematical game theory were strongly approved by these educationally sophisticated, lecture-experienced audiences. But the really funny results are in the "subjective" comments added to the questionnaire, that is, in what listeners wrote as prose responses to the invitation to comment (the following comments are from a number of different respondents). "No respondent [in the first group]," Dr. Naftulin and his co-authors wrote, "reported having read Dr. Fox's publications. [But] subjective responses included the following: 'Excellent presentation, enjoyed listening. Has warm manner. Good flow, seems enthusiastic. What about the two types of games, zero-sum and non-zero-sum? Too intellectual a presentation. My orientation is more pragmatic.'" From the largest group of subjects for this experiment, the substantive comments were, if possible, even funnier: "Lively examples. His relaxed manner of presentation was a large factor in holding my interest. Extremely articulate. Interesting, wish he had dwelled more on background. Good analysis of subject that has been personally studied before. Very dramatic presentation. He was certainly captivating. Somewhat disorganized. Frustratingly boring. Unorganized and ineffective. Articulate. Knowledgeable."[3]

We highly recommend this article. It should still be possible to find it in any university, especially one with a good medical or education library. The "educational seduction" of the title refers to what "Dr. Fox" did for (and to?) his listeners. This result and many others like it should have affected all schools of education, if not teachers generally. However, such was not the case. The possibility, indeed the likelihood, of intellectual "seduction" in circumstances such as these is probably increasing as specialization increases. Countless clones of Dr. Fox tread the academic and public policy boards today, as always. Readers familiar with the now-universal practice in higher education of using end-of-course student evaluations as key evidence in faculty promotion and tenure decisions will know this: evaluations by students, who lack the requisite knowledge but are called on to judge their professors' expertise in their disciplines, can determine the academic fate of nontenured faculty and the possibility of merit raises for tenured ones. Intellectual seduction by substantive ("content") nonsense, offered to audiences who want or like to hear what they are being told, or who simply assume that what they don't understand must be correct if it *sounds* scholarly, is nearly universal.

This book is about a current, national, intellectual seduction phenomenon, not in mathematical game theory, but close enough to it. It is a case, at least formally, not much different from the Dr. Fox lecture, except that the lecturers here actually believe what they are lecturing about, or at least they want very much to believe it, or are convinced that they *must* believe it. And they are not actors, but *executors* of a real and serious political strategy. The "audiences" in this case are large; they consist of decent people: students, parents, teachers, public officials across the length and breadth of the United States (and now in other countries of the "developed world")—people who don't, in most cases, know much about science, especially the modern biological sciences. But they are people who are deeply and justifiably concerned about their religious faith, the state of their society, and the education of their children. They include some people for whom "fairness" and openness to the ideas of "the other side" have become the cherished, even the indispensable, characteristics of our civilization. Their insistence on the equal worth of *all* earnestly held opinions—whether or not those opinions are well founded—makes them relativists whether they know it or not. This book is about the newest form of *creationism,* named by its proponents "intelligent design" (ID); but it is, especially, about the organization of the system of public and political relations that drives the movement. That system operates on a very detailed plan—a set of well articulated goals, strategies, and tactics—named "The Wedge" by its executors. It offers an upgraded form of the religious fundamentalist creationism long familiar in America.

Neo-creationism

Creationism has been a perennial nuisance for American science education. Despite the persistent fecklessness of creationist arguments and their continued failure in the courts since 1925, the creationists refuse to go away. The attempts to insert religion into public elementary and secondary science education are unceasing, and they now include direct efforts to influence college students as well. Efforts to force it into curricula—especially those having anything at all to do with biology and the history of Earth—have been unremitting since the late nineteenth century, and they have continued into the present. The most notorious recent, nearly successful, attempt was the 1999 deletion of evolution and all immediately relevant geology and cosmology from the Kansas public school science standards, by action of the state board of education. Scientific integrity was restored to those defaced standards only after a protracted political effort to defeat creationist board members and replace them with moderates—who eventually undid the damage to science teaching and to the state's reputation.

The defeated have not given up, however; today they are more active than ever in the politics and public affairs of Kansas and other states. And increasingly it appears that pro-evolution (pro-science) victories are secure only until the next election, when old battles may be revived by "stealth" candidates who do not disclose their anti-evolution agenda until *after* they are elected to office. Soon after the restoration of the integrity of science standards in Kansas, new efforts, even more forceful and better organized than those in Kansas, were mounted in Ohio. More are brewing in several other states, gaining added impetus from the Wedge's efforts in the United States Congress. Nor is the phenomenon likely to remain limited to the United States; similar efforts are in progress or being planned in a number of other countries.

This struggle is cyclic; there have been short periods of relative quiet after major creationist failures in the courts. But the effects of the struggle are being felt today far beyond pedagogy in the schools. They are everywhere visible, and except for a few conscientious media outlets, they also threaten to lower the already variable and uncertain standards of science journalism. Contrary to the perception of most scientifically literate people, creationism as a cultural presence has in the recent past grown generally stronger—even as its arguments, in the face of scientific progress, have grown steadily weaker and more hypocritical. Despite the intense activity of creationists, no faction, nor any individual advocate of one, and no modern creationist "research" program has as yet come up with a new, verifiable, fruitful, and important fact about the mechanisms or the history of life or the ancestral relationships among living things on Earth. For that reason, the scorecard of scientific successes for any form of creationism, including ID theory, is blank.

Creationists, including the newest kind—the neo-creationist "intelligent design theorists" who are the subject of this book—offer an abundance of theories. These theories are often decorated with open or only thinly disguised religious allusions, and they always include the now-standard rejection of naturalism, which is, in these circumstances, the indirect admission of supernaturalism. Their contributions to ongoing science consist of nit-picking and the extraction of trivialities from the vast literature of biology and of unsupported statements about what—they insist—*cannot* happen: "Darwinism"—organic evolution shaped by natural selection and reflecting the common ancestry of all life forms. In the face of the extraordinary and often highly practical twentieth-century progress of the life sciences under the unifying concepts of evolution, their "science" consists of quote-mining—minute searching of the biological literature—including outdated literature—for minor slips and inconsistencies and for polemically promising examples of internal arguments. These internal disagreements, fundamental to the working of all natural science, are then presented dramatically to lay audiences as evidence of the fraudulence and impending collapse of "Darwinism." How are such audiences to know that modern biology is *not* a house of cards, *not* founded on a "dying theory"?

Intelligent Design

Until a few years ago, "scientific" creationism was led by biblical literalists like Duane Gish and Henry Morris, whose Bible-thumping and logic-chopping were easy to discount, even for ordinary (nonscience) journalists, by exposing the obvious errors of fact and logic—independently of the gross errors of actual science. But those old-timers have now been eclipsed by a new brand of creationists who have absorbed a part of their following: the new boys are intelligent design promoters, mainly those associated with the Discovery Institute's Center for the Renewal of Science and Culture (now Center for Science and Culture), based in Seattle, Washington. This group operates under a detailed and ambitious plan of action: "The Wedge." Through relentlessly energetic programs of publication, conferences, and public appearances, all aimed at impressing lay audiences and political people, the Wedge is working its way into the

American cultural mainstream. Editorials and opinion pieces in national journals, prime-time television interviews, and other high-profile public appearances, offhand but highly visible negative judgments on evolution or "Darwinism" from conservative politicians and sympathetic public intellectuals (assisted in their anti-science by a scattering of "feminist epistemologists," postmodernists, and Marxists)—all these contribute to a rising receptiveness to ID claims by those who do not know, or who simply refuse to consider, the actual state of the relevant sciences. In documenting and analyzing the political and religious nature of the Wedge, and bringing together expert comment on the ID "science" claims, we show that such grateful reception of the glad tidings of intelligent design is entirely unjustified by either the scientific, the mathematical, or the philosophic weight of any evidence offered.

THE WEDGE'S HAMMERS

Under cover of advanced degrees, including a few in science, obtained in some of the major universities, the Wedge's workers have been carving out a habitable and expanding niche within higher education, cultivating cells of followers—students as well as (primarily nonbiology) faculty—on campus after campus. This is the first real success of creationism in the formerly hostile grove of academe. Furthermore, the Wedge's political alliances reach into a large, partisan elite among the nation's legislators and other political leaders. Armed thus with a potentially huge base of popular support that includes most of the Religious Right, wielding a new legal strategy with which it hopes to win in the litigation certain to follow insertion of ID into public school science anywhere—and lawyers ready to go to work when it does—the Wedge of ID creationism is, indeed, intelligently designed. To be sure, its science component is *not*. But in a public relations–driven and mass-communications world, that is not a disadvantage. In the West, opinions, perceptions, loyalties, and, ultimately, votes are what matter when the goal is to change public policy—or for that matter, cultural patterns. Serious inquiry and questions of truth are often a mere diversion.

This newly energized, intellectually reactionary enterprise will not fade quietly away as the current team of ID promoters ages. It is already too well organized and funded, and the leading Wedge figures have invested too much of themselves for that to happen. Moreover, there is every reason to think that religiously conservative, anti-science agitation will increase, especially as the life sciences and medical research continue to probe the fundamentals of human behavior. As that happens, the general public uneasiness with evolutionary biology and the underlying genetics and cell biology becomes simple hostility, not just on the political right. Some of the far-left intelligentsia help to fuel the hostility, at least in academia. Therefore, we have undertaken to document very thoroughly, largely but not exclusively by means of the Wedge's own announcements and productions, its steadily increasing output of anti-evolution and more broadly anti-science materials.

The Discovery Institute's creationists are younger and better educated than most of the traditional "young-earth" creationists. Their public relations tricks are up to date and skillful; they know how to manipulate the media. They are very well funded, and their commitment is fired by the same sincere religious fervor that characterized earlier and less affluent versions of creationism. This combination makes them crusaders, just as inspired as, but much more effective than, the old literalists, whose pseudo-science was easily recognized as ludicrous. And the Wedge carries out its program as a part of the evangelical Christian community, which William Dembski credits with "for now providing the safest haven for intelligent design."[4] The welcoming voices within this community have all but drowned out those of its many members who are honest in their approach to science, sincere in their Christian faith, and appreciative of the protection afforded to *both* by secular, constitutional democracy. Dembski admits that the Wedge's acceptance among evangelicals is not "particularly safe by any absolute standard."[5] Yet in our survey of this issue, we see that the evangelical voices most prominently heard, with a few notable exceptions, support the Wedge.

FOCUS ON EDUCATION

Unfortunately, ID, by now quite familiar among scientifically qualified and religiously neutral observers as the recycled, old-fashioned creationism it is, drapes its religious skeleton in the fancy-dress language of modern science, albeit without having contributed to science, at least so far, any data or any testable theoretical notions. Therefore, ID creationism is most unlikely in the short term to change genuine science as practiced in industry, universities, and independent research laboratories. But the Wedge's public relations blitz (intended to revolutionize public opinion); its legal strategizing (intended as groundwork for major court cases yet to come); and its feverish political alliance-building (through which the Discovery Institute hopes to shape public policy) all constitute a threat to the integrity of education and in the end to the ability of the public to judge scientific and technological claims. This last threat is not just a secondary, long-term worry. Competent, honest scientific thinking is critically important *now*, not only to the intellectual maturation of our species, especially of its children, but also to optimal management of such current, urgent policy problems as environmental preservation and improvement, energy resources, management and support of scientific research, financing medicine and public health (including human heredity and reproduction), and, in general, the support and use of advanced technology.

Led by Phillip Johnson, William Dembski, Michael Behe, and Jonathan Wells—the four current top names of the Discovery Institute's Center for Science and Culture—with a growing group of like-minded fellows and co-workers, this movement seeks nothing less than to overthrow the system of rules and procedures of modern science and those intellectual footings of our culture laid down in the Enlightenment and over some 300 years. If this sounds overwrought, we ask our readers to proceed at least a little way into the following chapters to judge for themselves. In any case, the Wedge *admits* that this is its aim. By its own boastful reports, the Wedge has undertaken to discredit the naturalistic methodology that has been the working principle of all effective science since the seventeenth century. It desires to substitute for it a particular version of "theistic science," whose chief argument is that nothing about nature is to be understood or taught without reference to supernatural or at least unknowable causes—in effect, to God. The evidence that this is a fundamental goal follows within the pages of this book. No matter that these creationists have produced not even a research *program*, despite their endlessly repeated scientific claims. Pretensions to the contrary, this strategy is not really aimed at science and scientists, whom they consider lost in grievous error and whom they regularly accuse of fraud (as we will demonstrate), of conspiring to hide from a gulled public the failures of modern science, especially of "Darwinism." It is aimed, rather, at a vast, mostly science-innocent populace and at the public officials and lawmakers who depend on it for votes.

A Neo-creationist's Progress

In April 2001, ID movement founder Phillip Johnson released on the creationist Access Research Network website "The Wedge: A Progress Report."[6] There he reviewed the Wedge's goals: "to legitimate the topic of intelligent design . . . within the mainstream intellectual community" and "to make naturalism the central focus of discussion [meaning "of attack"] in the religious world." He cited the establishment of a "beachhead" in American journalism, exemplified by articles in major newspapers. He declared that "the Wedge is lodged securely in the crack" between empirical science and naturalistic philosophy, which he calls "the dominant naturalistic system of thought control." According to Johnson, "the [Wedge] train is already moving along the logical track and it will not stop until it reaches its destination. . . . The initial goals of the Wedge strategy have been accomplished. . . . [I]t's not the beginning of the end, but it is the end of the beginning."[7]

There is some justification for this aggressive show of confidence. As Johnson says, ID has won significant coverage in major U.S. newspapers and, more recently, abroad as well. In the *New York Times,* James Glanz wrote that "evolutionists find themselves arrayed not against traditional creationism, with its roots in biblical literalism, but against a more sophisticated idea: the intelligent design theory." On the front page of the *Los Angeles Times*, Teresa Watanabe wrote that "a new breed of mostly Christian scholars redefines the old evolution-versus-creationism debate and fashions a movement with more intellectual firepower, mainstream appeal, and academic respectability."[8] And Robert Wright (author of *The Moral Animal: Evolutionary Psychology and Everyday Life*, Vintage Books, 1994) points out in a critical *Slate* article that while ID presents no new ideas of any significance, the *New York Times* article "has granted official significance to the latest form of opposition to Darwinism." Wright concludes that although ID is just a new label, a marketing device for an old product, it is also an effective one.[9]

The admirable, but in this particular case misguided, concern of most Americans to be fair, "even-handed," to consider both sides of a dispute respectfully, especially the side claiming to suffer discrimination, creates a fertile field for ID activists. They have enough financial backing and self-righteous zeal to outlast what little effectively organized opposition to them presently exists, especially in the higher education community, which one would quite reasonably expect to be in the forefront of opposition to the Wedge. There is, of course, the further—and very real—possibility that the demographics of the judiciary will shift toward creationism should there be appointments of judges with strong doctrinal or emotional ties to the Religious Right, where one's views on evolution are once again, as they were in the 1920s, a "litmus test." There is no doubt that the Wedge's immediate goal is to change what is taught in classrooms about the basics of biology and the history of life, as we show here from its own documents, sources of support, and productions. But based on our demonstration in chapter 9 of the religious foundation of the intelligent design movement and the importance of this foundation to the Wedge's goal of "renewing" American culture, we also believe that its ultimate goal is to create a theocratic state, which would provide a protective framework for its pedagogical goals. In an important respect, the Wedge is another strand in the well organized Religious Right network, whose own well documented but poorly understood purposes are strongly antagonistic to the constitutional barriers between church and state.

As of March 2001, creationists had launched programs to change public school curricula in one out of five states across the nation. During the writing of this book, creationists were causing significant problems in Ohio, Washington, Idaho, Montana, Kansas, Missouri, Alabama, Georgia, Michigan, Pennsylvania, and Wisconsin.[10] At present, there are renewed rumblings in New Mexico, where a hard-fought battle was presumably resolved. These programs have not yet attained their broadest goals, but they continue to divert precious educational resources, time, and energy from the real problems of public education in the United States toward the work of responding to creationist attacks. Even in the small, rural state of Louisiana, ID advocates seem to be waiting in the wings to initiate a sequel to recent attempts by Representative Sharon Weston-Broome to declare the idea of evolution "racist."[11] In Kansas, where creationist changes to the state's science standards have finally been reversed, the Discovery Institute is nevertheless actively assisting a satellite group, the Intelligent Design Network (IDnet), in pushing ID more aggressively than ever. In June 2001, IDnet held its Second Annual Symposium, "Darwin, Design, and Democracy II: Teaching the Evidence in Science Education," featuring three key Wedge campaigners—Phillip Johnson, William Dembski, and Jonathan Wells.[12] The great public universities are now a main target of wedge efforts: a Discovery Institute fellow, Jed Macosko, taught ID in a for-credit course at the University of California-Berkeley; his father, Chris Macosko, has been doing the same at the University of Minnesota.[13]

Concern about the Wedge is building, very late but finally, in scientific and academic quarters. The American Geophysical Union considered ID a problem serious enough to require scheduling at least six presentations on it at the spring 2001 conference.[14] Philosopher Robert Pennock's eye-opening book, *Tower of Babel: The Evidence Against the New Creationism* (MIT, 1999), analyzed and recounted the philosophical and scientific flaws of ID creationism. It is followed by his anthology, *Intelligent Design Creationism and Its Critics: Philosophical, Theological, and Scientific Perspectives* (MIT, 2001). These books seem to be making a contribution in awakening academics to the need for an effective counter-strategy. Similar books are on the way; and in book reviews and a spate of recent writings, distinguished scientists are at last taking the trouble (and it *is* troublesome, *and* time-consuming, *and* costly!) to rebut, point by point, the new creationist claims. Of course, those claims are not really new. They are rather pretentious variants of the ancient, and discredited, argument from design (aptly renamed for our era, by Richard Dawkins, the argument from personal incredulity).

So far, however, no book has documented the genesis, the support, the real goals, and the remarkable sheer volume of Wedge activities. We have come to believe that such a chronicle is needed if people of good will toward science and toward honest inquiry are to understand the magnitude of this threat—not only to education but to the principle of separation of church and state. The chapters that follow are our effort to supply the facts: as complete an account, within the limits of a single volume and the reader's patience, as can be assembled—and checked independently—from easily accessible public sources. To convince those with the indispensable basic knowledge who are in a position to act, that they must do so, we must first make the case that (1) a formal intelligent design strategy, apart from and above the familiar creationist carping about evolutionary and historical science, does exist, and (2) it is being executed successfully in all respects *except* the production of hard scientific results—data. To accomplish these aims, we have had to accumulate the evidence, which consists of the massive schedule of the Wedge's own activities in execution of the strategy, together with the actual pronouncements of Wedge members. We have allowed them to speak for themselves here at length and as often as possible.

The Wedge's busy schedule of ID activities and its increasing public visibility have been accompanied by a steadily evolving public relations effort to present itself as a mainstream organization. In August 2002, the CRSC changed its name, now calling itself simply the "Center for Science and Culture." This move parallels the Wedge's low-key phase-out of the overtly religious banners on its early web pages: from Michelangelo's God creating Adam, to Michelangelo's God creating DNA, to the current Hubble telescope photo of the MyCn18 Hourglass Nebula.[15] But despite the attempt to alter its public face, the Wedge's substantive identity remains. Thus, we refer henceforth to the Center for Science and Culture by the name under which it has been known during the period covered in this book: the Center for the *Renewal* of Science and Culture (CRSC).

The readers' patience may well be tried at times by the repetitiousness of Wedge activities: conferences, websites, trade book and media publications and appearances, testimony before legislative bodies and education committees, summonses to religious and cultural renewal predicated on anti-science. The Wedge's efficient and planned repetitiousness is itself one of our main points. *In fact, it is one of the most remarkable examples in our time of naked public relations management substituting successfully for knowledge and the facts of the case*—substituting for the truth. For that reason alone, it is both interesting and important. It must be known and understood if there is to be recognition—among scientists as well as the literate nonscientist public—of current anti-evolutionism and its aims.

The Issue

The issue, then, is *not*—as ID creationists insist it *is*, to their increasingly large and credulous audiences nationwide—that the biological sciences are in deep trouble due to a collapse of Darwinism. The issue is that the public relations work, but not the "science," of the Wedge and of ID "theorists" is proving all too effective. It is not refutations or technical dismissals of ID scientific claims that are needed. The literature of science and the book review pages of excellent journals are already replete with those: expert reviews of ID books and other public products are readily available to anyone. We provide here what we hope is an adequate sampling of those technical dismissals and expert scientific opinions, and we document the sound science and the ID anti-science as needed. But in the past few years, very detailed disproof has been provided, again and again, by the commentators best qualified to speak to the substance: some of the world's most honored evolutionary and physical scientists, as well as some of the most distinguished philosophers of mind and science. Rather, what is needed now is documentation of the Wedge *itself*, from its own internal and public relations documents, so that the public may understand its purposes and the magnitude of its impact, current and projected. The issue is not Darwinism or science: the issue is the Wedge itself.

Providing the necessary documentation, including the minutiae that can turn out to be important, is always a writer's strategic problem when the intended audience is broader than a small group of specialists. Even scholars who demand and are accustomed to copious documentation can find it off-putting. Others, members of the most important audience of all—curious, able, and genuinely fair-minded general readers—who rarely if ever read with constant eye and hand movement between text and references, are strongly tempted to give up when confronted with profuse supporting data and the necessary but distracting scholarly apparatus of notes and references. We do not have a *good* solution to this problem. The endnotes can be taken, however, as running commentary, supplementary to, *but not* essential for, the main text. Our references to literature include, whenever possible and therefore in abundance, pointers to sites on the World Wide Web.

No reader needs to use the notes to apprehend the argument and to judge its broad justifications—or lack of them. The main text can usefully and properly be read for itself alone. But for those readers who decide that this argument is to be taken seriously, and who feel the need to arm themselves with facts, they are here; *or* there is a pointer to them, immediately serviceable for anyone with access to a computer and an Internet connection. Initially, we envisioned a much shorter response than this book to the Wedge's campaign. We have delayed work on other projects to write it, even though we would have preferred not to have found it necessary. The more we examined the situation, the more expansive and invasive the Wedge's program proved to be, and the greater, therefore, was the need we saw for full public examination and for a proper response to it. We have watched and waited for the coalescence of an appropriately organized counter-movement, and, indeed, a few small organizations and individual members of the scientific and academic communities, as well as concerned citizens, have recently mounted admirable efforts, with only a minute fraction of the resources available to the Wedge. But those active people *are* few, and they need the help of everyone who has a stake in the high quality of our civic, scientific, and educational cultures.

1

How the Wedge Began

If we understand our own times, we will know that we should affirm the reality of God by challenging the domination of materialism and naturalism in the world of the mind. With the assistance of many friends I have developed a strategy for doing this. . . . We call our strategy the wedge.

Phillip E. Johnson, *Defeating Darwinism by Opening Minds*

I was an establishment figure when I was young, but now I have become a cultural revolutionary.

Phillip Johnson,
Silicon Valley Magazine

Inquiry is the search for knowledge, whether in the work of a theoretical physicist, an automobile mechanic, or any other honest student of physical reality. Fanaticism—religious, political, or cultural—is the eternal enemy of inquiry. Fanatics have always been preoccupied with controlling education, especially that of children. Freedom of speech, especially in the schools, is their traditional foe and target. In the West, at least, the irresistible compulsion of ideologues to control teaching is well recognized—even in the recurrent periods, like the present, of ideological vigilantism. Vigilantes of one ideology are the keenest watchers for intrusions of the next ideology. People who favor the growth of knowledge and intellectual freedom are usually able to see and willing to oppose fanaticism, even when it lurks under a facade of religious or socio-political rectitude. Alertness to strong ideology masquerading as education has been the main obstacle to the spread of dogma in a democratic society.

There is now, however, a new variant of the old (anti)scientific creationism—a no-holds-barred commitment to particular, parochial religious beliefs about the history and fabric of the world and the place of humanity in it. This variant has eliminated brilliantly the obstacle of rational opposition to ideology substituted for education. The new strategy is wonderfully simple. Here is how you implement it: exploiting that modern, nearly universal, liberal suspicion of zealotry, you accuse the branch of legitimate inquiry whose results you hate, in this case the evolutionary natural sciences, of—what else?—zealotry! Fanaticism! Crying "viewpoint discrimination," you loudly demand adherence to the principle of freedom of speech, especially in teaching, insisting that such freedom is being denied your legitimate alternative view. You identify your (in this case, religious) view of the world as the victim of censorship by a conspiracy among most of the world's scientists, whom you label "dogmatic Darwinists" or the like.

This bold strategy is working, not just with religious fundamentalists, who do not need to be convinced anyway, but with people who have no such fundamentalist commitment and who are in principle well-enough educated to see what is happening. Among these increasingly susceptible persons are many politicians, who sense an opportunity to exploit for votes the cry of victimization, and many highly influential persons who have no selfish motives but who, like most of the population, lack the scientific knowledge needed to make an informed distinction between genuine science and pseudoscience.

This lusty new variant of creationism is advancing rapidly by means of a strategy called "The Wedge." We begin our account of its operations with its own (true) origin story. The Wedge is a movement with a plan to undermine public support for the teaching of evolution and other natural science supporting evolution, while at the same time cultivating a supposedly sound alternative: "intelligent design theory" (ID hereafter). The Wedge of intelligent design, which is simply a restatement of the ancient argument from design, did not arise in the mind of a scientist, or in a science class, or in a laboratory, or as a result of scientific research in the field. It appeared in the course of one man's personal difficulties after a divorce. Those led a middle-aged Berkeley professor of law, Phillip E. Johnson, into born-again Christianity. The Wedge movement, with its huge ambitions for revolutionizing all science and all culture (as will be shown), was the result of personal crisis and an epiphany in the life of a nonscientist, whose scientific knowledge is at the very most that of an untrained amateur.

In his own account, Phillip Johnson says that "the experience of having marriage and family life crash under me, and of achieving a certain amount of academic success and seeing the meaninglessness of it, made me . . . give myself to Christ at the advanced age of 38. And that aroused a particular level of intellectual interest in the question of why the intellectual world is so dominated by naturalistic and agnostic thinking."[1] Nancy Pearcey, a fellow of the Center for the Renewal of Science and Culture and a Johnson associate in this creationist section of the Discovery Institute (see later), links Johnson's religious conversion and his leadership of the intelligent design movement in two recent publications. In an interview with Johnson for *World* magazine, Pearcey says, "It is not only in politics that leaders forge movements. Phillip Johnson has developed what is called the 'Intelligent Design' movement. . . . Mr. Johnson is a Berkeley law professor who, spurred by the crisis of a failed marriage, converted to Christianity in midlife."[2] In *Christianity Today*, she identifies the causal relationship between Johnson's new beliefs and his deep animosity toward evolution: "The unofficial spokesman for ID is Phillip E. Johnson, a Berkeley law professor who converted to Christianity in his late 30s, then turned his sharp lawyer's eyes on the theory of evolution."[3]

Johnson's search for meaning in his life set the stage for another epiphany during a sabbatical leave in England: "In 1987, when UC Berkeley law professor Phillip Johnson asked God what he should do with the rest of his life, he didn't know he'd wind up playing Toto to the ersatz winds of Darwinism. But a fateful trip by a London bookstore hooked Mr. Johnson on a comparative study of evolutionary theory."[4] Johnson purchased Richard Dawkins's *The Blind Watchmaker* and "devoured it and then another book, Michael Denton's *Evolution: A Theory in Crisis*." Says Johnson, "I read these books, and I guess almost immediately I thought, *This is it. This is where it all comes down to, the understanding of creation*."[5] The Wedge's gestation had begun.

According to Johnson, the Wedge movement, if not that name for it, began in 1992: "The movement we now call The Wedge made its public debut at a conference of scientists and philosophers held at Southern Methodist University [SMU] in March 1992, following the publication of my book *Darwin on Trial* [1991]. The conference brought together as speakers some key Wedge figures, particularly Michael Behe, Stephen Meyer, William Dembski, and myself."[6] Johnson had established contacts with a "cadre of intelligent design (ID) proponents for whom Mr. Johnson acted as an early fulcrum. . . . Mr. Johnson made contact, exchanged flurries of e-mail, and arranged personal meetings. He frames these alliances as a 'wedge strategy,' with himself as lead blocker and ID scientists carrying the ball behind him."[7] In 1993, a year after the SMU conference, the Wedge held another meeting (June 22–24, 1993), "The Status of Darwinian Theory and Origins of Life Studies": "the Johnson-Behe cadre of scholars met at Pajaro Dunes. . . . Here, Behe presented for the first time the seed thoughts that had been brewing in his mind for a year—the idea of 'irreducibly complex' molecular machinery."[8] This idea has come to serve as something of a joke among evolutionary biologists (which Behe is not) and other scientists, but it seems to be the gladdest of glad tidings for the scientifically naive.

When the July 1992 issue of *Scientific American* published Stephen Jay Gould's devastating review of Johnson's *Darwin on Trial*, in which Gould described the book as "full of errors, badly argued, based on false criteria, and abysmally written," Johnson's supporters formed the "Ad Hoc Origins Committee" and wrote a letter (probably in 1992 or 1993) on Johnson's behalf: "This letter was mailed to thousands of university professors shortly after Gould wrote his vitriolic analysis of . . . *Darwin on Trial*. Included with it was Johnson's essay 'The Religion of the Blind Watchmaker', replying to Gould, which *Scientific American* refused to publish."[9] Among the thirty-nine signatories were nine (listed here with their then-current affiliations), who a few years later became Fellows of the Center for the Renewal of Science and Culture:

Henry F. Schaefer III, Ph.D.
Chemistry
University of Georgia
Stephen Meyer, Ph.D.
Philosophy
Whitworth College
Michael Behe, Ph.D.
Biochemistry
Lehigh University
William Dembski, Ph.D.
Philosophy
Northwestern University
Robert Kaita, Ph.D.
Physics
Princeton University
Robert Koons, Ph.D.
Philosophy
U[niversity of] T[exas], Austin
Walter Bradley, Ph.D.
Mechanical Engineering
Texas A & M University
Paul Chien, Ph.D.
Biology
University of San Francisco
John Angus Campbell, Ph.D.
Speech
University of Washington

These names recur, as we shall see, throughout the subsequent history of the Wedge. The signers describe themselves as "a group of fellow professors or academic scientists who are generally sympathetic to Johnson and believe that he warrants a hearing. . . . *Most of us are also Christian theists who like Johnson are unhappy with the polarized debate between biblical literalism and scientific materialism. We think a critical re-evaluation of Darwinism is both necessary and possible without embracing young-earth creationism*" (emphasis added). Notre Dame philosopher Alvin Plantinga was also a signatory to this letter, which is early evidence of his continuing support of and continued, active participation in the intelligent design movement. Nancy Pearcey refers to Plantinga as a "design proponent."[10] Thus, a critical mass of religiously committed supporters had already begun to coalesce around Johnson. None of those named had significant, professional credentials in evolutionary biology, nor had any of them published scientific peer-reviewed research on, or criticism of, evolution. In fact, not one of them has done so to this day.

But by 1995, Johnson's mission had crystallized, and he had a loyal contingent of associates to help carry it out. That summer they held another conference, "The Death of Materialism and the Renewal of Culture," which served as a matrix for the "Center for the Renewal of Science and Culture," organized the following year.[11] Johnson produced another book, *Reason in the Balance: The Case Against Naturalism in Science, Law and Education* (InterVarsity Press, 1995), in which he positioned himself as a "theistic realist" fighting against "methodological naturalism":

> First, here is a definition of MN [methodological naturalism], followed by a contrasting definition of my own position, which I label "theistic realism" (TR). . . . 1. A methodological naturalist defines science as the search for the best naturalistic theories. A theory would not be naturalistic if it left something (such as the existence of genetic information or consciousness) to be explained by a supernatural cause. Hence all events in evolution (before the evolution of intelligence) are assumed to be attributable to unintelligent causes. The question is not *whether* life (genetic information) arose by some combination of chance and chemical laws . . . but merely *how* it did so. . . .
>
> The Creator belongs to the realm of religion, not scientific investigation.
>
> 2. A theistic realist assumes that the universe and all its creatures were brought into existence for a purpose by God. Theistic realists expect this "fact" of creation to have empirical, observable consequences that are different from the consequences one would observe if the universe were the product of nonrational causes. . . . God always has the option of working through regular secondary mechanisms, and we observe such mechanisms frequently. On the other hand, many important questions—including the origin of genetic information and human consciousness—may not be explicable in terms of unintelligent causes, just as a computer or a book cannot be explained that way.[12]

This superficially reasonable opposition between (what he defines as) naturalism and "theistic realism" became the hallmark of Johnson's persuasive technique with legally, philosophically, and scientifically lay audiences. Now that the metaphysical terrain of ID was mapped, Johnson and his allies needed a formal strategy for executing the mission of toppling "naturalism" from its pedestal in Western culture, and necessarily thereby, of putting modern science in its proper (in their view) place. By 1996, the most crucial of preliminary developments had been achieved: the Center for the Renewal of Science and Culture was established under the auspices of the Discovery Institute (DI), a conservative Seattle think tank that had itself been established in 1990.[13] The Wedge had found a home. In its summer 1996 *Journal*, "a periodic publication that keeps DI members and friends up to date on Discovery's programs and events," the Institute announced the CRSC's formation, which "grew out of last summer's [1995] 'Death of Materialism' conference."[14] According to DI president Bruce Chapman, "The conference pointed the way and helped us mobilize support to attack the scientific argument for the 20th century's ideology of materialism and the host of social 'isms' that attend it." (That list of social "isms" includes, of course, everything that religious conservatives see as evil in contemporary culture and in the modern world.) Larry Witham's December 1999 *Washington Times* column reveals the CRSC's topmost position on the roster of its parent organization's priorities:

> The eight-year-old Discovery Institute is a Seattle think tank where research in transportation, military reform, economics and the environment often takes on the easygoing tenor of its Northwest hometown. But it also sponsors a group of academics in science affectionately called 'the wedge.' . . . The wedge is part of the institute's four-year-old Center for Renewal of Science and Culture (CRSC), a research, publishing and conference program that challenges what it calls an anti-religious bias in science and science education. "I would say it's our No. 1 project," said Bruce Chapman, Discovery's president and founder.[15]

With formation of the CRSC, the Wedge's core working group was in place: Stephen Meyer and John G. West, Jr., as co-directors; William Dembski, Michael Behe, Jonathan Wells, and Paul Nelson as 1996–1997 full-time research fellows; and Phillip Johnson as advisor.[16] Once the movement was securely housed within DI, execution of the Wedge strategy began to pick up speed. In November 1996, Johnson and his associates convened the "Mere Creation" conference at Biola University in California.[17] The importance of this conference for the subsequent development of the Wedge cannot be overestimated. Indeed, in the foreword to the book issued from it, its importance was made explicit by Henry Schaefer, a Georgia chemist and a signer of the Ad Hoc Origins letter, who had defended Phillip Johnson against the destructive analysis in *Scientific American* of *Darwin on Trial*: "An unprecedented intellectual event occurred in Los Angeles on November 14–17, 1996. Under the sponsorship of Christian Leadership Ministries, Biola University hosted a major research conference bringing together scientists and scholars who reject naturalism as an adequate framework for doing science and who seek a common vision of creation united under the rubric of intelligent design."[18] (Christian Leadership Ministries, the "Faculty Ministry of Campus Crusade for Christ, International," has continued actively to assist the Wedge, both logistically and by way of its provision of "virtual" office space to Wedge members on its "Leadership University" website.)[19]

Unfortunately, Dr. Schaefer's description of the Mere Creation conference as "a major research conference" was either simple hyperbole or wishful thinking. It did *not* in fact produce any original, peer-reviewed scientific research.[20] It did, however, yield a badly needed and eventually very effective public relations strategy. The movement's goal at this conference was already clear to third-party observers such as Scott Swanson, who wrote about the conference for *Christianity Today*:

> The fledgling "intelligent-design" movement, which says Darwinian explanation of human origins are inadequate, is aiming to shift from the margins to the mainstream. . . . The first major gathering of intelligent-design proponents took place in November at Biola University in La Mirada, California.[21] . . . If the turnout at the conference is any indication, intelligent design is gaining a following. More than 160 academics, double what organizers had envisioned, attended from 98 universities, colleges, and organizations. The majority represented secular universities.[22]

Although, according to Swanson, the organizers "chose not to use the conference as a forum to develop a statement of belief for the movement," he learned that "leaders are planning a spring conference at the University of Texas and have begun publishing a journal, *Origins and Design*, edited by Paul Nelson." This "spring conference" materialized as the "Naturalism, Theism, and the Scientific Enterprise" meeting, held at UT in February 1997 and organized by CRSC fellow Robert Koons, a philosopher and UT faculty member.[23] With a core of supporters who had now been able to convene and strategize, the Wedge's remarkably short embryonic period was over: "Prior to the [Biola] conference, the intelligent-design movement was a loose coalition of academics from a wide variety of disciplines. The conference brought together like-minded people, potential activists, 'to get them thinking in the same range of questions,' says . . . Phillip Johnson."[24]

William Dembski edited a book of conference presentations entitled *Mere Creation: Science, Faith and Intelligent Design* (such books, like the conferences themselves, being a centrally important component of the Wedge strategy). Henry Schaefer wrote its foreword, in which he revealed that the Wedge strategy had now solidified in important ways, as indicated by the adoption of very specific goals for disseminating the Wedge's message "both at the highest level and at the popular level":

> Preparing a book for publication, with chapters drawn from the conference papers (this goal has been met with the publication of the present volume);
>
> Planning a major origins conference at a large university to engage scientific naturalists;
>
> Outlining a research program to encourage the next generation of scholars to work on theories beyond the confines of naturalism;
>
> Exploring the need for establishing fellowship programs, and encouraging joint research (Seattle's Discovery Institute is the key player here . . .);
>
> Providing resources for the new journal *Origins & Design* as an ongoing forum and a first-rate interdisciplinary journal with contributions by conference participants (see www.arn.org/arn);
>
> Preparing information usable in the campus environment of a modern university, such as expanding a World Wide Web origins site . . . and exploring video and other means of communication.

Schaefer also lists the members of the steering committee for the conference:

Michael Behe
Walter Bradley
William Dembski
Phillip Johnson
Sherwood Lingenfelter
Stephen Meyer
J. P. Moreland
Paul Nelson
Pattle Pun
John Mark Reynolds
Henry F. Schaefer III
Jeffrey Schloss[25]

The activities Schaefer lists in his foreword prefigure most of the activities now being carried out, and the steering committee metamorphosed into some of the Wedge's most active members. In fact, *all steering committee members except Johnson, who is the CRSC's advisor, and Sherwood Lingenfelter, Biola University provost who hosted the conference, have become CRSC fellows.*

By 1997, Johnson was talking publicly about the Wedge strategy in his book *Defeating Darwinism by Opening Minds* (dedicated "To Roberta and Howard [Ahmanson], who understood 'the wedge' because they love the Truth").[26] Johnson devotes chapter 6 to "The Wedge: A Strategy for Truth," calling on the familiar metaphor of a splitting wedge employed to widen a small crack, which can then split a huge log: "We call our strategy 'the wedge.' A log is a seeming solid object, but a wedge can eventually split it by penetrating a crack and gradually widening the split. In this case the ideology of scientific materialism is the apparently solid log."[27] Johnson's 1998 book *Objections Sustained: Subversive Essays on Evolution, Law and Culture*, is dedicated "To the members of the Wedge, present and future."[28] One of his recent books is *The Wedge of Truth: Splitting the Foundations of Naturalism* (InterVarsity Press, 2000).

Of course, without money, a multifaceted and determined strategy like the Wedge would have been no more than a pipe dream. The money was forthcoming, however. CRSC was soon funded quite generously by benefactors, the most munificent of whom is Howard Ahmanson (through his organization, Fieldstead and Company). Ahmanson's award, along with that of the Stewardship Foundation, is acknowledged in DI's announcement of the CRSC's establishment in its August 1996 *Journal*:

> For over a century, Western science has been influenced by the idea that God is either dead or irrelevant. Two foundations recently awarded Discovery Institute

> nearly a million dollars in grants to examine and confront this materialist bias in science, law, and the humanities. The grants will be used to establish the Center for the Renewal of Science and Culture at Discovery, which will award research fellowships to scholars, hold conferences, and disseminate research findings among opinionmakers and the general public. . . . Crucial, start-up funding has come from Fieldstead & Company, and the Stewardship Foundation which also awarded a grant.[29]

Financial security—money in the bank, with which to get things done—having been assured for the Wedge, at least for a number of years, the CRSC could now proceed: it could focus its resources and its undivided attention on strategic planning and implementation on behalf of its ultimate purpose—to divest contemporary natural science of its intellectual legitimacy and public respect and to replace it, insofar as circumstances allow and wherever possible, especially in education, with a rigorously God-centered view of creation, including a new "science" based solidly in theism.

2

The Wedge Document: A Design for Design

Discovery Institute's Center for the Renewal of Science and Culture seeks nothing less than the overthrow of materialism and its cultural legacies.

"The Wedge Strategy," a.k.a.
"The Wedge Document"

Although Phillip Johnson has talked openly about the Wedge strategy, he has not elaborated publicly all of its detail and logistics. The whole plan is exceedingly ambitious. The full particulars can be found in a paper that surfaced on the Internet in March 1999 and has come to be known as the "Wedge Document." This is a five-year plan (1999–2003) for the Center for the Renewal of Science and Culture, although it also represents goals stretching into the next twenty years. The CRSC obviously takes its Wedge strategy as a long-term commitment. Entitled "The Wedge Strategy," with the name of the organization, "Center for the Renewal of Science and Culture," beneath the title, this document elucidates current activities of the CRSC, as well as the intentions and hopes for the future that underlie them. It is important in three respects: (1) it confirms the existence of a formal strategy, (2) it provides insight into how the Wedge views its program, and (3) it provides a way to measure the Wedge's advance. Although no longer recent news among those who follow the creationism issue, the document remains an informal reference point in discussions of the Wedge. Therefore, in light of the remarkable political (not scientific) successes its adherents have already achieved, its provenance, contents, and style are worthy of close examination. That is our purpose in this chapter.

Although the Wedge Document's history and function as the original plan of operations for the Wedge program have never explicitly been acknowledged by the Discovery Institute, the case for its authenticity seems unshakable to all who have examined it and who are familiar with the rhetoric issuing from the CRSC before and after the document's appearance. It is obviously of the first importance for our account of the Wedge that its authenticity be established, even though major sections of it are used in more recent and clearly official statements of its promoters. (In fall 2002, DI belatedly admitted owning the document—a year *after* publication of an article by one of us citing identical wording on an early DI website.) Beyond the usual reasons for establishing the genuineness of such a document, there is another, rather unusual one: since it surfaced on the Web, the Wedge Document's explanations of what it presents as the depraved and moribund condition of Western culture, especially through the twentieth century and now into the twenty-first, might be taken by people acquainted with the hyperbole of the extreme Religious Right for an elaborate spoof—a sophomoric parody of the moral thunderbolts periodically flung by creationists and other religious zealots. "Biological evolution," according to one such formulation, is the trunk of a tree of evil that bears the foliage of "philosophical evolution," which in turn produces the rotten fruits of secularism, crime, dirty books, "homosex," relativism, drugs, sex education, communism, genetic engineering, abortion, hard rock, inflation, and others.[1] One might therefore interpret as deliberate comic excess the Wedge Document's announcement that one of the CRSC's tasks is to "brief policymakers" (e.g., members of the U.S. Congress) on the "opportunities for life after materialism"—if it were not utterly clear that this offer is not meant in jest. There is no evidence of a sense of humor *anywhere* within the Wedge.

In the Wedge Document, all the world's evil is traced to "materialism"; and the most insidious of all the materialist forces, indeed the source of them all, is taken without hesitation to be "Darwinism," along with such other science as might support it or call into question the accepted truths of religious doctrine. Sadly, the Wedge Document is not a joke. It is taken with utmost seriousness by its authors, and it is meant to encourage and cultivate the financial and political support needed to sustain an ambitious, expensive, and relentless attack on evolutionary science. Here and elsewhere in the book, we quote from it selectively.

Judging from statements in the document, it was written about 1998, as indicated by several examples:

> We believe that, with adequate support, we can accomplish many of the objectives of Phases I and II in the next five years (1999–2003). . . .
>
> InterVarsity will publish our large anthology, *Mere Creation* (based upon the Mere Creation conference) this fall, and Zondervan is publishing *Maker of Heaven and Earth: Three Views of the Creation-Evolution Controversy*, edited by fellows John Mark Reynolds and J. P. Moreland. . . .
>
> During 1997 our fellows appeared on numerous radio programs (both Christian and secular) and five nationally televised programs, TechnoPolitics, Hardball with Chris Matthews, Inside the Law, Freedom Speaks, and Firing Line. The special edition of TechnoPolitics that we produced with PBS in November elicited such an unprecedented audience response that the producer Neil Freeman decided to air a second episode from the "out takes."[2]

Verification of the quoted dates helps not only to date the document but to establish its authenticity. A number of facts ascertained independently of the document are consistent with its contents. The copyright date of the book *Mere Creation: Science, Faith & Intelligent Design* is 1998.[3] The book *Maker of Heaven and Earth: Three Views of the Creation-Evolution Controversy*, by Reynolds and Moreland, was published by Zondervan in March 1999.[4] The TechnoPolitics broadcasts referred to aired on November 15 and December 19, 1997, as listed on the creationist website, Access Research Network.[5] In addition, DI president Bruce Chapman recently acknowledged using the document for fundraising in 1998, but immediately added a nonsensical hedge: "I don't disagree with it. . . . but it's not our program."[6] Whatever he may mean here, our study points to the Wedge Document as a precise reflection of DI's program.

Beyond such consistency of dates, two kinds of information add to the bona fides of the original Wedge Document: correspondence between Jay Wesley Richards, program director of CRSC, and James Still of the Secular Web; and comparison of the Wedge Document's language and concepts with those today employed regularly and emphatically on the website of the CRSC.

Correspondence with Jay Wesley Richards

Acording to James Still, former editor of the Secular Web, the Wedge Document surfaced on the Internet on March 3, 1999.[7] When he and others became aware of it and its contents, Still made contact with Richards. Richards's responses to Still's inquiries (as related by Still) leave little doubt of its genuineness:

> I remember when the Wedge paper first started making its rounds on the Internet at the beginning of March 1999. People were speculating about its authenticity, what it might mean, and whether the wedge strategy should be taken seriously. So I wrote a story on it for the Secular Web and asked Jay Richards, the CRSC's Director of Program Development, whether or not the paper was indeed authored by the CRSC. He didn't want to confirm its authenticity

> outright, of course, but he admitted that it was an "older, summary overview of the 'Wedge' program." Out of politeness I didn't press him on it. For the rest of the conversation, we both treated it as authentic and he was kind enough to explain in great detail the policy behind the three phases outlined within it. If anyone doubts whether or not the paper represents the true position of the CRSC, all that person has to do is visit the CRSC's website where large portions of the paper are reprinted for all to read.[8]

Still's recollection is confirmed in an e-mail he received from Richards after he sent Richards the web address to the article. In this message, Richards simply thanked Still for quoting him accurately, acknowledged that scientific naturalists would disagree with CRSC's program, and for the sake of scientific progress expressed his wish for honest debate rather than personal attacks.[9] Richards answered Still's questions in a way that reveals his recognition of the document itself: "When asked if he worried that Phase II [of the Document: "Publicity & Opinion-making"] will seem like a heavy-handed spin and that no one will take seriously the work accomplished in Phase I ["Scientific Research, Writing & Publicity"], Richards said that the publicity will not drive the scholarship but that the scholarship will come first and foremost."[10]

Richards's reply to Still is as significant for what he does not say as for what he does: he does not disavow the authenticity of the Wedge Document or of any of its contents. His e-mail message to Still after reading Still's article on the Secular Web was the second opportunity Richards had to disavow the document. He had every chance to declare it a fabrication if indeed it was that, but he did not on either occasion. Indeed, Richards makes no critical comments at all. Moreover, the Wedge Document, along with Still's article about it, had been posted on the Secular Web, as well as the American Humanist Association website, since early 1999 without protest from the CRSC or any individual member of it. Kansas Citizens for Science, a group formed in 1999 to counteract creationist activities in Kansas, has even used the Wedge Document as a flier in its activities to inform the public of the existence and nature of the Wedge.

Still recounts Richards's comments in his Secular Web article: "The white paper created quite a buzz among many skeptics after it was widely circulated on the Internet. However, CRSC Senior Fellow and Director of Program Development Jay Richards said that the mission statement and goals had been posted on the CRSC's web site since 1996."[11] Richards's characterization of the document as an older summary may indicate that the document written in 1998 is an updated version of an earlier prospectus for the CRSC's program, since Still recalls that Richards had made some connection between the document and a 1996 press release.[12]

There is independent support for the 1998 document's being an updated version. In the February/March 1998 *SBC* (Southern Baptist Convention) *Life* (before the document surfaced on the Internet in 1999), Hal Ostrander, an ardent Wedge supporter, outlines "the wedge strategy of the design theorists." He refers to four rather than three phases:

> The first part is that of research and publicity—where leading scholars are enlisted for the cause, where trailblazing books are written and published, and where considerable attention is drawn to these matters in the scholarly and popular press. . . . The second part of the program is that of recruitment and alliance building. In this stage the next wave of theistic scholars begins their cutting-edge work. . . . The third part is that of academic breakout conferences. At this point the firestorm debates begin. . . . The fourth and final part of the program has a great deal to do with the popular media—where educational video projects, high school textbooks, public TV documentaries, and educational materials for various religious communities become the order of the day.[13]

Ostrander says that implementation of the Wedge strategy is "slated for 1996 through 2001," indicating that there indeed was an earlier version of the 1998 document, which would have been first drawn up for implementation by the CRSC at its founding in 1996.

The August 1996 issue of Discovery Institute *Journal* announces the establishment of the CRSC as a new arm of the institute:

> For over a century, Western science has been influenced by the idea that God is either dead or irrelevant. Two foundations recently awarded Discovery Institute nearly a million dollars in grants to examine and confront this materialistic bias in science, law, and the humanities. The grants will be used to establish the Center for the Renewal of Science and Culture at Discovery, which will award research fellowships to scholars, hold conferences, and disseminate research findings among opinion makers and the general public.[14]

The CRSC website became accessible early in 1996.[15] Richards's remarks are strong evidence that the Wedge Document is genuine. And, although these separate considerations suggest that it does represent an older version of their program, it is abundantly clear that the Wedge Document's contents are anything but outdated in the view of its promoters. Virtually the entire plan—with the striking exception of the "scientific research" on intelligent design offered as the key to the rest of the program—is in full execution as we write.

Comparison of Language and Concepts

James Still's comment that large portions of the Wedge Document were on CRSC's website is correct. The case for the authenticity of the Wedge Document as the original plan could be made entirely by inspection of the CRSC's official announcements of goals, objectives, and strategies. The most convincing evidence for the Wedge Document's authenticity therefore comes from the CRSC website, on pages that contain verbatim wording from the circulated online document and on other pages with similar wording and identical concepts. Many such pages appear to be early ones, dating from the CRSC's establishment, and are no longer accessible on the site. The most important of these pages was entitled "What is The Center for the Renewal of Science & Culture All About?" When found, this page was not directly retrievable from the main CRSC website; it was stored in a directory that was then, but is no longer, accessible. This early page in its entirety is virtually identical to the introduction of the online Wedge Document and confirms the latter's authenticity. An excerpt from this early page is sufficient illustration:

> The proposition that human beings are created in the image of God is one of the bedrock principles on which Western civilization was built. . . .
>
> Debunking the traditional conceptions of both God and man, thinkers such as Charles Darwin, Karl Marx, and Sigmund Freud portrayed human beings . . . as animals or machines who inhabited a universe ruled by chance and whose behavior and very thoughts were dictated by the unbending forces of biology, chemistry, and environment. This materialistic conception of reality eventually infected virtually every area of our culture. . . .
>
> Discovery Institute's Center for the Renewal of Science and Culture seeks nothing less than the overthrow of materialism and its damning cultural legacies.[16]

An excerpt follows from the online Wedge Document's introduction; with the exception of a few words, it is the same as the CRSC's early web page:

> The proposition that human beings are created in the image of God is one of the bedrock principles on which Western civilization was built. . . .
>
> Debunking the traditional conceptions of both God and man, thinkers such as Charles Darwin, Karl Marx, and Sigmund Freud portrayed humans . . . as animals or machines who inhabited a universe ruled by purely impersonal forces and whose behavior and very thoughts were dictated by the unbending forces of biology, chemistry, and environment. This materialistic conception of reality eventually infected virtually every area of our culture. . . .
>
> Discovery Institute's Center for the Renewal of Science and Culture seeks nothing less than the overthrow of materialism and its cultural legacies.

Readers may view the entire Wedge Document directly; it is online at both http://www.public.asu.edu/~jmlynch/idt/wedge.html and http://www.antievolution.org/features/wedge.html. In short, the introduction to the online Wedge Document and the entire early page from the CRSC website itself ("What is The Center for the Renewal of Science & Culture All About?") are the same, with only insignificant changes. But there is more authenticating evidence that predates the Wedge Document itself.

In summer 1995, the DI sponsored a "Death of Materialism" conference, out of which the CRSC was born.[17] John G. West, Jr. (now associate director of the CRSC) delivered an address entitled "The Death of Materialism and the Renewal of Culture," in which his opening paragraphs are identical to the online version of the Wedge Document's introduction, save for roughly a dozen words as shown in this excerpt:

> The proposition that human beings are created in the image of God is one of the bedrock principles on which Western civilization was built. . . .
> Debunking the traditional conceptions of both God and man, thinkers such as Karl Marx, Charles Darwin, and Sigmund Freud portrayed human beings . . . as animals or machines who inhabited a universe ruled by chance and whose behavior and very thoughts were dictated by the unbending forces of biology, chemistry, and environment. This materialistic conception of reality eventually infected virtually every area of our culture.[18]

Listed on one of CRSC's now-inaccessible web pages, West was one of its first research fellows (1996–1997).[19]

Beyond the evidence predating the Wedge Document, there are plentiful contemporary traces of its existence and continued operation. For example, the spring/winter 1998 Discovery Institute *Journal*, DI's annual report, contains an unsigned article that restates all the major arguments in the Wedge Document's introduction:

1. The harmful effects of "scientific materialism" on "politics, medicine, the welfare system, law, and the arts";
2. CRSC's goal of undermining "scientific materialism"; and
3. CRSC's desire to "*bring about nothing less than a scientific and cultural revolution*." (emphasis added)

The similarities are also obvious in the opening statements of the *Journal* article:

> During the past century, human beings have been treated increasingly as the products of their genes and their environment. The cultural consequences of this scientific materialism can be seen in virtually every area of human endeavor, including politics, medicine, the welfare system, law, and the arts. . . . Discovery Institute's Center for the Renewal of Science and Culture is devoted to the overthrow of scientific materialism. . . . [T]he Center hopes to bring about nothing less than a scientific and cultural revolution.[20]

A brochure entitled "Exploring a Designed Universe" that was available on the CRSC site in pdf also contains wording identical to the introduction of the Wedge Document. This brochure appears to have been produced no earlier than September 1998, since it advertises William Dembski's book *The Design Inference*, also published in 1998.[21]

The CRSC continues to repeat the assertions on which the Wedge Document stands. The most striking current evidence for authenticity of the original Wedge Document is the similarity between its language and concepts and the language and concepts used regularly on CRSC's website today, specifically, on the key page "Life After Materialism?"[22] The following are some of these similarities:

Wedge Document	CRSC: "Life After Materialism?"
Thinkers such as Charles Darwin, Karl Marx, and Sigmund Freud portrayed humans . . . as animals or machines who inhabited a universe ruled by purely impersonal forces and whose behavior and very thoughts were dictated by the unbending forces of biology, chemistry, and environment.	Marx, Freud, and Dewey . . . portrayed humans . . . as mere animals or machines controlled by impersonal forces of biochemistry and environment.
This materialistic conception of reality eventually infected virtually every area of our culture, from politics and economics to literature and art.	This materialistic conception . . . infected almost every area of Western thought and culture . . . politics and law . . . literature and personal mores.
The center awards fellowships for original research, holds conferences, and briefs policymakers about the opportunities for life after materialism.	Can there be life after materialism?

The genuineness of the Wedge Document as a statement of the Wedge's strategy is therefore not a matter of speculation. Yet, even if it were not the authentic foundational document, the astonishing increase of CRSC activities enunciating and claiming to implement that document's stated principles demonstrate the existence of a well-orchestrated strategy for inserting intelligent design creationism into the American cultural mainstream and for securing to it a permanent, and if possible, dominant, place throughout American education. Today, with its program of action spelled out in the Wedge Document and the official, published successors to it, and with ample funding secured, the Wedge is at work and gaining power, despite occasional setbacks. Having begun with only four research fellows, the CRSC as of early 2003 consists of at least forty-three fellows, fourteen of whom have senior status. Phillip Johnson is still the advisor, along with George Gilder.[23] Their pursuit of the Wedge's goals continues largely, if not entirely, unopposed by seriously organized political effort and certainly undeterred, even welcomed, by those who share the Wedge's broader political and religious aims. The Wedge is now making excellent progress even in the halls of Congress—we present evidence later.

The Wedge-forced split in the log of "materialist" science and culture widened with the establishment, in October 1999, of the Michael Polanyi Center at Baylor University, although that center no longer bears the original name and has been absorbed into Baylor's Institute for Faith and Learning. The eventual destiny of that part of the Wedge's agenda embodied in the Polanyi Center—of which much more later—remains to be determined. But with the DI having provided the CRSC, and therefore the Wedge, a proper, functional home, it has grown in a very few years from infancy to robust adolescence and is racing toward adulthood.

3

Searching for the Science

Materialistic thinking dominated Western culture during the 20th century in large part because of the authority of science. The Center for the Renewal of Science and Culture seeks, therefore, to challenge materialism on specifically scientific grounds. Yet Center fellows do more than critique theories that have materialistic implications. They have also pioneered alternative scientific theories and research methods that recognize the reality of design and the need for intelligent agency to explain it.

"Design: A New Science for a New Century," Center for the Renewal of Science and Culture website, January 2001

It is the empirical detectability of intelligent causes that renders Intelligent Design a fully scientific theory, and distinguishes it from the design arguments of philosophers, or what has traditionally been called 'natural theology.' The world contains events, objects, and structures which exhaust the explanatory resources of undirected natural causes, and which can be adequately explained only by recourse to intelligent causes. Scientists are now in a position to demonstrate this rigorously. Thus what has been a long-standing philosophical intuition is now being cashed out as a scientific research program.

William Dembski, "The Intelligent Design Movement," *Cosmic Pursuit*, March 1998

In an article entitled "Shamelessly Doubting Darwin," William Dembski makes a bold claim: "A growing movement of scientists known as 'design theorists' is advocating a theory known as 'intelligent design.' Intelligent design argues that complex, information-rich biological structures cannot arise by undirected natural forces but instead require a guiding intelligence. *These are reputable scientists who argue their case on strictly scientific grounds and who are publishing their results in accepted academic outlets. This includes my own work and that of Jonathan Wells, Siegfried Scherer, and others*"[1] (emphasis added).

The Center for the Renewal of Science and Culture portrays itself as constantly involved in scientific research; indeed, the first item in the organization's title—the "Renewal of Science"—suggests that scientific renewal will precede cultural renewal. As the assertion from the Wedge Document makes clear, *science*—a science "consonant with Christian and theistic convictions"—is the fuel that will drive the engine of renewal for American culture. The CRSC's description of itself on its website assures readers, with astounding self-confidence (and no facts), that the organization has revealed the collapse of "strictly materialistic thinking in science":

> During recent decades, evidence from many scientific disciplines has suggested the bankruptcy of strictly materialistic thinking in science and the need for new explanations and perspectives. Consider: . . . In molecular biology, the presence of information encoded along the DNA molecule has suggested the activity of a prior designing intelligence.[2] . . . Predictably, many defenders of the status quo have refused to address the new evidence. . . . The Center for the Renewal of Science and Culture takes a different view, and that's why we are supporting scientists who aren't afraid to follow the evidence where it leads.[3]

Given such intense self-congratulation about CRSC achievements in science, the most urgent question raised in the mind of an objective observer must be this one: *What* scientific research have CRSC fellows in fact produced to support intelligent design? The Wedge Document, after all, outlining the major phases of its plan, begins in Phase I with the output of scientific research—"the essential component of everything that comes afterward"—to support intelligent design:

> The Wedge Projects
>
> Phase I. Scientific Research, Writing & Publicity
> - Individual Research Fellowship Program
> - Paleontology Research program (Dr. Paul Chien et al.)
> - Molecular Biology Research Program (Dr. Douglas Axe et al.)[4]

The Wedge Document makes clear that CRSC does not consider the chronological order of these phases unchangeable: "The Wedge strategy can be divided into three distinct but interdependent phases, which are roughly but not strictly chronological. We believe that, with adequate support, we can accomplish many of the objectives of Phases I and II in the next five years (1999–2003), and begin Phase III." This built-in flexibility has proven very convenient in light of the Wedge's scientific track record.

Thus, by the CRSC's own description, production of scientific research results, along with writing and publicity, is the foundation of the Wedge strategy. In support of "significant and original research in the natural sciences, the history and philosophy of science, cognitive science and related fields," the CRSC mounts a generous fellowship program, providing "full-year research fellowships between $40,000 and $50,000" and "short-term research fellowships between $2,500 and $15,000 for either summer research, release time from teaching or book promotion activities, or other research-related activities."[5] During CRSC's first year of operation alone, it awarded more than $270,000 in research grants.[6] Such lucrative support should have enabled industrious young scientists to develop scientific research programs and compile data at least in *support* of intelligent design, if not to prove it—*if ID is anything like as obvious a fact about the living world as its promoters claim*.

Yet in this most important of all the CRSC's goals—the only one that can truly win them the intellectual credibility they crave—their record is conspicuously one of failure. Thus, the CRSC boasts of its scientific research program and achievements, on one hand, while, on the other, Phillip Johnson admits that the needed scientific accomplishments have yet to be realized. For example, the CRSC's website declares, "[Center Fellows] have also pioneered alternative scientific theories and research methods that recognize the reality of design. . . . This new research program—called 'design theory'—is based upon recent developments in the information sciences[7] and many new evidences of design."[8] Yet in 1996, when the Wedge strategy was being formalized at the Mere Creation conference at Biola University, Phillip Johnson acknowledged that intelligent design proponents did *not* have the science they needed to accomplish their goals:

> What we need for now is people who want to get thinking going in the right direction, not people who have all the answers in advance. In good time new theories will emerge, and science will change. *We shouldn't try to shortcut the process by establishing some new theory of origins until we know more about exactly what needs to be explained*. Maybe there will be a new theory of evolution, but it is also possible that the basic concept will collapse and science will acknowledge that those elusive common ancestors of the major biological groups never existed. If we get an unbiased scientific process started, we can have confidence that it will bring us closer to the truth.
>
> For the present I recommend that we also put the Biblical issues to one side. The last thing we should want to do, or seem to want to do, is to threaten the freedom of scientific inquiry. Bringing the Bible anywhere near this issue just

raises the "Inherit the Wind" stereotype, and closes minds instead of opening them.

We can wait until we have a better scientific theory, one genuinely based on unbiased empirical evidence and not on materialist philosophy, before we need to worry about whether and to what extent that theory is consistent with the Bible. *Until we reach that better science, it's just best to live with some uncertainties and incongruities,* which is our lot as human beings—in this life, anyway.[9] (emphasis added)

Despite the significant absence of "some new theory of origins," CRSC fellow Nancy Pearcey also wrote in 1997 that "the design movement offers more than new and improved critiques of evolutionary theory. . . . Its goal is to show that intelligent design also functions as a positive research program." She asserts that "the design movement shows promise of winning a place at the table in secular academia, while uniting Christians concerned about the role science plays in the current culture wars."[10] But this coveted acceptance into secular academia is impossible without a sound scientific research program, buttressed by peer-reviewed publication of new results in the worldwide scientific literature. In the natural sciences, at any rate, this remains a truism: *no publishable results, no recognition.* Yet in the same article, Pearcey says that, according to Johnson, "The key . . . is not to defend a prepared position so much as to promote critical thinking." That laudable but hackneyed sentiment follows Johnson's 1996 remarks acknowledging the *absence* of any scientific position from which to counter the evil of evolutionary theory.

The dearth of scientific results in support of ID was confirmed in George W. Gilchrist's 1997 survey of the scientific literature (to which we refer later). He reports that "this search of several hundred thousand scientific reports published over several years failed to discover a single instance of biological research using intelligent design theory to explain life's diversity."[11] The situation has not changed since 1997.

The CRSC's Scientific Output

Phase I: Scientific Research, Writing, and Publicity

Although DI never publicly released the Wedge Document, with its confident predictions of scientific research to advance ID, in its 1999 *Journal* article entitled "The Promise of Better Science and a Better Culture," DI touts the scientific achievements of its Center for the Renewal of Science and Culture:

> Today . . . Darwinist dogma is being challenged by new science. It isn't easy getting a hearing, but it is happening more and more. Science's grand tradition of self-examination is leading to new theories based on better evidence, and pointing away from materialism.

> Defenders of Darwinian orthodoxy are quarreling among themselves as never before as disturbing evidence against Darwinism appears in such fields as Big Bang cosmology, paleontology (especially in Cambrian era fossils) and molecular biology. Moreover, an alternative to Darwinism—within science—is emerging in the theory of "Intelligent Design." The Center for the Renewal of Science and Culture at Discovery Institute is a major factor in the new scientific debate and the examination of its implications for culture and public policy.[12]

This statement is remarkable, even coming from a politically driven "think" tank, for its insouciant disregard of well-known facts. It is *not* hard to get a hearing in regular science journals for ideas like ID—provided that they meet the minimum standards of objectivity and technical competence. Anybody who offered even merely interesting, and not necessarily conclusive, evidence for ID would become a scientific celebrity overnight,[13] as happens regularly to *competent*, articulate dissidents who argue against the standard model (whatever that may be), in evolution or any other area of natural science. Nothing has happened in science—yet—that "points away from materialism," despite unceasing, indeed desperate, efforts of many good scientists over a few hundred years to find such evidence. "Defenders of Darwinian orthodoxy," whatever that means, do indeed quarrel among themselves, sometimes bitterly, about contemporary technical issues and on matters of emphasis,[14] as well as about professional politics. That is how science works: it is a social, as well as an intellectual, process. But those "defenders" are all "Darwinians," who reject outright or simply pay no attention to the vapors of religiously obsessed anti-Darwinism, promoted without evidence or compelling theoretical argument.

So, despite their public boasts about scientific accomplishment, the only Phase I goals in which CRSC fellows have been regularly and unequivocally successful are in writing (popular general or religious journal pieces and trade books) and public relations.[15] They have produced no original scientific data, not even a genuine scientific research plan, to mark progress or successful accomplishment in this most crucial phase. Michael Behe's book, *Darwin's Black Box*, which CRSC advertises as one of the most important contributions to its scientific agenda, has resulted in no scientific research program, not even by the person from whom one would most reasonably expect it—Dr. Behe himself.

At the end of 1997, in the CRSC's "Year End Update" published on its website, the CRSC chronicled its activities for that year. Among those activities was a "Consultation on Intelligent Design," bringing together "CRSC fellows and friends from around the world." Featured at this consultation, as "highlights of the weekend," were the scientists named in the Wedge Document as those whose work promises confirmation of the scientific truth of intelligent design: "Paul Chien, Professor of Biology at the University of San Francisco and a new fellow of the CRSC"; "Doug Axe, Postdoctoral Research Associate of Cambridge

University's MRC [Medical Research Council] Unit for Protein Function and Design"; and "Michael Behe, Associate Professor of Biological Sciences at Lehigh University."[16] Axe is the molecular biologist of the group and therefore a specialist. Molecular biology is the modern science of macromolecules (DNA, RNA, proteins, polymeric carbohydrates) and their functioning in cells, that is, within the subcellular level—what Behe mockingly calls "Darwin's black box." Behe and others of the ID group insist that this is the level at which the truth of ID is most startlingly obvious and the existing proofs for it the most powerful. In a subsequent chapter, we will look for specific scientific contributions to ID of Drs. Chien and Behe. Here, we examine first and very briefly the contributions to ID of Dr. Axe and then, broadly, of the entire CRSC science contingent.

MOLECULAR BIOLOGY

Douglas Axe, a research scientist formerly at the Centre for Protein Engineering at Cambridge University's Medical Research Council (MRC) and now with the Babraham Institute, is listed in the Wedge Document as advancing intelligent design through his "Molecular Biology Research Program." One of that document's highlighted activities is to supply "front line research funding at the 'pressure points,'" one of which is "Doug Axe's research laboratory in molecular biology." The National Institute for Research Advancement online database, "World Directory of Think Tanks" (1999), using information provided in 1998 by DI's Jay Richards, includes Axe as one of DI's "Chief Researchers."[17] He is listed in the spring 1998 DI *Journal* as a fellow. As of March 3, 2000, Axe's name was still on the list of fellows on CRSC's website, though by May 29, 2000, it had been removed. His biographical sketch, which was on the CRSC website as of May 29 but has also since been removed, listed his credentials: a Ph.D. in chemical engineering from the California Institute of Technology, postdoctoral research in molecular biology at the MRC at Cambridge University in England, and "numerous technical articles" in journals such as *Proceedings of the National Academy of Sciences* (U.S.), *Biochemistry*, and *Journal of Molecular Biology* (*JMB*).[18]

These are solid credentials in a technically sophisticated subdiscipline of modern biology. Although no longer listed on the CRSC website as a research fellow, Axe reported in June 2000 that he was still affiliated with the organization. The CRSC has paid for part of his research concerning "the relationship between amino acid sequence and function in proteins," which has "potential relevance" to several fields of science, including molecular evolution.[19] Axe said that he knows very little about the organizational features of CRSC and that funding is his only connection with it.[20] But at this writing (spring 2003), he is still formally affiliated with DI according to his curriculum vitae, where he lists himself as a "Senior Fellow, Discovery Institute (1998–present)."

Although his work in behalf of the CRSC has apparently been limited, Axe has indeed engaged in some ID activity, according to CRSC literature. The CRSC's November/December 1997 "Year End Update" reports that he served on a panel of biologists during its Consultation on Intelligent Design—one of the "highlights of the weekend"—making a presentation on "the growing challenge to Darwinian biology."[21] Along with others, he is also acknowledged by William Dembski in his book, *Intelligent Design: The Bridge Between Science and Theology*, as having "contributed significantly to the book."[22] The Wedge Document also lists as one of its goals "an active design movement" in the United Kingdom, where Axe lives and works. (According to his c.v., he is a native of the United States but a naturalized British citizen.) But we have found no evidence of his involvement in any such effort so far.

Despite his receipt of a CRSC research fellowship, Axe has published nothing in the scientific literature that supports intelligent design. A May 2000 SciSearch survey without date restrictions using the term "Axe D" yielded five titles that Axe acknowledges as his, but that provide no form of support for ID. One of us[23] has read with care his August 2000 *Journal of Molecular Biology* article (cited by William Dembski; see later), a workmanlike analysis of functional constraints on the amino acid sequence in a particular protein (and hence on its gene, the encoding DNA). There is *nothing* in that article—certainly nothing explicit by any stretch of the verb's meaning—to "support" ID. No working molecular or cell biologist, among the several colleagues we have consulted, has reported otherwise.

A February 5, 2001, Google search yielded only three hits not directly related to the Wedge Document, in which Axe's name appears. One was a course website at California State University-Fullerton, which contains a link to one of Axe's articles in a section entitled "Molecular Evidence."[24] The linked article is "An Irregular β–Bulge Common to a Group of Bacterial RNases Is an Important Determinant of Stability and Function in Barnase," co-authored by Douglas D. Axe, Nicholas W. Foster, and Alan R. Fersht (*JMB* 1999, 286, 1471–1485). This is one of the papers Axe acknowledged as his. Of the other two hits, one was a message from William Dembski posted by Larry Arnhart to a moderated public "evolutionary-psychology" discussion group on December 4, 2000. In that message, Dembski asserts that Axe's 2000 *JMB* article supports ID:

> Below is an abstract by a colleague of mine at the MRC at Cambridge. . . . This paper is indeed relevant to ID and the author is firmly in the ID camp. Note the sentence toward the end of the abstract, "Contrary to the prevalent view, then, enzyme function places severe constraints on residue identities at positions showing evolutionary variability, and at exterior non-active-site positions,

in particular." The "prevalent view" referred to here follows directly from neo-Darwinism. Axe's results are more clearly consistent with ID, though he doesn't say as much.[25]

The article to which Dembski refers is "Extreme Functional Sensitivity to Conservative Amino Acid Changes on Enzyme Exteriors" (*JMB* 301, 2000, 585–595). Dembski clearly has great confidence in this article and in Axe's work in general, as he indicated in a March 2002 *Newsday* story:

> The research, by Douglas Axe of the Centre for Protein Engineering in Cambridge, England, introduces a concept called "extreme functional sensitivity" that relates a protein's specialized function to the changes permitted in its amino acid sequence. Axe's premises are hinted at in an article published two years ago in the Journal of Molecular Biology, but Dembski and others say Axe plans to go public with his findings soon and "shake things up."
>
> "As these things get nailed down," Dembski says, "the Darwinian stories about them will become increasingly implausible."[26]

Axe's responses to inquiries about his own view of whether his work supports ID are much less definitive than Dembski's public statements. When asked to clarify his position regarding his work to date with respect to intelligent design, he gave the following response:

> These three statements summarize my position:
>
> I remain open-minded with respect to the possibility that a sound argument can be made for intelligent design in biology.
>
> I have not attempted to make such an argument in any publications.
>
> Since I understand that Bill Dembski has referred [in his book *No Free Lunch*] to my work in making such an argument, I shall remain open to the possibility that my published findings may support such an inference until I have had a chance to see his argument.[27]

We discuss Dembski's use of Axe's work in chapter 5. Suffice it to say, however, that if there is any support for ID in the *JMB* 2000 article, which Dembski does cite in *No Free Lunch*, we have not found it; and it is not reflected in a scientific database survey conducted on February 6, 2001. Although the article is indexed in Biological and Agricultural Index, Medline, BIOSIS, and SciSearch, a SciSearch survey using the keywords "Axe" and "intelligent design" yielded no results. Axe's reference list does not include any known ID proponents. The key words accompanying the abstract, which Axe himself supplied, do not include "intelligent design" or anything remotely suggestive of it.[28] An October 2001 search for articles by Axe in Medline and BIOSIS likewise turned up nothing new; nor did a simultaneous general search for articles on ID. So, judging from the careful, noncommittal response received directly from Axe, our survey of the scientific literature, and our own study of his otherwise competent molecular biological research, we find no significant scientific support for ID in results he has so far produced.

Intelligent Design in the Broader Scientific Literature

In the May/June 1997 issue of the *Reports of the National Center for Science Education,* George Gilchrist reports his survey of five massive, computerized databases up to 1997 for any scientific publications, among hundreds of thousands indexed, on intelligent design as a biological theory. He searched BIOSIS, the Expanded Academic Index, the Life Sciences Collection, Medline, and the Science Citation Index (which SciSearch also searches). His survey yielded a total of only thirty-seven references, of which "none report[s] scientific research using intelligent design as a biological theory."[29]

A similar search conducted in May and June 2000 for this study, supplementing Gilchrist's survey by looking for ID articles published *since* 1997, had the same result: no scientific research supporting ID as a biological theory had been published. The cited SciSearch surveys conducted for publications by Dr. Chien, Dr. Axe, and Dr. Behe yielded no intelligent design publications. In order to survey other databases for anything the SciSearch survey might have missed, we conducted surveys with no date restrictions of the BIOSIS and Medline databases using both "intelligent design" and "design theory" as key words. The "intelligent design" search in BIOSIS yielded four articles, only one of which was about intelligent design: "Redundant Complexity: A Critical Analysis of Intelligent Design in Biochemistry," by Niall Shanks and Karl H. Joplin, in *Philosophy of Science,* June 1999, *which is a strong critique of Dr. Behe's notion of "irreducible complexity."* The authors maintain that "recent work on self-organizing chemical reactions calls into question Behe's analysis of the origins of biochemical complexity." (Behe's response to this article appears in the March 2000 issue of *Philosophy of Science.*) With "design theory" as the key word, a BIOSIS survey yielded sixteen articles, none of which was about design theory as it relates to evolution or human origins. The Medline search using both "intelligent design" and "design theory" yielded fourteen articles, none of which is about intelligent design creationism. A SciSearch survey using "intelligent design" as a key word yielded sixty-one titles. All except four were concerned with industrial technology, engineering, computers, shipbuilding, and so on. Of the remaining four, only two were articles on ID as a biological theory: the Shanks-Joplin article and Behe's *Philosophy of Science* reply to it, already cited. The other two were letters entitled "Intelligent Design" in *Geotimes* and "Intelligent Design Reconsidered" in *Technology Review.* These titles are ambiguous because articles with "intelligent design" in their titles were also listed but were clearly not about intelligent design as a biological theory (e.g., "HyperQ Plastics: An Intelligent Design Aid for Plastic Material Selection"). Table 3.1 provides a summary of the 2000 findings:

Table 3.1 Summary of ID Publications, Year 2000.

Keywords	Index	Date restrictions	Total references	Intelligent design/ design theory references (biological only)
Intelligent design	BIOSIS	None	4	1 Shanks-Joplin
Design theory	BIOSIS	None	16	0
Intelligent design/ design theory	Medline	None	14	0
Intelligent design	SciSearch	None	61	2 Shanks-Joplin and Behe

Table 3.2 Gilchrist's Summary of ID Articles.

Keywords	Index	Years	Intelligent design (total references)	Intelligent design references (biological only)
Intelligent design	BIOSIS	1991–1997	1	0
Intelligent design	Medline	1990–1997	1	0
Intelligent design	Science Citation Index	1992–1995	4	0

These data can be compared to the relevant parts of Dr. Gilchrist's findings, in table 3.2, which stop at 1997.[30]

Surveys conducted in SciSearch and Medline in October 2001 revealed no change in these results.

Bibliometry

These survey results show that there is as yet no published research in scientific journals that supports, in any unequivocal way, the ID theory of the Wedge. Even in the venues where its adherents do argue for and publish their ideas, the ID promoters make relatively little use of up-to-date scientific results. This is not trivial. In the life sciences, especially in molecular biology, cell biology, genetics, evolution—that is, where scientific progress is both rapid and worldwide—failure to be aware of the current science, as revealed by the bibliography of a submitted manuscript, guarantees its rejection. Dr. John Lynch, lecturer at Barrett Honors College and affiliated professor with the Institute of Human Origins at Arizona State University, has done a bibliometric survey of the frequency with which ID proponents cite the most recent science. He explains that "within bibliometrics, a standard measure is the ratio of papers cited that are seven years [of age] or younger to those that are twenty years [of age] or older, with higher ratios being expected in 'fresh, vibrant, new, cutting edge' science."[31] Because of the method of calculation, even small differences in this ratio can be highly significant. Lynch's figures show the relative frequency of citations of "cutting edge science" by ID theorists in comparison to the frequencies of such citations in other types of presentations and publications in the same broad areas of concern:

Publication type	*Recent /old citation ratio*
Popular Science (evolution books by such writers as Richard Dawkins)	2.209
Popular Anthropology (Johanson, Eldredge, Leakey, et al.)	3.968
Scientific Journal (*Systematic Biology*)	2.964
Philosophy of Biology (*The Philosophy of Biology*, D. L. Hull and Michael Ruse, eds., Oxford University Press, 1998)	3.036
Creationist Conferences (Proceedings of the International Conference on Creationism)	0.864
Impact, from the Institute for Creation Research	1.302
Miscellaneous Individual Anti-Evolutionists (Duane Gish, Henry Morris, Francis Hitching, Walter Brown, et al.)	0.743
A newly developing *scientific* field (morphometrics conference proceedings)	4.005
Intelligent design theory	1.379

To derive his figure for ID theorists, Lynch surveyed a variety of ID publications: every book published up to that time by Phillip Johnson and CRSC fellows Michael Behe and William Dembski; the book *Of Pandas and People*; Leonard Brand's *Faith, Reason and Earth History: A Paradigm of Earth and Biological Origins by Intelligent Design* (Andrews University Press, 1997); and CRSC fellow Siegfried Scherer's *Typen des Lebens*, (Pascal Verlag, 1993).[32]

Lynch's figures reveal that the frequency of current science citations by ID proponents, who claim that ID is based on new and exciting scientific research, is less than half that of popular anthropology and differs little from the low frequency of such citations by biblical literalist, young earth creationists (YECs) of the Institute for Creation Research—from whom ID proponents try very hard to distance themselves (as we shall see). The ID-theorist citation rate is not even twice that of the two lowest creationist rates, the International Conference on Creationism Proceedings and miscellaneous YECs, including Duane Gish and Henry Morris. Even if, as Lynch points out, "the IDTers argue that the initial stages of their 'revolution' is philosophical," their citation rate for up-to-date work is less than half that of typical publications in the philosophy of biology, fewer of which were actually surveyed by Lynch in comparison to the extent of his survey of IDT publications. Lynch points out that

> what is interesting here is that the IDT stuff is not following the citation pattern for a "cutting-edge" science, relying on less "fresh" material than popular scientific writings and having a similar pattern to ICR's *Impact*. The relatively high value for popular works in physical anthropology can obviously be explained by the nature of the constant discovery in the field, and the [popular appeal] of the material, which leads to many new works appearing for the public describing the "latest" finds. In short, intelligent design theorists are no different from young earth creationists in that they tend to use out-dated material and do not follow the patterns of "normal" science or philosophy [of science].[33]

Scientific Data Supporting Intelligent Design?

John G. West, Jr., CRSC associate director, has identified "notable" scientists who support intelligent design: "In addition to [Michael Behe, Walter Bradley, and Jonathan Wells], notable scientists espousing design include Dean Kenyon, . . . San Francisco State University; Paul Chien, . . . University of San Francisco; Jeffrey Schloss, . . . Westmont College; Scott Minnich, . . . University of Idaho; Pattle Pak-Toe Pun, . . . Wheaton College; Henry Schaefer III, . . . University of Georgia; and Robert Kaita, . . . Princeton University."[34]

The Wedge is thus, in principle, amply supplied with scientific expertise. But the reality is this: one of the "Five Year Objectives" given in the Wedge Document is "one hundred scientific, academic and technical articles by our fellows"; but it is nothing more than wishful thinking. As the surveys reveal, the Wedge strategy has failed in its most important goal: the production of scientific research that supports intelligent design creationism, and the publication of such data in scientific journals. Publication in journals would indicate also that the data had been presented at actual scientific conferences, where it can be examined and discussed by other scientists, particularly by real peers—experts in the relevant fields. But it is clear that CRSC fellows are not making such presentations. Not only have Chien, Axe, and Behe failed to produce such work, but so has every other CRSC fellow—indeed, so has every known (professional or academic) proponent of ID in the world. Nothing of genuine scientific experimental or theoretical significance has been written on intelligent design *as a biological theory*—neither by Jonathan Wells nor by Paul Nelson, the CRSC fellows who presented papers at the Chengjiang fossil symposium in China. (see chapter 4).

Contrary to William Dembski's assertion in his "Shamelessly Doubting Darwin," nothing of scientific significance, nothing on ID that is evoking interest or concern in a professional scientific discipline, has been published by Siegfried Scherer or by Dembski himself. If there were such work, it would inevitably have shown up in the survey of the international databases and be under discussion in the frequent conferences of evolutionary biologists around the world. In the Wedge strategy's most crucial goal—indeed, its very foundation—the CRSC has fallen disastrously short.

Such consistent failure—if one can even call it failure, for no real attempt to produce relevant, supporting scientific data seems to have been made—is evidence that scientific research, the first stated goal in Phase I of the Wedge strategy, is not now, and perhaps never was, a primary goal of the CRSC. Indeed, Phillip Johnson, quoted by Laurie Goodstein in the *New York Times*, indicates that "getting into the mainstream," however it is to be accomplished, is in fact the primary goal:

> Another ally of Mr. Johnson is Michael Behe, a biochemist at Lehigh University who contends that the molecular machinery of cells is so complex and interdependent that this is proof of purposeful design. Mr. Behe's book, *Darwin's Black Box: The Biochemical Challenge to Evolution*, was chosen as 1997 Book of the Year *by the evangelical monthly Christianity Today.* (emphasis added)
>
> Entering the fray with a recent article in *Commentary* is David Berlinski, a philosopher,[35] who asserts that after more than 140 years the Darwinists have failed to prove their case. . . .
>
> This triumvirate has been duly picked apart by mainstream scientists. Kenneth Miller, a biologist at Brown University . . . skewered Mr. Behe's book in a recent review.[36] But that the book was even reviewed is progress in Mr. Johnson's view: "This issue is getting into the mainstream. People realize they can deal with it the way they deal with other intellectual issues like whether socialism is a good thing. My goal is not so much to win the argument as to legitimate it as part of the dialogue."[37]

Phillip Johnson's is a clear statement of what would really be a worthy goal for a Hollywood public relations agent: as generally agreed in that world, there is no such thing as bad publicity; all publicity is good; it builds "name recognition." It does not say much, though, for the purposes and the dedication to intellectual quality of ID science. Such ID-based science and thinking is supposed to be "revolutionizing" science and culture, driving it toward truth. Nevertheless, surveying the endeavors that the Wedge *is* executing successfully shows that the CRSC is indeed winning in the game of getting into the academic and cultural mainstream, even as it fails on the substance—that is, in the production of new and serious science. This success is a testament to the Wedge's religious passions and to its remarkable energy. It ought also to be a distressing reminder to readers who care about honest inquiry: that public interest in an idea, and its acceptance of ideas generally, has nothing necessarily to do with the truth, or even the general merit, of those ideas.

4

Paleontology Lite and Copernican Discoveries

From our modern vantage, it's hard to realize what an assault on the senses was perpetrated by Copernicus and Galileo. . . .

Things got steadily worse over the years. . . . Darwin shook the world by arguing that the familiar biota was derived from the bizarre, vanished life over lengths of time incomprehensible to human minds. Einstein told us that space is curved and time is relative. . . .

Now it's the turn of the fundamental science of life, modern biochemistry, to disturb.

Michael Behe, *Darwin's Black Box*

Paleontology Lite

Dr. Paul K. Chien is described in the Wedge Document as carrying out the CRSC's research in paleontology. He is identified in Wedge publicity as an important "paleobiologist" who has led others to see that the fossil record of the late Precambrian and early Cambrian periods (roughly 550 to 500 million years ago, or MYA) not only fails to support Darwinism but in fact refutes it. That would be a tall order and a world-class scientific accomplishment, but it is wildly unlikely: neither claim is new or true. But Chien *has* labored assiduously to nurture connections between key

CRSC personnel and Chinese paleontologists in Kunming, China, where the now well-known Chengjiang phosphate-rock fossils, representing the Precambrian and the long-known Cambrian radiation (sometimes called the "Cambrian explosion") have aroused great paleontological interest. In fact, he was one of the organizers of a June 1999 symposium in China on the "Origins of Animal Body Plans and Their Fossil Records," to which scientists from around the world were invited for discussion of the Chengjiang material. By that time, Chien had been fostering relationships with the Chinese paleontologists for several years.

In its November/December 1997 "Year End Update," DI reported that "Paul Chien, Professor of Biology at the University of San Francisco and a new fellow of the CRSC, gave an exciting slide presentation of rare, Cambrian-era fossils unique to the Chengjiang region of mainland China." This update boasts that "Dr. Chien has an extraordinary opportunity to work with leading Chinese scientists on the interpretation of these crucial fossils." And DI continues to make heavy public relations use of Chien's connection and collaborations with those scientists. The most important example is their argument for teaching ID in the nation's public schools. It is found in typical form in an article entitled "Intelligent Design in Public School Science Curricula: A Legal Guidebook," by CRSC fellows David DeWolf and Stephen Meyer, with Mark DeForrest:

> In recent years the fossil record has also provided new support for design. Fossil studies reveal a "biological big bang" near the beginning of the Cambrian period 530 million years ago. At that time roughly fifty separate major groups of organisms or "phyla" (including most of the basic body plans of modern animals) emerged suddenly without evident precursors. . . . Moreover, the fossil record shows a "top-down" hierarchical pattern of appearance in which major structural themes or body plans emerge before minor variations on those themes. . . . Not only does this pattern directly contradict the "bottom-up" pattern predicted by neo-Darwinism, but as University of San Francisco marine paleobiologist Paul Chien and several colleagues have argued, . . . it also strongly resembles the pattern evident in the history of human technological design, again suggesting actual (i.e., intelligent) design as the best explanation for the data.[1]

Among these ostensibly scientific assertions, including the *recent* discovery of the "Cambrian explosion" and a supposedly rigid Darwinian requirement for "bottom-up" evolution, not one is better than careful *mis*-interpretation (known in the news media as "spin"). They are issued, presumably, for consumption by the scientifically naïve, since the scientifically informed would laugh at them all. Yet the Wedge is making the "Cambrian explosion" the pedagogical centerpiece of its plan to place ID in the school *science* class. Details of those valuable fossil finds at Chengjiang are indeed scientific data.[2] But before proceeding to their discussion and to the Wedge's alleged research program in paleontology under Chien's leadership, a brief review of key issues surrounding Precambrian and Cambrian fossils is necessary.

Recognition of the remarkable speed and productivity of the Cambrian radiation—the *relatively* "sudden" appearance of new animal body plans in the fossil record of the Cambrian period, over the course of a few tens of millions of years—is not at all *recent*. The famous Burgess Shale Cambrian fossils were discovered long ago—in 1910, which in active scientific fields is very much *history*. There are common "Darwinian" explanations for it, recognized by most paleontologists as adequate or, at least, as the best explanations so far available for the apparent abruptness. But "abruptness" means *over ten or a few tens of millions* of years for the appearance of many basic animal body plans to which taxonomists (classifiers) now award the highest taxonomic status: "phylum." There was nothing like a *sudden* "explosion," even though a remarkable collection of fossil forms does appear for the first time, sooner or later, in lower (545–510 MYA) and middle (510–490 MYA) Cambrian rocks. *Pre*cambrian animals had trivially few or no hard body parts. The Cambrian was a time, however, when change and elaboration of body plan had apparently proceeded far enough for hard bodies to have emerged. Hard bodies—with an external or internal skeleton—make excellent fossils; soft ones, including nearly all the *Pre*cambrian life forms so far discovered, do not fossilize except in geologically rare, special, and fortunate circumstances. The Chengjiang phosphate deposits are one such special circumstance. (Certain small or microscopic hard *parts* of Precambrian animals, including spicules of very ancient sponges, *are* sometimes found.)

Nevertheless, Precambrian animal fossils do exist; they are fossils of animals living long *before* the Cambrian "explosion." The evidence shows that there was already a diverse fauna on Earth fifty million years or more before the Cambrian. Abundant Precambrian fossils were first discovered after World War II in the Ediacaran rocks of southern Australia. Most Ediacaran species were extinct by the time of the Cambrian radiation, although on the available morphologic evidence, some may well have been ancestral to Lower (early) Cambrian species. Suffice it to say that complex multicellular animals (metazoa) did *not* appear on Earth for the first time, all at once, in the Cambrian period: animals of the Cambrian had Precambrian ancestors.[3] There is no scientific reason to think that the Cambrian fauna sprang into being full-blown, without an evolutionary history—unless one wants a *special creation*. These facts of the matter of Precambrian metazoan life are in every respectable, modern evolution textbook.[4]

It is true that some professional paleontologists, including a few who wish to diminish (but not to deny!) the role of natural selection in the Cambrian radiation and evolution generally, who wish to suggest that other *natural* but less gradual forces played a role in it, relish the idea of an "explosion." But they remain vague about what, if anything, the word means, beyond "rather abrupt." Nevertheless, vague or not, use of the term by paleontologists who make specific, empirical points is legitimate. A glimpse of the terms of the current specialist argument, with specific empirical reference, appears in a recent issue of *Science* magazine, where two groups of paleontologists argue about whether *anything* that happened in the Cambrian radiation of animals deserves to be called an "explosion."[5] But remember that the implication of the ID proposals and literature, by contrast, is that in the Cambrian period a *real* explosion happened: it is that everything in the categories of animal life came into being suddenly, with no ancestors. This implies, of course, at least one point at which the Designer has intervened directly in the natural processes of descent and change.

Paleontologists agree that the Cambrian animal species were descended, one way or another, from metazoan (multicellular) ancestors. And they do not usually address the literate public with such journalistic grandiloquence as "Biology's Big Bang." Old-earth creationists, when commenting on the Cambrian radiation, routinely do. They often assert, moreover, that there has been a "top-down," as opposed to a "bottom-up" emergence of biological forms, meaning that all phyla (i.e., all the *major* animal groups) appeared abruptly during the Cambrian period and that since then there have been no, or only trivial, changes in body plans. They sometimes justify this claim by noting the occasional use of the terms "top-down" and "bottom-up" among professional evolutionists, even though scientists use these terms in different ways.

This top-down view of things remains weak (see later); but it is a legitimate scientific viewpoint. Yet the *scientific* argument between the predominant bottom-up view and the top-down view is, basically, an argument over the *order of descent*: through common descent from earlier forms, what developed first—*species* (a spectrum of smaller groups of more similar forms), or *phyla* (several groups of more disparate forms)? The debate is *not* about whether there were *any* ancestral forms. Both sides in the argument agree that there were, and both groups agree that Darwinian natural selection played either a significant or the major role in the Cambrian radiation.

The small subset of biologists who do speak of a top-down Cambrian radiation consider that the prevalent neo-Darwinian view (which they characterize as bottom-up) requires that different *species*, which are by definition (for classification purposes) at the "bottom" of the hierarchy of taxonomic (classificational) categories, appear *first* in the order of historical descent—before phyla, which are the highest, most inclusive categories. Thus, in this formulation, a bottom-up course of evolution would produce first, from one ancestor or a few, many originally very similar *species*, from which the higher categories—genera, orders, classes, phyla—would eventually emerge, as combinations of species, generated solely from the species already existing. (The taxonomic system itself need not reflect the order of descent, i.e., what life forms came first in time, but only where any given organism, either existing or extinct, fits in its genetic relationships to other organisms.)

On the most simple-minded idea of top-down evolution, on the other hand, *all the phyla as such—not species—appear first*, and from these phyla there subsequently descend, with only modest modification, through time and the great hierarchy of taxonomic categories, the eventual multimillions of Earth's different "species," extinct and living. The question then is what process produced these major, disparate phyletic groups so quickly from their relatively simple common ancestors. Biologists who actually argue top-down this way—and we stress that they are few—want to suggest that *in addition to* gradual change via natural selection and the like, some important, as yet unidentified *natural* process or processes operated in the Cambrian, and thereafter, too, somehow causing the *relatively* speedy diversification of metazoan body plans. Those disparate body plans are then interpreted in contemporary taxonomy, necessarily by hindsight, as representing most of the living animal phyla (and some that are extinct).

On the other hand, creationist *mis*interpreters of this argument, whether through willfulness or ignorance, make the illegitimate leap from it to an explicit or implied conclusion: that some form of special creation (or intelligent design) is the only sufficient explanation of top-down evolution—because, as they assert, no ancestors of the Cambrian fauna or transitional forms between them exist in the fossil record, and all the basic subsequent forms are already present in the Cambrian. But there are two problems with this creationist view of the top-down version of the debate. First, if there were in fact *no* evidence of ancestral or transitional life forms, or any hope of it, the abruptness would be irrelevant: there would *have* to have been some magical intervention anyway, regardless of the gradual or abrupt appearance of the Cambrian body plans! Second, ancestral and transitional forms *are* known and available for study, as all geologists and paleontologists today insist. Introducing an excellent summary on Precambrian and transitional forms, geologist Keith Miller (who is an evangelical Christian), complains about the serious misrepresentations of the fossil record that figure so prominently in evangelical Christian writings:

> The implication of much of the evangelical Christian commentary on macroevolution [change of body plan] is that the major taxonomic groups of living things remain clearly distinct entities throughout their history, and were as morphologically distinct from each other at their first appearance as they are today. There is a clear interest in showing the history of life as discontinuous, and any suggestion of transition in the fossil record is met with great skepticism. The purpose of this short communication is to dispel some of these misconceptions about the nature and interpretation of the fossil record.[6]

The simplest form of the erroneous top-down argument implies that there is more disparity of fundamental body plans early in the history of life on Earth than could ever be accomplished by ordinary Darwinian evolution; therefore, something else must have been at work and, in terms of geological time, at work for the merest instant of time. However, even if this top-down argument were correct (it is not, as we show), it would not necessarily point to any supernatural phenomenon. It would indicate merely that the fossil record is incomplete and that scientific interpretation of it is therefore incomplete—as is the scientific interpretation of nearly everything else. Good science is always incomplete. It is just the best we have for the moment, and it usually gets better.

The evolutionary steps taken by metazoa *before* the Cambrian "explosion" could themselves have been, and probably were, rapid—perhaps a few tens of millions of years in geological time, but they are lost to us because soft-bodied forms did not normally fossilize like the Cambrian fauna.[7] So, in the absence of a huge number of good fossils, we have no scientific warrant for saying that the body types found in Cambrian deposits "exploded" into being in the literal wink of an eye. And in any case, tens of millions of years are nothing like the wink of an eye. Metazoa, some of them precursors of the Cambrian forms, *were* around for at least that long. Earth was not sterile before the Cambrian: it was teeming with microbial (unicellular) life for at least two or three billion years; and at the end of the Proterozoic era (just before the Cambrian), there was an abundance of multicellullar life, too—chemically and genetically kin to present life forms but structurally simpler. Those readers who remain beguiled by the idea of a sudden and deeply mysterious explosion some 540 MYA should read the recent (1998), authoritative treatment of the Cambrian radiation by the distinguished explorer of the Burgess Shale, Simon Conway Morris,[8] who disputes the earlier (1989), and very influential, account of it by paleontologist Stephen Jay Gould (in Gould's *Wonderful Life*). Gould's eloquent prose seems to be the unintentional source of the popular but erroneous version of the idea of top-down evolution.[9] Needless to say, neither Morris nor Gould ever argued for the special creation of Cambrian fauna.

Finally, the top-down explosion scenario, as generally *meant* by creationists, who have co-opted the words but not their exact scientific intention, is structurally nonsensical. Miller's is the best recent, short, focused statement on this point, but the most accessible form of it is in Richard Dawkins's *Unweaving the Rainbow*. Top-down, at least as imagined by nonbiologists or other commentators who do not understand taxonomy, is a serious misunderstanding of how the taxonomic categories are assigned and of what they mean in practice. Taxonomic categories (taxa) are not themselves things: they have no independent existential status, nor do they represent absolute, unbridgeable degrees of (morphological) difference! They are the *names* of groupings assigned to organisms, for convenience, in a retrospective ordering of those organisms that have appeared on Earth in the course of geological time, of which we have adequate fossils or traces, and about whose relationships with similar forms we have some, not necessarily all, the desired evidence. Taxonomy ceased long ago to be the crude typology that it was in the early eighteenth century, when it began. There is *no* requirement, in the retrospective assignment of two Cambrian animals to different phyla, that those organisms at *that* time differed from each other as much as would randomly chosen representatives of, say, the oysters (phylum Mollusca) from your favorite Hollywood star (phylum Chordata)—*today*. Dawkins exemplifies the false top-down idea this way: "[I]t is as though a gardener looked at an old oak tree and remarked, wonderingly: 'Isn't it strange that no major new boughs have appeared on this tree for many years. These days, all the new growth appears to be at the twig level!'"[10]

Of course, some classes of genetic mutation are now well known that *do* change the body plan quite radically and relatively fast, in a single or a few mutational steps (although the likelihood of survival in the wild of *most* such macromutant forms would necessarily be small). The study of such mutant genes and their possible role in the generation of new body plans has been an interest of Neo-Darwinian genetics and embryology for decades. Again, however, all this is standard textbook material.[11] The evidence from the newer *molecular* systematics (taxonomy), the reliability of which is immensely strengthened by its consistency with the preexisting morphological and fossil evidence, indicates that common ancestors of phyla present in the Cambrian fauna, and surviving today, were there long before the Cambrian era dawned. As the evidence from molecular systematics grows firmer, this conclusion approaches general acceptance in the scientific community. Based on this emerging chemical evidence, a general comment on the new work, appearing in *Nature*, began with confidence, "If a recent analysis of animal evolution is correct, then the famed 'Cambrian Explosion' in the evolution of multicellular animals was not so much a Big Bang as simply the end of a long, slow, crawl."[12] The arguments among evolutionary biologists—and they can be bitter arguments, as elsewhere in frontier scholarship, including natural science—have nothing to do with any form of special creation or intelligent design. Readers who care about the issue should consult the professional (some quite readable) literature on the subject of top-down, bottom-up, and what they really mean.[13] There is a short but authoritative statement of the minority *scientific* top-down view, which tends to be coupled with arguments that macroevolution (more on this later) includes some processes distinct from those of microevolution. This is from Stephen Jay Gould himself, in the closing paragraph of one of the introductory essays for the massive new *Oxford Encyclopedia of Evolution*:

The explanation of trends by higher-level selection and the catastrophic basis of at least some mass extinctions . . . pose no threat or challenge to the importance or validity, but only to the exclusivity, of the conventional microevolutionary theory of Darwinian natural selection. But these and other examples do indicate that the historical contingencies of any complex chronology and the different rules and predictions of genuinely macroevolutionary principles must also play a major part of any fully adequate theory of evolutionary mechanisms and the pageant of life's stunning variety and history.[14]

Dr. Chien and the Chengjiang Fossils

The connection between Dr. Chien's involvement in gaining access to the Chinese scientists who discovered the Chengjiang fossils and the agenda of the CRSC is easy to detect. In an interview with Chinese Christian Mission for *Challenger* magazine, Chien's remarks display his essential (Chinese) patriotism and also his sincere anti-evolutionary fervor:

> The Chengjiang Biota is a "Treasure" discovered by our own scientists on our own land! . . .
>
> Many Chinese have been taught the wrong theory, namely Darwinism. When I told them about this new scientific finding, some were very angry because they had been told the wrong story all their lives. Of course some thought I was telling a lie. But after I showed them . . . the real fossils from Chengjiang, they . . . blamed the education they had received. . . .
>
> When I was younger, . . . I tried to stay away from the theory of evolution. I never believed it even when I was a non-Christian. But . . . I should have faced it straight on since I was very sure the theory of evolution would change drastically and dramatically.
>
> When a Christian gets into Biology, he naturally feels uncomfortable and alienated because he has to live with Darwinism every day. . . . However, in recent years, we do have a network of biologists and other scientists, historians, philosophers and theologians meeting regularly to talk about these things and support one another.[15]

Chien's last remark appears to be a reference to the CRSC, an appearance reinforced by his citation in the same interview of Phillip Johnson's report of an unnamed "Chinese professor" who said, "You Americans say that in China we cannot criticize the government, but you see we can criticize Darwin. In America you can criticize the government, but not Darwin." Indeed, it is abundantly clear in this interview that Chien shares the CRSC's implacable hostility toward "Darwinism," including the belief that "Darwinian theory" has become a "religion"[16] (whose death knell, nevertheless, has sounded):

> We will eventually . . . knock down the Darwinian paradigm. Other fields like biochemistry and developmental biology, even molecular genetics, will need to work together with the biologists to make an impact. . . .
>
> Darwinian theory is not just a scientific inquiry any more. . . . It has become a religion for many people although they do not confess . . . or recog-

nize it. Once it becomes a paradigm or reference for all other areas of life, the resistance to change is paramount. Many scientists build their life studies on . . . evolution; they can't see anything differently. But . . . some people, including myself, . . . are seeing big cracks in the Darwinian theory. . . . [W]e'll see it crumble like the Berlin Wall. We may see it happen in our lifetime.[17]

Chien's references to the CRSC are even clearer in the following remarks, since they reflect the CRSC's agenda and jargon: "[A] number of my friends in the network have begun to gain access to radios and TV programs all over the country to talk about real science (they are not the so-called 'creationists'). They are trying to help shape a better policy for science education in the States, which is: 'Teach Evolution, but ask tough questions.'"[18] This comment is an almost verbatim repetition of the title of Michael Behe's August 13, 1999, *New York Times* op-ed piece, "Teach Evolution—And Ask Hard Questions." In addition, Chien resorts in the *Challenger* interview to the CRSC's familiar conspiracy mongering to create the impression that important information is being withheld from the public: "Basically this is a hush-hush topic in scientific circles, particularly among the Darwinists. The only place you can find it is in Stephen Jay Gould's 1989 book, *Wonderful Life*, where he points out that things like the Cambrian Explosion are the 'trade secrets' of paleontology and the enigma of all enigmas. There has been a conspiracy among the scientists not to tell and talk about it. . . . Anybody speaking up against it will suffer professionally and personally. I have experienced it myself."[19]

A prime example of the conspiracy mongering from ID promoters can be found in Phillip Johnson's *Defeating Darwinism by Opening Minds*, in which he impugns the honesty of Niles Eldredge (a senior, distinguished paleontologist who has well-known arguments with "gradualist" Darwinism):

Eldredge also explains the pressures that could easily lead a forlorn paleontologist to construe a doubtful fossil as an ancestor or evolutionary transitional. Science takes for granted that the ancestors existed, and the transitions occurred, so scientists ought to be finding positive evidence if they expect to have successful careers. According to Eldredge, "the pressure for results, positive results, is enormous." This pressure is particularly great in the area of human evolution, where success in establishing a fossil as a human ancestor can turn an obscure paleontologist into a celebrity. . . . In light of these pressures and temptations, how confident should we be that fossils of "human ancestors" are really what they purport to be?

Despite Johnson's assurance at this point that he is not implying that "anybody is committing a deliberate fraud," he ends with the strong suggestion that this is exactly what Eldredge is doing:

Think how much pressure the other physical anthropologists are under to develop standards that will allow *some* fossils to be authenticated as human ancestors. A fossil field without fossils is a candidate for extinction. Keeping all that in mind, why do you think such a high proportion of the fossils used to prove "evolution" come from this one specialty? Why do you think Niles Eldredge, a specialist in marine invertebrates, uses hominid examples rather than the vast record of fossil invertebrates to argue the case for evolution? If anybody tries to tell you that questions like these are improper . . . your baloney detector should blow a fuse. A scientist who objects to scientific testing is like a banker who doesn't want the books to be audited by independent accountants. View such people with suspicion.[20]

In the Discovery Institute's spring 1999 *Journal*, the "fascinating and consequential dialogue" between Chien and Chinese scientists who discovered the Chengjiang Cambrian fossils is touted as a major development for the CRSC *which will result in a book "on the significance of the find and its relevance to the theory of evolution"* (emphasis added).[21] DI also announces in this issue that "at year's end [1998], several Discovery fellows were invited to a major international conference on the origin of animal body plans to be held in China" in June 1999 and that "Chinese scientists entertained William Lane Craig during the past year and such [CRSC] fellows as [William] Dembski, Robin Collins of the University of Texas, J. P. Moreland of Biola, Jonathan Wells, [and] Paul Nelson of the University of Chicago." This means that in the year prior to the International Symposium on the Origins of Animal Body Plans and Their Fossil Records, which Chien and the CRSC helped to organize (see later), six CRSC fellows had already been to China. Clearly, the CRSC has been nurturing the Chinese connection.

Chien's pursuit of relations with Chinese scientists working on the Cambrian fossils dates back to 1995, "when a number of publications caught my eye." Considering the great importance of these fossils to the study of paleontology, this was very late—seven years late—for them to have caught a competent paleobiologist's eye.[22] In March 1996, he "organized an international group to make a visit there" after hearing about the fossil discovery and having his interest "ignited" by the chapter on fossils in Phillip Johnson's *Darwin on Trial*.[23] This contact with Chinese scientists studying the Chengjiang fossils culminated in the June 1999 International Symposium on the Origins of Animal Body Plans and Their Fossil Records, of which Chien was a key organizing figure: he is a cosigner on both the first and second circulars put out by the Chinese Academy of Sciences; on the second, he is identified as one of the organizers.[24] On neither circular, however, is his affiliation with the CRSC or DI disclosed.

According to scientists who attended the Kunming conference, the involvement of DI in the conference became known only after the conference began. This involvement was first recognized by Dr. David Bottjer, professor of Earth Sciences (Paleobiology and Evolutionary Paleoecology) at the University of Southern California, who arrived in Kunming a few days early to arrange research projects on the Chengjiang fossils and to make a preconference field trip to Guizhou Province to see the site where the phosphatized Precambrian microfossils had been discovered.[25] Chien asked Bottjer to help edit a book of abstracts of the conference presentations to be handed out to participants. On reading the abstracts, Bottjer recognized creationist arguments and anti-evolutionists Jonathan Wells and Paul Nelson. He exclaimed to Chien, not knowing of the latter's connections with the CRSC, "There are creationists at the conference!" (At that point, Bottjer, like most professional scientists to this day, had no idea of the activities or purposes of the CRSC or its parent organization, the Discovery Institute, nor did he know that Chien was connected with such an organization.)

As the field trip to Guizhou progressed, it became increasingly clear to Bottjer that the aim of Chien and his CRSC colleagues was to produce and then to promote a book containing the conference papers of CRSC members *immediately juxtaposed to those written by respected scientists* in the relevant fields.[26] Since such a book would certainly have been used by the CRSC as evidence of an important, even a key, place for its ID creationist claims in mainstream evolutionary biology, or at least that ID is recognized in science as a very serious alternative to it, Bottjer became concerned. There were then and are to date no such ID contributions, and the experts do not see it as a serious alternative to anything in paleontology. His concerns and those of conference colleague Eric Davidson (Norman Chandler Professor of Cell Biology, California Institute of Technology) generated discussions among the attending scientists, including the host Chinese scientists. CRSC representatives at the symposium were fellows Paul Chien, Jonathan Wells, Paul Nelson, Michael Denton, and Marcus Ross, a graduate student who was also a CRSC fellow at the time. Rich McGee of Christian Leadership Ministries, who has worked closely with the CRSC in the pursuit of its ID agenda, also presented a paper. Fred Heeren, a free-lance reporter and old-earth creationist who also attended, interviewed conference participants.[27] Both Ross and Heeren told Bottjer that they were attending the conference with the aid of grants from DI.

Michael Denton and Rich McGee gave their talks on the first day of the conference; the offerings of Jonathan Wells and Paul Nelson came at the very end.[28] According to Nigel Hughes (associate professor of geology specializing in paleontology in the Department of Earth Sciences, University of California-Riverside), who also attended, it was during the presentations by Wells and Nelson, when the conference was at an end, that "the broader agenda of what was going on was apparent," meaning that the obviously creationist arguments and the real motivation for the book became clear at that point.[29] Eric Davidson dissected Wells's and Nelson's presentations during the sessions, identifying their errors. (According to Bottjer, the creationist presentations were "examples of 'cherry-picking'—picking things out of other people's publications and quoting them").[30] When at last the Western scientists at the conference understood how they were being used—at a supposedly scientific conference—they were displeased.

On the last day of the meeting, all participants attended a discussion of the expected publication. The Western scientists present unanimously agreed that there should be no book that included creationist papers alongside their own. According to Bottjer, one CRSC participant objected, insisting that their papers could pass peer review and should therefore be included. A CRSC participant also pointed out that "the Discovery Institute has been heavily involved in organizing this conference," implying that this involvement conferred the right to be represented in a conference publication.[31] Some of the scientists then argued that if DI's involvement in the conference had been made known to them, they would not have attended.

Bottjer's account is confirmed by Nigel Hughes, who remarks also, however, on the surprise of DI attendees on finding their arguments rejected. "Discovery Institute guys wanted their papers in [the book], subject to peer review. When it became clear that DI was involved in the basic organization of the conference, many scientists said that they would not have come if they had known. DI people appeared shocked to hear this. . . . They expressed shock that scientists did not want to be associated with them."[32] In the end, publishing plans were cancelled, and no book was published. As far as Bottjer knows at present, there are still no plans to publish.

Chien, in the *Challenger* magazine interview, offered his own explanation of why the book plans were cancelled:

> Before the international symposium on the "Origins of Animal Body Plans" last June, some scientists found a lot of areas not working right with neo-Darwinian theory. A Chinese professor suggested that we should have experts from other disciplines such as molecular genetics, developmental biology, even historians and philosophers, to join the discussions on the broader issues of this animal big bang. But during the meeting he was pressured by the American Darwinists not to allow open discussion, and the papers written from a non-Darwinian standpoint would not be published.[33]

According to this oblique and evasive version, no open discussion of the book plans was allowed because of pressure against it exerted on the Chinese by American scientists. But Bottjer's account of exactly such a discussion at the last session flatly contradicts Chien's, as does the recollection of Nigel Hughes:

> The plans for a book were discussed frankly in open session during the last day of the Chengjiang meeting. All persons present had the opportunity for input, and more than four people spoke up (Bottjer, Hughes, Nelson, . . . perhaps also Davidson, Wells . . . and others). The DI people expressed their interest in the book, and to his credit the DI's Paul Nelson expressed the wish that his

chapter should be externally reviewed. I said that the results I had presented in my paper were in press elsewhere and that I didn't feel that a book was necessary at this time. I'm amazed that Paul Chien apparently doesn't remember this discussion because it was the only time that the scientists publicly expressed the fact that we were unaware of the DI's involvement in the meeting prior to our arrival in China. In my view it was certainly the most spirited episode of the whole meeting.[34]

Hughes later wrote a review of the meeting, "The Rocky Road to Mendel's Play," in which he expresses pleasure in the fact that "evolutionary developmental biology now offers a route to link distant corners of evolutionary science," but also warns of "bandits that lurk along the way"—referring to the attempt by the CRSC to exploit the Chengjiang fossil discovery for its own purposes.[35] In his review, Hughes points out that the CRSC presenters made gross scientific errors, which were "candidly dispatched" by Eric Davidson. Questioning by other scientists as well revealed deep flaws in the positions the CRSC presenters were trying to defend—positions Hughes terms "the old Paleyan arguments wrapped in a variety of molecular guises," after which he remarks that the only *new* thing about these arguments was their presentation "at a meeting that was . . . billed as being *scientific*."

Given the CRSC's long, expensive involvement in what was advertised as a purely scientific meeting, it is not surprising that after it was over Phillip Johnson sought aggressively to exploit it. Hughes notes in his review that after the conference, commentary written by someone at DI (whom his review does not name) appeared in the August 16, 1999, *Wall Street Journal*, asserting that Chinese scientists have new evidence that casts doubt on the foundations of evolution. This commentary is in fact Johnson's own "The Church of Darwin." Responding to the Kansas Board of Education's deletion of evolution from its science standards, it is an example of the CRSC's exploitation of the Chinese connection—especially the Chinese scientists—to advance their own agenda: "A Chinese paleontologist lectures around the world saying that recent fossil finds in his country are inconsistent with the Darwinian theory of evolution. His reason: the major animal groups appear abruptly in the rocks over a relatively short time, rather than evolving gradually from a common ancestor as Darwin's theory predicts. When this conclusion upsets American scientists, he wryly comments: 'In China we can criticize Darwin but not the government. In America you can criticize the government but not Darwin.'"[36]

Hughes says that the CRSC has employed a clever strategy to get Chinese paleontologists—who may not fully understand the nature of the CRSC or its anti-evolution crusade—to take public positions on the Chengjiang fossil discovery that appear sympathetic to its arguments.[37] According to Hughes, who had himself pointed out numerous scientific errors made by CRSC presenters, the CRSC is not interested in scientific rigor; but he notes nevertheless that "they are dangerous, because they can 'talk the talk' enough to convince [even] reasonably well educated people, and claim affiliation with institutions like Berkeley and Chicago." More ominously, Hughes sees that the ultimate destination of the CRSC is public school science classrooms—by way of the U.S. Supreme Court. Nothing would accelerate the Wedge's intrusion into the science curriculum like having their claims published alongside the work of recognized and productive evolutionary biologists, in a book issuing from an international science symposium.[38]

But the most basic questions remain. Is Dr. Chien—charged with producing, out of his relationship with Chinese paleontologists, key scientific data that the CRSC must have in order to identify its statements on the "Cambrian explosion" and ID as respectable science—qualified in paleontology? Has he produced data or interpretations of his own in support of these claims? And if he has not done so thus far, is he likely to do so in the future? According to Kevin Padian, curator of the Museum of Paleontology and professor of paleontology and evolutionary biology at the University of California-Berkeley, "*Dr. Chien admits that he has no expertise or training in paleontology. He admits in interviews that he came into this issue believing that evolution is not true*. The real issues about the so-called sudden appearance of Cambrian faunas are not as complex as Chien portrays them, in my view and the view of experts on the subject" (emphasis added).[39] And Nigel Hughes says, "As far as I know, P. K. Chien is not a paleontologist and has published no peer-reviewed papers in paleontology. *He is not a 'player' in scientific issues related to the Cambrian radiation*" (emphasis added).[40] David Bottjer observes that "Chien has tried to produce straight science papers on Chengjiang fossils, but so far I don't believe that there have been any publications. He has a Chengjiang fossil collection . . . but even if he does have a lot of specimens, that is not proof that he has or can do anything scientific with them; lots of amateur (non-scientist) individuals have large fossil collections. *From my interactions with him in China, I can say that Chien knows nothing about the science. He did not give a talk at the conference. It is all a show to 'wedge in.'* He is interested in creationist goals" (emphasis added).[41] Chien's biosketch at the CRSC website lists his credentials: Ph.D. in biology, University of California-Irvine (Dept. of Developmental and Cell Biology); postdoctoral fellow in the Dept. of Environmental Sciences, California Institute of Technology (CIT), Pasadena; instructor in biology at the Chinese University of Hong Kong; consultant to CIT's Kerckhoff Marine Laboratory and to the "Scanning Electron Microscopy and Micro X-ray Analyst" at the Dept. of Biology, Santa Clara University (California); and professor and (former) chair in the Dept. of Biology at the University of San Francisco. Chien's bio also notes publications in "over fifty technical journals" and international lectures in connection with "cooperative research programs." (Also included are work "with leading Chinese scientists on the interpretation of crucial Cambrian-era fossils," plans to co-author a book with one of these (unnamed) Chinese scientists, and Chien's editing and translating Phillip Johnson's *Darwin on Trial* into Chinese.)[42]

Chien clearly has proper scientific credentials. However, paleontology—the one discipline indispensable for study of the Chinese Cambrian or any other fossils—is entirely absent from them. On the University of San Franciso's Department of Biology home page, Chien's degrees are listed: "B.S., Chung Chi College, N.T., Hong Kong, Chemistry, 1962; B.S., Chung Chi College, N.T., Hong Kong, Biology, 1964; Ph.D., University of California, Irvine, 1971." His Ph.D. research field is not listed, but as noted in his CRSC biosketch, it is biology. Chien's research interests, in his own statement of them, are listed on the University of San Francisco web page: "Prof. Chien is interested in the physiology and ecology of inter-tidal organisms. His research has involved the transport of amino acids and metal ions across cell membranes and the detoxification mechanisms of metal ions."[43] These are respectable but routine research interests in marine biology; they have nothing to do with fossils or the fossil record.[44] Chien has no standing in paleontology. Moreover, he is not interested in acquiring it, as he reveals in a 1997 interview published in *The Real Issue*, from Christian Leadership Ministries:

> RI: Do you intend to go back to Chengjiang, the Chinese Cambrian site?
> Chien: I would very much like to do that. Somehow I would like to get more involved in fossil work. Although I have lectured so many years in my own area of marine biology and pollution, I think I would like to concentrate on this aspect. This was an opportunity presented to me which nobody else has.
> RI: *Perhaps you could add "paleontologist" to your credentials.*
> Chien: *Not really; that's not my purpose. I am more interested in working on the popular level* (emphasis added).[45]

Despite this remarkable admission from an announced expert on the fossil record, who is here disavowing any wish to learn the relevant science and its methods, the introduction to the interview asserts that "what Chien found at the Chengjiang site, and what he has since learned about the Cambrian fauna, has changed the focus of his career. Today, Chien concentrates on further exploring and promoting the mysteries of the Cambrian explosion of life. Subsequently, he has the largest collection of Chinese Cambrian fossils in North America."[46]

So, while the Chengjiang fossils may have, in some way, "shifted the focus" of Chien's career, it has not shifted toward serious scientific work in paleontology. Rather, he sees his interest in the Cambrian explosion as "a hobby," and so states in this interview with *Challenger*:

> [Interviewer:] Do you plan to continue research on the Cambrian Explosion?
> [Chien:] I would like to continue to study it as a hobby and watch its development.[47]

A survey of the scientific literature reveals that, so far, Chien's involvement in the study of the Chengjiang fossils can indeed have progressed no farther than its adoption as a hobby. A survey in late May 2000 (*five years* after Chien became interested in the fossils and began his association with the Chinese paleontologists), using SciSearch, which contains citations dating to 1988, and using the name "P. K. Chien" with no date restrictions, yielded only six articles, none of which had anything to do with the Chengjiang fossils or ID. A Medline search on June 26, 2000, for "P. K. Chien" yielded only six articles, none of which was about the Chengjiang fossils or intelligent design. A combined search of Biological and Agricultural Index, Medline, and Zoological Record databases on June 26, 2000, for "P. Chien" (which also picked up anything by "P. K. Chien") yielded a total of forty-five articles, but none about either the Chengjiang fossils or intelligent design. A combined search of the Biological and Agricultural Index, Medline, and Zoological Record databases on June 26, 2000, using "Cambrian fossil" as the search term yielded a total of 128 articles, books, book chapters, and meeting papers, including those in different languages, *but not a single one was authored by Chien*. A combined search of Biological and Agricultural Index, Medline, and Zoological Record on June 26, 2000, using "Chengjiang" yielded fifty-five articles, book chapters, and meeting papers on the Chengjiang fossils, many in Chinese; but not one was even co-authored by Chien. A BIOSIS search on June 26, 2000, using "Chengjiang" and "Cambrian fossils" yielded a total of forty-seven articles, books (one), literature reviews, and what appeared to be a conference presentation; but none was authored by Chien. Needless to say, *none of the numerous references found in all these searches combined was about intelligent design*. Table 4.1 summarizes the findings from the 2000 searches.

The initial search of the databases was conducted at a time when Chien should have had ample time to produce at least one authentic paper on the Chengjiang fossils (or on their relationship to ID). A second search of the scientific databases in October 2001, like the June 2000 search, revealed no relevant publications by Chien. To date, therefore, Chien has published *nothing* in scientific journals on the Chengjiang fossils. Yet he is cited by DeWolf, Meyer, and DeForrest in note 23 of "Intelligent Design in Public School Curricula: A Legal Guidebook" as an authority on the Chengjiang fossils and referred to as a "marine paleobiologist." This citation is for a single article, co-authored with CRSC fellows Stephen Meyer, Marcus R. Ross, Paul A. Nelson, and ID disciple John Wiester, in an "expected publication by Michigan State University Press" edited by CRSC fellow John Angus Campbell, entitled *Intelligent Design, Darwinism, and the Philosophy of Public Education*.[48] This "expected publication" is not likely to have run the gauntlet of peer review by qualified paleontologists who have taken the trouble to judge, against expert contemporary standards, the scientific claims it makes. It is in-

Table 4.1. Summary of 2000 Literature Survey on Paul K. Chien

Keywords	Index	Date restrictions	Total references	Chengjiang fossil references	Intelligent design references
P. K. Chien	SciSearch	None Index covers 1988–present.	6	0	0
P. K. Chien	Medline	None	6	0	0
P. Chien, includes P. K. Chien	Biological and Agricultural Index and Medline and Zoological Record	None	45	0	0
Cambrian fossil	Biological and Agricultural Index and Medline and Zoological Record	None	128	0 Authored by Chien	0 Authored by Chien
Chengjiang	Biological and Agricultural Index and Medline and Zoological Record	None	55	0 Authored by Chien	0 Authored by Chien
Chengjiang and Cambrian fossils	BIOSIS	None	47	0 Authored by Chien	0 Authored by Chien

tended, almost certainly, as a celebration of ID, with, if possible, a few contributors who really are evolutionary biologists, in which the claims and debating techniques of CRSC regulars will be displayed juxtaposed to writings from a few recognizably mainstream scientists.[49]

Despite Chien's zero productivity in the area with which the CRSC has charged him, he *has* been active in promoting the publications of fellow creationists. According to the March/April 2000 newsletter of the American Scientific Affiliation, Chien helped get Phillip Johnson's *Darwin on Trial* and Michael Behe's *Darwin's Black Box* published in Chinese by the Chinese government, also serving as translator. The newsletter refers to "Paul Chien's new book":

> Chi-Hang Lee is excited about Paul Chien's new book, writing:
>
> Last week, excitedly I saw the newly published Chinese version of two books, both published by a government publishing house in Beijing. They are

Phil Johnson's *Darwin on Trial* and M. Behe's *Darwin's Black Box*. Both are in the official Chinese simplified script used in mainland China. These books resulted from Paul Chien's tireless efforts for several years. A team of four translators—among them three ASA members: Paul Chien, Pattle Pun, and Chi-Hang Lee—translated Johnson's book, which was previously published in the full (non-simplified) Chinese script. . . .

The Beijing version [of Johnson's book] has a new preface written by Chien, who hopes readers will keep an open mind. He especially mentions the fact that recent Yunnan fossils were headlined in the official Communist newspaper, the *People's Daily* on July 19, 1995, and that the Cambrian animal explosion "poses a challenge to traditional Darwinism."

Chien has become "quite an expert on these Yunnan Cambrian fossils; he has done research on site," Lee added. Of course, the Communist government press added their own disclaimer, saying that evolution is not compatible with the idea of creation, and warns the reader to question Phil's objectivity, since he admits in the book that he believes in a Creator!

To buy either of these books, contact Paul Chien by e-mail.[50]

The article cited by DeWolf, Meyer, and DeForrest and the translations of Johnson's and Behe's books appear to be, at present, the only publications Chien has produced that have any relation to the Chengjiang fossils. His work on these translations, while it does nothing useful as paleontology, is evidence of his desire to promote anti-Darwinism among the Chinese people, as he says quite forthrightly in the *Challenger* interview. This kind of work serves the CRSC's purpose of undermining evolution and advancing the cause of ID worldwide. The Wedge Document predicts, as part of the CRSC's Five Year Objectives, "an active design movement in Israel, the UK and other influential countries outside the US."[51] However, Chien's work does nothing, anywhere in the world, to provide genuine science in support of any such notion as intelligent design.

Now, Copernicus

Although not listed along with Chien and Axe in Phase I in the Wedge Document, Michael Behe is now the person whom the CRSC presents as its most formidable *biological* scientist. He has himself modestly compared the "discovery" of intelligent design—which, of course, includes his own contribution of "irreducible complexity" (IC)—with epochal discoveries of science in the past, that is, with "the assault on the senses . . . perpetrated by Copernicus and Galileo" and Einstein's theory of relativity. Now, according to Behe, "it is the turn of the fundamental science of life, modern biochemistry [Behe's discipline], to disturb."[52] His *Darwin's Black Box,* which the CRSC's Wedge Document lauds as having been published in paperback "after nine print runs in hard cover," is credited in the document with helping to increase the momentum of the Wedge strategy that began with Phillip Johnson's *Darwin on Trial*. CRSC cites his appearance on a 1997 *Firing Line* television debate in the Wedge Document as an example of its increasingly successful mass-audience public relations campaign. Behe also maintains a remarkably busy schedule of public appearances in behalf of intelligent design (and of his career as such).

In the CRSC's 1997 "Year End Update," Behe is named, along with Chien and Axe, as having made a presentation showing "the growing challenge to Darwinian biology" at its "Consultation on Intelligent Design." He is also cited for his participation in a CRSC dinner in Dallas for "ninety business, civic and academic leaders, and representatives from Christian non-profits." The update calls attention to the *Wall Street Journal* mention of "CRSC Fellow Michael Behe" in a December 24, 1997, article entitled "Science Resurrects God" and points out that

> Michael Behe and Phil Johnson have become inseparable from the increasing discussion of Intelligent Design throughout the world. In addition to the citations mentioned above, Behe and Johnson have recently been cited in newspapers from San Francisco, Portland, Minneapolis, Washington DC and Charleston, SC (among others). Internationally, they have appeared in print from Canada to France, (and Poland, where *Darwin on Trial* was recently translated). This attention followed the release of their respective books, and is an exciting foreshadowing of things to come.[53]

Yet the update does *not* provide any specific information about Behe's scientific accomplishments—likewise for Chien and Axe, despite their presence on the same 1997 program. The update *does* say that "a substantial amount of compelling evidence was presented at the meeting" and that "much of this scientific research cannot be continued without funding"—which is certainly the right thing to say in connection with fund-raising. But there is no indication of what said evidence *is*. The absence of any such information in a document intended to inform people of the progress of ID science is telling. Behe does indeed serve a vital purpose of the organization, but that purpose has not been, and is not yet, the production of new scientific knowledge.

An inspection of professional information about Behe on his departmental website at Lehigh University yields nothing that can reasonably be taken as research (other than, perhaps, some library research) on or supporting ID. On the Department of Biological Sciences home page, Behe is listed, along with others, under "faculty, research specialties, representative publication." His stated research specialty is "chromatin structure, nucleosome structure, DNA structure"; his "representative publication," which was posted with his research interests in June 2000, but which has since been removed, was "'Randomness or Design in Evolution' in *Ethics & Medics* 23, 3–4 (1998)."[54] *Ethics & Medics*, published by the National Catholic Bioethics Center, "makes every effort to publish articles consonant with the Magisterial teachings of the Catholic Church," although "views expressed are those of individual authors and may advance positions that have not yet been doctrinally settled."[55]

Behe's "Randomness or Design in Evolution" is a note on the problem of randomness for the Catholic view of evolution. He cites Cardinal Joseph Ratzinger's claim that "life is the result of intelligent purpose, not ontologically random events," noting that the cardinal "implies that biochemistry . . . provides particularly strong support for this view." Behe argues that "the critical question [for Catholics] is whether life is an unintended accident or the purposeful work of a Creator" and that, following Cardinal Ratzinger, Catholics must not hesitate to believe and to acknowledge publicly that "modern science has yielded physical evidence which points strongly to an intelligent designer."[56] The latter works, so long as the unhesitating need not specify what the "physical evidence" is and how it "points." So far, it has not been identified for them to acknowledge.

Behe's own faculty page is about his professional research, and here he says nothing about his work promoting ID. He has posted a picture of himself at what appears to be a bookstore, holding up a copy of *Darwin's Black Box,* but lists four representative publications, *none* about ID. He lists his favorite links, one of which is the creationist website, Access Research Network, where he maintains a schedule of speaking engagements.[57] Other than these very low-key references to what must therefore be interpreted as an avocation, there is nothing related to intelligent design in his professional postings.

On the Access Research Network site entitled "Michael J. Behe Online Articles," Behe has posted links to several articles, three of which appear at first glance to be scientific.[58] An examination reveals, however, that they are *not* articles that have come through peer review by the relevant competent scientific community:

1. "Experimental Support for Regarding Functional Classes of Proteins to Be Highly Isolated from Each Other." This article was not published in a scientific journal, but rather "presented at the 1992 SMU Symposium and printed in the conference proceedings of 'Darwinism, Science or Philosophy?'" The conference was one of the earliest ID gatherings, attended by Behe, Phillip Johnson, William Dembski, and Stephen Meyer (among others).[59] It was co-sponsored by the Foundation for Thought and Ethics, which sponsored the writing of and holds the copyright for the creationist textbook *Of Pandas and People.*[60] Seven years after its offering at the meeting, the paper was printed on June 7, 1999, in the "Books & Arts" section of *The Weekly Standard,*[61] where a doctrinaire, neo-conservative anti-Darwinism seems to be company policy.

2. "Molecular Machines: Experimental Support for the Design Inference." According to the copy of the paper posted on Access Research Network, Behe presented it "in the Summer of 1994 at the meeting of the C. S. Lewis Society, Cambridge University."[62] A printed version of a

speech, "Evidence for Design from Biochemistry" (which is also posted), delivered by Behe at DI's 1996 God and Culture conference, appears to be a version of this 1994 paper.[63]

3. "Histone Deletion Mutants Challenge the Molecular Clock Hypothesis." The link with this title on Behe's online article page at Access Research Network is not actually to the article itself (which appears in the October 1990 issue of *Trends in Biochemical Sciences*), but rather to an anonymous, unreferenced review of the article.[64] Moreover, even though Behe's article in this journal is only two pages long and therefore not a book of any kind, the review is incorrectly labeled "Book Reviews." None of Behe's references in the *Trends* article itself is to his own work. He recounts the work of others *and never mentions intelligent design*. Behe may have intended for there to be implied relevance to ID, since he refers to the very old work of E. Zuckerkandl and L. Pauling in their *Evolutionary Divergence and Convergence in Proteins* (Academic Press, 1965), which stated a version of the "molecular clock hypothesis" to the effect that "random mutations in the genomes of all species are fixed at a rate that is constant, with respect to absolute time (as opposed to species generation time)."[65] Behe then cites work of others that supposedly casts doubt on this hypothesis.[66] This is the only article referred to on Behe's ARN page that is actually printed in a scientific journal. But it is twelve years old—antediluvian in contemporary molecular evolution research—and nothing more recent is posted.

Dr. Thomas Wheeler, biochemist at the University of Louisville School of Medicine, points out that the subject of Behe's article does *not* support intelligent design:

> As summarized in the web site, the article points out a problem with histones [unusual, smallish, and very ancient proteins that are tightly associated with the DNA in eukaryotic chromosomes] as a molecular clock. Certain regions are highly conserved in sequence (suggesting critical function), but can be deleted in yeast and still leave viable organisms. . . . The article has nothing to do with ID (*it seems it could be used as an argument against design: if these regions are not essential in yeast, why are they there with almost the same sequences as in humans?*). I can see two reasons this article might be [referred to] on the ARN site. First, it is a publication of Behe in a scientific journal (though not an original research paper) and thus lends credibility to him as a spokesman for ID. Second, since it casts doubt on one tool used in analyzing data related to evolution, maybe the reader is supposed to think that it casts doubt on evolution itself.[67] (emphasis added)

A survey of the literature shows that Behe, like Chien and Axe, has to date published no scientific, peer-reviewed research on any aspect of ID in any scientific journal. *Darwin's Black Box*, published by a mainstream company, The Free Press, and aimed at a popular audience, has been reviewed and criticized exhaustively by scientists; but the new ideas in the book have not been published in a scientific journal because, as most of those reviews show, the key arguments for "irreducible complexity" (hence for "intelligent" design of that complexity) are bad science or not science at all. (We will return to this.)

A May 2000 SciSearch survey using "Behe, M." yielded ten articles in scientific journals, none about intelligent design. The only titles attributed to Behe referring either to evolution or to ID were five letters: "Embryology and Evolution," *Science*, 1998, vol. 281, no. 5375; "Defining Evolution," *Scientist*, 1997, vol. 11, no. 22; "Defining Evolution," *Scientist*, vol. 11, no. 12; "Darwinism and Design," *Trends in Ecology and Evolution*, 1997, vol. 12, no. 6; and "Understanding Evolution," *Science*, 1991, vol. 253, no. 5023. Not one of these five letters is more than a page long. A new search in October 2001 of the Medline and SciSearch databases for articles on ID (which would have picked up anything by Behe) revealed no new scientific publications by Behe (or anyone else) on either ID or IC.

Behe did write a response to an article in the July 1999 issue of *Philosophy of Science*, "Redundant Complexity: A Critical Analysis of Intelligent Design in Biochemistry," by Niall Shanks and Karl H. Joplin.[68] This response, "Self-Organization and Irreducibly Complex Systems: A Reply to Shanks and Joplin," appeared in the March 2000 issue of *Philosophy of Science*. However, the only work of his own cited in the response is his popular book, *Darwin's Black Box*, published in 1996. He cites also fellow CRSC member William Dembski's *The Design Inference* (Cambridge University Press, 1998). He cites articles and books written by others, but none in scientific journals and bearing his own name—the only kind of citation appropriate for an intended technical rebuttal to the technical critique of Shanks and Joplin.

Darwin's Black Box, then, is the prime locus of his statement and promotion of IC. IC is a theoretical characteristic of a complex system, such that it cannot function unless *all* its parts are available and working from the beginning. Absence or loss of any part leaves the system nonfunctional. Such a system cannot, *by definition*, come into being gradually by slow accumulation of working parts. It is either there with all its parts and functioning, or it is not. Behe's homely example of this for laymen is a spring-operated mousetrap. But he claims that many or most of the critical pathways of chemical reaction and structure building *inside living cells* are irreducibly complex and therefore could not have arisen by Darwinian "gradualism." They must have been "designed."

Not unexpectedly, this claim has generated a tremendous and highly critical response from the scientific community and joy and praise among creationists and some political conservatives (including neo-conservatives), who for social or political reasons dislike everything to do with "Darwinism." Important responses to *Darwin's Black Box* have been written by scientists; one of the best known is Kenneth Miller's *Finding Darwin's God* (Cliff Street Books, 1999). Miller, who is professor of biology in the Department of Molecular Biology, Cell Biology, and Biochemistry at Brown University, also published a review of the book in *Creation/Evolution* (vol. 16, 36–40, 1996).[69] In addition to Miller's work, a comprehensive website devoted to Behe's views on ID has been constructed by Dr. David Ussery, associate research professor at the Center for Biological Sequence Analysis, Institute of Biotechnology, Technical University of Denmark. Ussery, too, has written critical reviews of Behe's work.[70] A well-known website by John Catalano, "Behe's Empty Box," provides abundant links to criticisms of IC by qualified scientists and philosophers, *as well as* comprehensive links to Behe's writings.

In his introduction to the pages he devotes to Behe, Catalano highlights the absence of scientific work by Behe himself: "Yes, Michael Behe is a scientist, but is 'Intelligent Design' science? If so, it will be the first science established without a single technical paper published for peer review, including *zero* by Behe himself. For some reason he has decided to completely bypass professional review and go directly to a Darwin-doubting public."[71] The Talk.Origins Archive has also compiled numerous links to criticisms of Behe's theory of IC, such as the devastating critique by Yale biologist Robert Dorit in *American Scientist* magazine.[72]

Despite the absence of convincing research and publication in support of ID by Behe, he maintains an energetic defense of it in the popular media, including articles, op-ed pieces, and interviews. He authored two op-eds in the *New York Times*: October 29, 1996, and August 13, 1999.[73] He was interviewed about *Darwin's Black Box* on the Wisconsin Public Radio program "To the Best of Our Knowledge" in January 1997.[74] Behe uses religious and creationist venues extensively to publicize his ideas on ID and evolution. In an article entitled "Dogmatic Darwinism" in *Crisis*, a Catholic magazine, he asserts that "Darwinism is dying of the same affliction that has killed other discarded theories—the progress of science itself."[75] The Access Research Network website markets *Irreducible Complexity: The Biochemical Challenge to Darwinian Theory*, which was "videotaped before a live audience at Princeton University" in 1997. This sounds like a university lecture under the auspices of Princeton University faculty, including its distinguished scientists; however, it appears to have been only a lecture Behe delivered at a gathering sponsored by Christian Leadership Ministries[76]—not at all the same thing. Access Research Network has also posted the "Molecular Machines Museum," a series of graphics claimed to illustrate IC.[77] Apologetics.org., a website maintained by Trinity College of Florida and the C. S. Lewis Society, sells a videotape entitled *Opening Darwin's Black Box: An Interview with Dr. Michael Behe*.[78] (C. S. Lewis, a foremost twentieth-century Christian apologist, was a literary scholar of Oxford University, later in life at Cambridge, who, after an early and intense conversion to Christianity, wrote prolifically in praise and defense of it throughout the remainder of his life.) This has been just a sampling of the pro-ID activities in which Behe engages.

Perhaps the most telling evidence that Behe does not intend to remedy the scientific deficiencies for which he is criticized consists of remarks he made at the April 2000 "Nature of Nature" conference at Baylor University, as reported by Glenn Morton, who attended the presentation: "It was sad to see Behe not answer a question seriously when he was asked what he would want science to do differently. (He was asked what he would do if he had control of all the funding. [Answer]: keep it himself. And then he did say that he wanted someone else to do research in a lab to support his theory. Why wouldn't he want to do his own research?)"[79]

In light of Behe's failure to publish anything about ID in scientific journals—publication that would gain him some measure of professional respect, if not agreement, in evolutionary and modern molecular biology—it is ironic that he issues this warning to proponents of "the theory of Darwinian molecular evolution": "'Publish or perish' is a proverb that academicians take seriously. If you do not publish your work for the rest of the community to evaluate, then you have no business in academia and, if you don't already have tenure, you will be banished."[80] The irony of this, from one who has neither published nor perished, is almost too heavy.

Copernican Counterclaims

Two counterclaims made on Dr. Behe's behalf must now be addressed directly. One is his own unwavering charge that the scientific community has produced no scientific results supporting the ability of "Darwinian mechanisms" (such as natural selection) alone to produce complex (that is, potentially "irreducibly complex") biochemical systems at the cellular level. He insists that there is *and can be* no plausible Darwinian explanation for the existence of complex biochemical systems such as metabolic pathways, the assembly of organelles such as cilia and flagellae, and the densely interlocked chemical steps of blood clotting. For Behe, "unexplained" means "unexplainable." The second counterclaim, made by an apparently pseudonymous defender, one "Mike Gene," is that Behe's work *has* influenced mainstream science through scientific publication.

NO DARWINIAN EXPLANATIONS FROM SCIENCE?

Behe accuses the scientific community itself of failing to publish anything that addresses adequately the evolution of molecular function at the subcellular level of organization. He has made such claims on at least four documented occasions over a period of years, and no doubt he has done so at other times in other places.

In 1994, in the paper presented to the C. S. Lewis Society at Cambridge University, Behe attacked evolutionary biology as follows:

> Other examples [besides cilia] of irreducible complexity abound, including aspects of protein transport, blood clotting, closed circular DNA, electron transport, the bacterial flagellum, telomeres, photosynthesis, transcription regulation, and much more. Examples of irreducible complexity can be found on virtually every page of a biochemistry textbook. But if these things cannot be explained by Darwinian evolution, how has the scientific community regarded these phenomena of the past forty years? A good place to look for an answer to that question is in the *Journal of Molecular Evolution*. . . . In a recent issue of *JME* there were published eleven articles; of these, all eleven were concerned simply with the analysis of protein or DNA sequences. None of the papers discussed detailed models for intermediates in the development of complex biomolecular structures. In the past ten years *JME* has published 886 papers. Of these, 95 discussed the chemical synthesis of molecules thought to be necessary for the origin of life, 44 proposed mathematical models to improve sequence analysis, 20 concerned the evolutionary implications of current structures, and 719 were analyses of protein or polynucleotide sequences. There were zero papers discussing detailed models for intermediates in the development of complex biomolecular structures. This is not a peculiarity of *JME*. No papers are to be found that discuss detailed models for intermediates in the development of complex biomolecular structures in the *Proceedings of the National Academy of Science*, *Nature*, *Science*, the *Journal of Molecular Biology* or, to my knowledge, any journal whatsoever.[81]

In 1996, at DI's God and Culture conference, Behe made the same charge, along with an assessment of what should happen to theories about which nothing is published in the scientific literature:

> No papers are to be found that discuss detailed models for intermediates in the development of complex biomolecular structures in the *Proceedings of the National Academy of Science*, *Nature*, *Science*, the *Journal of Molecular Biology* or, to my knowledge, any science journal whatsoever. . . . If a theory claims to be able to explain some phenomenon but does not even generate an attempt at an explanation, then it should be banished. *Despite comparing sequences, molecular evolution has never addressed the question of how complex structures came to be. In effect, the theory of Darwinian molecular evolution has not published, so it should perish.*[82] (emphasis added)

John Catalano, on his "Behe's Empty Box" website, points out that Behe makes these charges in *Darwin's Black Box* (1996): "There has never been a meeting, or a book, or a paper on details of the evolution of complex biochemical systems" (Behe, 179). But Catalano lists hundreds of scientific publications whose very existence refutes Behe's charges. In 1998, Behe made a similar claim in "Molecular Machines" (*Ethics & Medics*, 23(6), June 1998), where he migrates gracefully from the charge of *zero* publications in DBB to "only a handful" here: "Such systems [as cilia] are so recalcitrant to Darwinian explanations that few scientists even try to account for them. For example, in the past twenty years there have been approximately ten thousand papers published in scientific journals investigating how cilia work, yet only a handful that try, even in general terms, to understand how cilia might have been produced by gradual Darwinian processes. In reviews of my book by scientists the lack of Darwinian explanations for complex cellular systems has been freely admitted."[83]

Again, in June 2000, at the Design and its Critics conference at Concordia University in Mequon, Wisconsin, Behe accused the scientific community of failure to publish anything that could account for the evolution of a complex biochemical system. This accusation was recorded by Kenneth Miller, who was present at the conference. In a letter to the *Pratt Tribune* [Kansas], Miller relates this incident: "[Michael Behe] argued that evolution cannot explain the complexity of the cell, and offered as proof the contention that no scientific journal had ever published a detailed, step-by-step account of the evolution of a complex biochemical system. That argument sounded compelling, until the conference was presented with a list of published papers that had done exactly that."[84]

Catalano and Miller, among others, have produced lists of the publications that Behe says do not exist. Catalano's published list was last updated in February 2000. He observes that "[Behe's] claim is false since papers *do* in fact exist that attempt to flesh out the details of the evolution of various biochemical systems and structures."[85] Miller produced his list at the conference at which Behe made the charge. He has made this short list available for our study, with annotations relating them to Behe's claims and to a few of the specific questions the papers address.

> Question: Do partial biochemical systems have functions of their own that can be favored by natural selection?
>
> Yes. Behe contends that complex biochemical systems must be fully assembled before they can have function. However, this is not true. Musser and Chan in 1998 described the 6-part cytochrome C oxidase proton pump found in eukaryotic cells, and showed that parts of the complex consisting of as few as 2 of the 6 proteins could form a complex with full biochemical function. That function, of course, is a different one from the complete complex, but that's the point. Complex biochemical systems evolve by natural selection working on the functions of their component parts. See: SM Musser & SI Chan (1998) Evolution of the cytochrome C oxidase proton pump. J. Mol. Evol. 46: 508–520.
>
> Question: Has anyone published a paper showing the detailed, step-by-step evolution of a complex biochemical system?
>
> Yes. In the very same year that Michael Behe's book was published (1996) a research team published just such an explanation for a very complex biochemical system: The Krebs Cycle [one of the best known, most complex, and most important of the metabolic machines common to all aerobic cells, known also as the citric acid cycle]. There is no reason, of course, to fault Dr. Behe for not knowing about a paper that appeared after his book went to press. However, the appearance of this paper (and a 1999 follow-up) do demonstrate that his oft-stated claim that the literature is silent on the evolution of such systems is not correct. [See the following.]
>
> Meléndez-Hevia et al (1996) The puzzle of the Krebs citric acid cycle: Assembling the pieces of chemically feasible reactions, and opportunism in the design of metabolic pathways during evolution. J. Mol. Evol. 43: 293–303.

A more recent paper analyzing the way in which this complex pathway evolved appeared in 1999:

Huynen, Dandekar, and Bork (1999) Variation and evolution of the citric-acid cycle: A genomic perspective. Trends in Microbiology 7: 281.

Furthermore, these are not the only papers describing the way in which complex biochemical systems might have evolved. In 1998 a paper appeared describing a way in which some of the most fundamental have evolved. These authors wrote: "We describe a sequential (step by step) Darwinian model for the evolution of life from the late stages of the RNA world through to the emergence of eukaryotes and prokaryotes." [See the following.]

Poole, Jeffares, and Penny (1998) The Path from the RNA world. J. Mol. Evol. 46: 1–17.

Similar papers describing other complex biochemical systems are not at all difficult to find:

Alberti (1999) Evolution of the genetic code, protein synthesis, and nucleic acid replication Cell. Mol. Life Sci. 56: 85–93.

Zhang et al (2000) Evolution of the rodent eosinophil-associated RNase gene family by rapid gene sorting and positive selection. PNAS 97: 4701–4706.

Finally, one of the most interesting recent papers that bears on this claim is a summary of work describing the evolution of a complex, multi-step pathway that detoxifies a chemical pesticide (pentachlorophenol). As the author wrote: "Because PCP was only introduced into the environment in 1936 (Ref. 9) and is not known to be a natural product, it is likely that this pathway has been assembled during the past few decades. *Thus, we have a rare opportunity to observe the early events in the evolution of a new metabolic pathway*" [emphasis added].

SD Copley (2000) Evolution of a metabolic pathway for degradation of a toxic xenobiotic: The patchwork approach. Trends in Biochemical Sciences, 25: 261–265.[86]

This short list of scientific publications, Miller adds at the end, "is why the scientific community has been so thoroughly unimpressed by Dr. Behe's claims . . . because their refutation is almost too obvious." To be sure, some of the papers cited here by Professor Miller had not yet appeared when Behe made his charge *for the first time* as early as 1994. The fact that he *continues to make these accusations*, however, indicates either that (1) Behe has not reviewed the scientific literature seriously, (2) he has done so and does not recognize the significance of it, or (3) he is aware of the literature and ignores it in order to maintain his point in its purity for the nonspecialist, biochemically naïve audiences to which it is addressed. And of course, as every competent evolutionary biologist and geneticist knows, and as has been shown in the spate of reviews of Behe's book, considerations in the scientific literature of biochemical evolution extend back in time at least to the insights of the Nobel Prize–winning geneticist Herman J. Muller in the 1930s.[87] There is thus no excuse for a patently false claim that the literature of evolution is silent on these questions.

Shortly after the DAIC conference, Behe posted in rapid succession on the DI website (a safe enough venue) his responses to criticisms of "intelligent design theory" and to his failure to publish in the relevant scientific literature:

1. "'A True Acid Test'": Response to Ken Miller" (July 31, 2000);
2. "In Defense of the Irreducibility of the Blood Clotting Cascade: Response to Russell Doolittle, Ken Miller and Keith Robison," (July 31, 2000);
3. "A Mousetrap Defended: Response to Critics," (July 31, 2000);
4. "Philosophical Objections to Intelligent Design: Response to Critics," (July 31, 2000);
5. "Irreducible Complexity and the Evolutionary Literature: Response to Critics," (July 31, 2000); and
6. "Correspondence with Science Journals: Response to Critics Concerning Peer-Review," (August 2, 2000).[88]

His assiduous issuing of such ad hoc arguments serves only to make the absence of ordinary experimental or theoretical research on ID stand out against the claims of its power as a biological explanation. If ID were truly generating new scientific inquiry, which is what a decent scientific hypothesis must do, Behe would surely identify it among his professional activities at Lehigh University; he could list it among his research interests on the faculty website instead of relegating it to an obscure creationist website like Access Research Network. It surely is not the case, after all, that his colleagues at Lehigh are in the dark about what he does and says about science when away from the campus on a speaking engagement. (Nor is it likely, academics being what they are, that most of them care much.) He could apply for research support to the National Science Foundation, in the competition for funds in which all serious scientists must engage. The foundation would be bound by its rules to present his arguments to scientific peers for review, and Behe would be informed in writing, by the rules, of how that review had gone and of what was said in the course of it. Such a review would be indispensable and actionable hard evidence of bias—if there were any bias detectable. Behe has said that "what I'm really eager to do is write grant proposals to do research on some of the ideas I have as a result of intelligent design theory."[89] *That was in 1996.*

In a December 2001 article in *The Chronicle of Higher Education,* in which ID was accorded a high level of visibility in the cover story, Behe as much as admitted that he has no intention of pursuing through formal scientific channels the Wedge's announced goal of scientific credibility. Kenneth Miller, also interviewed in the article as a leading ID critic, pointed out that as a member of the American Society for Biochemistry and Molecular Biology (ASBMB), Behe has a right to present his ideas at ASBMB's annual conference: "If I thought I had an idea that would completely revolutionize cell biology in the same way that Professor Behe thinks he has an idea that would revolutionize biochemistry . . . I would be talking about that idea at every single meeting I could possibly get to." Indeed, the ASBMB web page states that "the ASBMB Meetings Committee oversees the structure of the meetings and assures that the scientific interests *of the entire Society membership* are included in the program" (emphasis added).[90] In response, however, Behe replied, "I just don't think that large scientific meetings are effective forums for presenting these ideas."[91]

But Behe knows the peer-review process. After all, he *does* have legitimate science publications in peer-reviewed journals. An August 2002 search of the *Biological Abstracts* and *Zoological Record* databases revealed that he has published at least five articles—one in 1996, the same year as *Darwin's Black Box*—in *Biochemical and Biophysical Research Communications, Journal of Molecular Biology, Nucleic Acids Research,* and *DNA Sequence*.[92] These articles are proof that he has not suffered from discrimination. His work has been rewarded with publication when it is genuinely scientific, even though two of these publications appeared in 1997 and 1998, *after* he published *Darwin's Black Box*. The point here is that when Behe could have been writing and submitting for peer review any original ID research he might have done, he chose not to do so.

Behe is certainly aware of all the formal and informal ways of making his case to his peers. If he were truly a victim of entrenched Darwinism in the editorial systems of science journals, as he claims in "Correspondence with Science Journals: Response to Critics Concerning Peer Review," he could simply take advantage of CRSC's ample financial resources to send his ID findings directly, *as preprints*, to selected leading scientists in the field, enabling them to judge his work despite the "orthodoxy" he claims prevails "when a group (such as the editorial board) gets together." The cost would be trivial, and the necessary mailing lists are readily available. Yet he has done none of these things.

So, prodigious, prosy efforts in defense of ID notwithstanding, Behe's responses to his professional critics, published on the website of a political/religious think tank, do *not* qualify as scholarly scientific publication. There is finally, however, good reason to sympathize with his avoidance of the regular scientific literature. In it, the fate of one of his primary claims for irreducible complexity—that it is perfectly exemplified in complex subcellular machines such as the bacterial flagellum—would be sealed: outright rejection. Within the vast and relevant scientific literature, there are publications that Behe says do not exist but in fact *do*, with compelling evidence, for example, that the contemporary flagellum is descended from earlier, simpler systems, some of which were (or are still in some bacteria!) used for a completely different function. A demonstration of this, with current bibliography, has been provided online by Ian Musgrave.[93]

DR. BEHE'S "ENTRY" INTO THE PEER-REVIEWED SCIENTIFIC LITERATURE

Access Research Network, run by CRSC fellows and their supporters,[94] has an online discussion forum, the "ARN Intelligent Design Discussion Forum," in which there has been much discussion of Behe's concept of irreducible complexity, for which, incidentally, despite its mathematical sound, *he* has never attempted any formal or mathematical justification. That is left to William Dembski, apparently. And, of course, the core idea of irreducible complexity is not Behe's at all. It is original to traditional creationism and one of the most favored of all creationist arguments: that life is just too complex, is much too "improbable," to have arisen by any sort of "chance" process. Recently, a discussant (using the apparent pseudonym "Mike Gene") reported, with happy excitement, "It's official. Behe's concept of irreducible complexity (IC) has found itself in the peer-reviewed scientific literature" and "is being taken seriously by scientists."[95] This announcement refers to an article by Richard H. Thornhill and David W. Ussery, "A Classification of Possible Routes of Darwinian Evolution" (*Journal of Theoretical Biology*, 203, 2000), which addresses directly Behe's claim that natural selection cannot operate at the cellular level and that therefore subcellular structures and systems are irreducibly complex.[96]

Mr. "Gene," in a display of argumentative ingenuity, represents the Thornhill-Ussery paper as *supportive* of IC: "First of all, this article shows that Behe's work has indeed contributed to science. . . . Behe's skepticism [about the possibility of molecular specificity arising via natural selection] has served as an impetus for these scientists to develop a classification that did not exist before. Therefore, Behe has indeed contributed in an indirect way by serving as the stimulus for the creation of such a classification." Gene reinforces this discovery by asserting that "Behe has contributed to science by forcing non-teleologists to once-and-for-all lay the various Darwinian pathways on the table. This is progress as we can now look to the data to determine if there is any evidence that these pathways apply to an [*sic*] particular IC system in question. Behe's notion of IC does indeed help us to effectively rule out some of the Darwinian pathways, as admitted by T & U. . . . Without realizing it, T & U have made a significant contribution to ID."[97]

The problem with this claim is that IC has "found itself in the peer-reviewed scientific literature," *but only in the form of sharp dismissals.* It is not in the scientific literature because of any positive or noteworthy scientific results or theories offered by Behe. The Thornhill and Ussery citations of Behe's work are solely to *Darwin's Black Box* and his responses (on the Access Research Network website) to postings on the Talk.Origins Network. The Thornhill-Ussery paper addresses and refutes Behe's thesis by offering "possible routes of Darwinian evolution" of complex, subcellular structures:

> It was recently suggested that many biological structures are irreducibly complex, and therefore inaccessible by Darwinian evolution. Thus far, this is merely a restatement of the (fallacious) popular creationist argument about organs such as the eye. However, the new departure was to argue that the components of biochemical systems, unlike those of supramolecular structure, are single molecules, which are often functionally indivisible. The conclusion was that irre-

ducibly complex structures of functionally indivisible components are inaccessible by Darwinian evolution. . . . (Behe, 1996a).

The above thesis is unsound, as it is not certain either that any biological structures are irreducibly complex, or that their component molecules are functionally indivisible (Coyne, 1996; Doolittle, 1997; Fulton, 1997; Ussery, 1999). . . .

One factor hampering examination of the accessibility of biological structures by Darwinian evolution is the absence of a classification of possible routes. A suggested classification is presented here.[98]

Given the nature of the Thornhill-Ussery paper and the absence of any scientific citations of Behe's work, Thornhill-Ussery cannot be interpreted as strengthening Behe's concept of IC or as recognizing any merit in his argument for ID. Ussery (speaking only for himself) responds to Gene's comments:

Mike Gene writes:

"First of all, this article shows that Behe's work has indeed contributed to science."

My [Ussery's] reply—well, yes, but you have to realize that this is kind of in the context of something like my grandma told me something that encouraged me today and so she "contributed to science." "Contributing to science" and saying that Mike Behe's "Irreducible Complexity" (IC) is actually SCIENTIFIC are two very different things![99]

Mike Gene writes:

"Thus, before we go on, let's consider that despite all the expressed incredulity that is so common among Behe's critics, he has indeed contributed to science by forcing scientists to classify routes of evolution and by showing that 50% of the possible routes can't generate IC machines. This is progress. Without Behe, for example, many would probably still think that classic evidence of RM&NS [random mutation and natural selection] allows us to think the bacterial flagellum evolved by the same mechanism."

My reply—See my "grandma analogy" above. Yes, Mike Behe is absolutely right to point out that Darwinian gradualism can't fully explain the origins of complex systems—and that is indeed a "contribution"—but this does not mean that his conclusion—"Intelligent Design"—is scientific. The whole point of the article is that there are many other valid scientific options to describe this observation (for that matter, you could claim that the world sits on the back of a turtle as well—this is an "option" . . . not really very satisfactory from a scientific point of view, but about as valid as saying that the IC systems point to the 19th-century Victorian view of God creating the world as a complex, designed machine with humans as the pinnacle of this wonderful creation, and that this was done in six 24-hour days). . . .

Mike Gene says:

"Behe has contributed to science by forcing non-teleologists to once-and-for-all lay the various Darwinian pathways on the table."

My reply—See "grandma's analogy", above. Yes, Behe is right that gradualism can't explain everything—but that doesn't mean IC should be admitted as an equally viable alternative.[100]

Ussery rejects Gene's identification of Behe's work as a contribution to science. Articles published outside the scientific literature—on intelligent design generally and irreducible complexity specifically, with neither compelling theory nor data to support them—simply do not qualify. Judgments of the quality and suggestiveness of such extra-scientific publications seem normally to be inversely proportional to the scientific sophistication of the judging audience. The way in which one's ideas influence science does matter, as Ussery points out: "[W]e *did* indeed cite Mike Behe's book, and so now we have actually 'helped ID' get into the scientific literature. (?) All I have to say is that if someone thinks getting trashed is good recognition, then they don't understand science. Maybe 'getting your name out' is good in politics, but one's reputation is important in science, and to be debunked does not mean that your idea was necessarily 'scientific.'"[101]

Articles such as the Thornhill-Ussery paper, which (1) is published in the *Journal of Theoretical Biology*, (2) offers possible routes of Darwinian evolution in a venue where it may be read and evaluated by other scientists, and (3) is supported by the work of other scientists who have conducted original research and published it in scientific journals, *do* qualify as contributions, even if they generate dispute. Behe has not presented and defended in the scientific literature his hypothesis of IC and its incompatibility with "Darwinism"; however, scientists *are* using that venue—the only legitimate one—to refute it.

Finally, there are quite fundamental flaws in Behe's recycled version of the argument from design. Perhaps the most important is that it takes no proper account (and continues, in his unending "refutations" of his critics, to take no proper account) of the common DNA replication process with which, given his stated research specialties, he must certainly be familiar—gene duplication. The genomes of animals (including humans) carry huge quantities of a kind of DNA of no function or unknown function. This is clearly a graveyard of slavishly self-replicating DNA, of evolutionary accidents, residing and going along for the ride in every cell nucleus; it is referred to familiarly as "junk DNA." Whether or not all of it is truly junk, a large part of it clearly consists of what were once copies of *functioning* genes. Most of the now altered, imperfect, inactive copies—"pseudogenes"—have lost the original function or all function whatsoever through the accumulation over time of deleterious mutations.

But does gene duplication lead *only* to accumulating junk? No, not at all. Some kinds of duplicate (or multiplicate) genes are indeed used—very much so, as students learn in elementary molecular genetics. There are striking examples of such use in the literatures of comparative biochemistry, developmental biology, and molecular biology, one example being the structures and functions of the *family* of globin genes. These genes contain the codes for the family of closely related proteins that carry oxygen in the blood from the lungs to the rest of the body, where it is given up and used in metabolism. This is a textbook case of increasing complexity in a biochemical system, in the course of Earth history, via gradual incorporation of parts that were not there at the start. It is not, and cannot be, irreducibly complex in any biologically meaningful sense.

The chromosomal mechanics and the chemistry of gene duplication are well understood. Gene duplication is a frequent accident of normal DNA replication.[102] The presence in the cell of duplicates of functioning genes allows for several possibilities. One is that the duplicates go on accumulating errors (mutations), stop working altogether, and, after long intervals, are lost. But another is that simply because they are *there* for a long time and accumulate changes, they acquire, via the opportunism of somatic existence and natural selection, new functions, which are either supportive of or supplementary to the original function. Or they acquire altogether different and newly useful functions in other systems. There are plenty of examples of single proteins with *multiple* functions, including one of the enzymes (thrombin) of the blood-clotting cascade upon which Behe focuses his claims of IC.

Now suppose that one or more *supplementary* functions have appeared in this way, through duplication of the gene for an original key protein in, say, a biochemical pathway or an extended subcellular structure (such as a cilium or flagellum). Suppose that in the course of time, the *original* gene and its protein products are lost due to mutation. The original, complete function (or structure) may then be lost; that mutation will then be detrimental or lethal. But the original complete function *need not* be lost. It might be slowed. Or it may be otherwise changed (becoming, perhaps, a bit more efficient) because the product(s) of the once "supplementary," duplicate gene(s) turn out to be capable of taking over the function(s) of the original product. This is not snatching possibilities from thin air. Redundancy of protein function is common in higher animals and plants. Then, if over time there are multiple supplements, *the whole process becomes more complex*—it evolves. It gains information. Not only is this sort of scheme (with its many possible variants) plausible, but there is robust evidence of it, even in some of the systems Behe insists could not have arisen cumulatively, over time, by natural selection. All this, too, is in the literature; most has been there for a long time.[103] Evolutionist Jerry Coyne puts the matter in a nutshell:

> [B]iochemical pathways did not evolve by the sequential addition of steps to pathways that became functional *only at the end*. Instead, they have been rigged up with pieces co-opted from other pathways, duplicated genes, and early multifunctional enzymes. Thrombin, for example, is one of the key proteins in blood clotting, but also acts in cell division, and is related to the digestive enzyme trypsin. (emphasis added)[104]

A simple logical flaw in the irreducible molecular complexity/intelligent design argument is its assumption that the evolution and embryonic development of macroscopic, multicellular tissues and organs (e.g., glands, eyes, brains) are relatively simple and divisible—by comparison with the evolution or assembly of subcellular, molecular processes or structures. So, without *quite* conceding that evolution through natural selection and recombination actually work anywhere at all, Behe admits that the standard accounts of evolution at least *within* major groups of plants and animals (the neo-creationist's "microevolution") are probably correct. Simply put, he admits by implication, in *Darwin's Black Box*, that Darwinism is an acceptable theory for some but perhaps not all of evolution. However that term is used, and it is used differently in different venues (see chapter 5), "microevolution" refers to a vast range of phenomena: from the rapid acquisition of drug resistance in viruses, bacteria, and uni- and multicellular parasites, to the appearance of tolerance among insects to new pesticides, to the comprehensive remodeling of foot, leg, and skull bones in the lineages of higher vertebrates. But, Behe's argument implies, the making of macroscopic structures, even the course of "microevolution" as he and other creationists define it, is a *simple* process compared with the complexity of building a biochemical pathway in evolutionary time, which process, he says, must be irreducibly complex. Therefore it is the product of an intelligent designer. That is his claimed Copernican breakthrough.

If the evolution of macroscopic structures *is* simple compared to the evolution of biochemical pathways, Behe has nowhere shown it. He has not counted the steps that must be accomplished simultaneously in the embryological emergence of some such "simple" structure as the vertebrate (or the squid's, or the insect's) eye, with its neural connections in the brain. "Complexity" means different things to different experts in the mathematical sciences of complexity,[105] but nowhere in Behe's book and its follow-ups does he take the trouble to define complexity any more precisely than it is used in colloquial speech. Thus, he means by it the opposite of "simplicity," period. But since complexity in this usage sooner or later means number of different component elements, such counting would be essential if one were trying to compare the complexity of a macroscopic process such as eye-building with the complexity of, say, the Krebs cycle or blood clotting or cilia-making inside a single cell.

Behe does count steps—the chemical reactions—in some complex biochemical processes—the focus of his IC argument. But any competent embryologist can do likewise with development of the eye, the heart, or the brain. The number of steps or elements in Behe's examples of specific biochemical processes inside the cell is a hundred or fewer for each process. The number of steps that must occur simultaneously or sequentially, and interactively, to make an eye is *vastly* larger. For one thing, there are usually involved in it hundreds to multiple thousands to millions of individual cells (*each* with its own complex subcellular systems), with many subpopulations of those cells *doing different things* simultaneously, at different locations in the embryo. And the whole process differs quantitatively and in the details of outcome for every species. If there is such a thing as "irreducible complexity" for subcellular objects, then there is necessarily irreducible complexity, raised to a higher power, in the uncountable details of the processes of embryonic determination, induction, differentiation, and cell-to-cell communications, in the formation and function of morphogenetic fields, by which the bodies of plants and animals are constructed.[106] So if, according to Behe's reasoning, biological structures below the level of the whole cell are inexplicable by Darwinian evolution, then any "microevolution" of multicellular life above that level is all the more inexplicable. Evolution by natural selection and other "Darwinian" processes must be impossible at every level.

There is solid, unequivocal evidence for physically and temporally intermediate structures and functions at all levels of plant and animal anatomy, hence for their descent with modification—their evolution. IDers cannot have it both ways: they cannot concede, as Behe and most of the others do, that what they call "microevolution" is (more or less) acceptable, but that processes within the individual cell cannot have evolved. Again, Coyne puts it crisply:

> [T]here is ample evidence for the evolution of morphology and anatomy from studies of paleontology, embryology, biogeography, and vestigial organs. Such evolution must, of course, be based on the evolution of molecules and biochemical pathways. Second, we have plenty of evidence for the evolution of molecules. This includes the remarkable congruence between phylogenies based on anatomy and those based on DNA or protein sequence.[107]

"Irreducible complexity," as ID advocates use it, is therefore an empty formalism as applied to the history of life. Without specification of the origins *and history* of all parts of a structure or system claimed to be "irreducibly complex," one cannot know that it *is* irreducibly complex. A claim of irreducible complexity amounts to saying that there cannot exist, or ever have existed, *simpler* forms of a contemporary cellular pathway or structure. But that is clearly not true. Simpler forms of complex structures in our cells can and do exist in the cells of other creatures. Behe's claim of the irreducible complexity of subcellular machines assumes, incorrectly, that their parts have always been exactly as they are now, having always served the same purposes in the living cell. Even further, it assumes that all of them were planned and in place from the beginning of their existence ("creation"), already encoded in the very first living cell, to work together—mostly in the distant future, even 3.5 billion or more years in the future! The very first cells on earth would have had to contain at least all the genes, if not all the DNA, of your cells and mine! That is clearly not the case; there are not nearly enough genes in the simplest of existing cells (a few hundred). Even if Behe's descriptions of cell chemistry are sound, Behe's positions on its meaning and evolutionary implications make no sense at all, scientific or otherwise.

At this writing, IC has suffered still another check. After defining IC in *Darwin's Black Box* as "a single system composed of several well-matched, interacting parts that contribute to the basic function, wherein the removal of any one of the parts causes the system to effectively cease functioning," Behe has had to admit a fundamental error in his definition. He argues that an IC system cannot be produced by "slight, successive modifications of a precursor system [by natural selection], because any precursor to an irreducibly complex system that is missing a part is by definition nonfunctional" (39). Robert Pennock shows in *Tower of Babel* that Behe cannot derive the *empirical* conclusion he reaches from his *conceptual* premise that precludes *by definition*—that is, *in principle*—the functionality of a supposedly irreducibly complex system from which a part is removed. To show Behe's error, Pennock provides the example of a clock that fulfills Behe's definition but that nevertheless can function (in a different way) despite losing a part; this makes the clock *without* the part a functional precursor to the clock *with* the part and thus a possible stepping-stone for natural selection, in just the manner Darwin proposed. Granting Pennock's counter-example, Behe admits in his 2001 *Biology and Philosophy* article that "there is an asymmetry between my current definition of irreducible complexity and the task facing natural selection. I hope to repair this defect in future work."[108] Two years later, he has not yet done so.

Meanwhile, of course, science progresses. While creationists announce that one or another feature of life is inexplicable in principle, or simply not demonstrable by natural science, the features of life continue to be explained and so demonstrated. Pennock himself has provided a follow-up to his earlier criticism of Behe. Along with Richard E. Lenski, Charles Ofria, and Christoph Adami, Pennock has published an article in *Nature* that models and demonstrates with astonishing effectiveness the evolution of an IC system—in this case in digital organisms. Moreover, comparative molecular biology has dealt yet another blow to IC: a very recent discovery in the physiology of cyanobacteria (blue-green algae) represents beautifully a fully "Darwinian" pathway to a physiological function that is eventually dependent upon an apparently "IC" molecular machine. This is the device that controls the circadian rhythm—the "biological clock"—of these ancient and still ubiquitous microorganisms. Volodymyr Dvornyk, Oxana Vinogradova, and Eviatar Nevo have shown convincingly how the three-gene circadian control system in the modern prokaryotes has "evolved in parallel with the geological history of the earth, and that natural selection, multiple lateral transfers, and gene duplications and losses have been the major factors shaping . . . [that] evolution."[109]

5

A Conspiracy Hunter and a Newton

Father's [the Reverend Sun Myung Moon's] words, my studies, and my prayers convinced me that I should devote my life to destroying Darwinism, just as many of my fellow Unificationists had already devoted their lives to destroying Marxism. When Father chose me (along with about a dozen other seminary graduates) to enter a Ph.D. program in 1978, I welcomed the opportunity to prepare myself for battle.

Jonathan Wells, "Why I Went for a Second Ph.D.," Unification Church Sermon

What intelligent design is attempting to advance now is the positive research program; a program of positive vision for biology which incorporates legitimate insights from Darwinism throughout the sciences.

William Dembski, Templeton *Research News*, November 2001

My thesis is that the disciplines find their completion in Christ and cannot be properly understood apart from Christ. . . . The point to understand here is that Christ is never an addendum to a scientific theory but always a completion.

William Dembski, *Intelligent Design: The Bridge Between Science and Theology*

Despite sponsorship of such tracts as the ironically titled book *Mere Creation*,[1] spokesmen for the intelligent design movement bridle at being called "creationists." They insist that they are not creationists or, rarely but more honestly, that they are not creationists of the biblical-literalist kind. They distance their movement from scriptural literalism and from antique forms of the argument from design, holding adamantly that their version, "intelligent design," is *not* introspection or revelation, *not* theology or apologetics, but simply excellent new science. And they announce that the new knowledge, obtained by means of (their) researches on "design" and informed analysis of naturalistic science, proves that contemporary evolutionary biology is a house of cards, a system of evasions. Good science, their science, makes it clear that neo-Darwinism cannot explain life—not even its structures, much less its history. Darwinism is therefore a materialist "creation myth." On the other hand, the discovery of intelligent design, they insist, demonstrates that the physical world, including its precious cargo of life, must be the work of a maker. Handiwork implies a hand: agency, *telos*—a purpose. This they want everyone, especially children, to understand.

Their argument has failed utterly to impress the overwhelming majority of the world's scientists. But the Wedge strategy does not require that *scientists* be impressed. It does depend critically, however, on convincing the public that mainstream science *is* impressed by ID (but running scared, avoiding debate). What the Wedge public relations campaign and the writings of its executors must accomplish is to convince the general public, including especially the political powers, that ID is a body of solid scientific research, sufficiently large and growing, with enough qualified practitioners and adherents to qualify as a credible and intellectually significant alternative to Darwinism. If it is anything less than that, a reasonable public, even a public the majority of whose members are theists, may (and certainly should) resist giving ID "equal time" or, indeed, any time at all in the already overburdened school *science* curriculum. ID theory must be treated as at least the equal of mainstream science. Otherwise the Wedge fails to penetrate.

We have already seen that this early and often-announced eruption of new science demonstrating ID has been, so far, nothing like volcanic. The Discovery Institute's continuing fuss over the Chengjiang fossils and the Cambrian "explosion" has nothing to do with the real scientific questions about that geological era; nor is there anything in the hyperbolic "Biological Big Bang" of the Cambrian period that poses any legitimate challenge to standard evolutionary biology as a whole. ID advocates have not raised substantive scientific arguments about the Precambrian and Cambrian biota—although there is plenty to argue about, as in all current and active science. Rather, the real scientific arguments are all technical disputes among Darwinists, who are concerned not with *whether* those biota had evolved, but with exactly *how and at what rates* they did evolve, from predecessors in earlier periods of Earth's history.[2]

The writings of Michael Behe, taking the debater's affirmative for the irreducible complexity of life and the consequent necessity of a designer, are influential, but for reasons entirely apart from their scientific content. In fact, although it contains well-written popularizations of some facts of elementary cell biology and biochemistry, Behe's book, *Darwin's Black Box,* contains no new science; and he has reported no new science of his own in this area since its publication. The main conclusions he draws from the standard science he uses in his version of irreducible complexity, as we have seen and continue to see in Behe's attempts to discredit expert critics, are tendentious. The task of rescuing the argument from oblivion has now been taken up, not by Behe, but by his "design theorist" colleague, William Dembski.

The researches on amino acid composition and the genetic encoding of proteins and protein structure, and the constraints imposed by catalytic function—the work of Douglas Axe—are sound enough protein engineering; but they imply neither an obvious challenge to Darwinism nor any visible support for ID theory. Nor, as already noted (chapter 3), do Axe's publications in refereed journals, describing this work, make any reference to ID. Vague suggestions by other DI members, who are not themselves competent in molecular biology, that Axe's work generates painful questions for Darwinism (as in William Dembski's references to it, discussed later) are just whistling in the dark. In fact, the exponentially growing science of genomics offers precisely the opposite message: the genetics of protein structure and function has proved to be a rich new contribution to evolutionary biology.[3]

Yet the writings and speeches of Phillip Johnson, Paul Chien, and Michael Behe insinuate broadly and repeatedly that there is a conspiracy afoot among the world's biologists—that the syndics of science are covering up the crumbling of Darwinist foundations, and therefore the collapse of modern (naturalistic) biology, whose foundation is organic evolution. Sometimes this is suggested, slightly more generously, as testifying merely to a blind, doctrinaire naturalism, or materialism, of scientists. But the implied accusations of fraud are always there. At least one of the ID movement's purportedly scientific thinkers, Jonathan Wells, does not merely *suggest* scientific deceit—he claims instead to have laid it bare: fraud and active conspiracy in science, and indeed in the whole system of science education. His allegations, to which we shortly turn, are that the evidence for biological evolution taught in our schools and colleges is half-truths and outright lies. Wells is therefore the CRSC's full-time conspiracy theorist. His theory is said by promoters to be grounded in expert knowledge of biology (in which he received a second Ph.D., his first having been in theology). But as we shall see, this expertise is less than conspicuous, even in his own field of interest—developmental biology. His widely marketed claims of scientific conspiracy have included to date *no* significant evidence of fraud and *no* foundational troubles for evolutionary biology.

Most, but not all, of the stars of ID train their rhetorical fire on what they portray as the stiff necks and the cover-up proclivities of establishment scientists and sometimes on the accommodationist theistic evolutionists who are content with both creation and evolution. One of these stars, fast becoming the intellectual protagonist of the movement, with print output exceeding even that of the redoubtable Phillip Johnson, is William Dembski—Christian apologist and mathematician. Broadcast charges of fraud and conspiracy are not a part of Dembski's role, and that sets him apart. His is, at least on the surface, a more decorous method, enough to have led to a book publication by a normally reputable university press.[4] He claims to have proven by mathematical means (the "positive research program" referred to in our epigraph) that certain natural phenomena are inexplicable by natural law or by chance alone and must therefore be the results of design. Dembski believes that for explaining human origin and that of our physical universe, only the ID alternative fits.

For this argument, one of Dembski's academic admirers (who is neither a mathematician nor a physicist) has dubbed him "the Isaac Newton of information theory."[5] In this chapter, we complete our sketch of the ostensibly scientific work generated to date by ID theorists with synopses of the main claims of Drs. Wells and Dembski (and a brief but necessary recollection of Michael Behe) and point to rebuttals of their claims by experts. All deserve more comprehensive scrutiny than can be given here. But Wells and Dembski have already received such scrutiny elsewhere, in print and online, at the hands of many accomplished reviewers, some of whom we cite.

For our purpose the sketches that follow suffice to show that ID has not sent mainstream science cowering in retreat and that, by strong scientific consensus, ID and its conspiracy mongering are *not* a viable or honestly competing alternative to contemporary evolutionary biology. The works of Wedge members and their admirers do not now offer a body of respectable opposing science, equal in merit to the mainstream consensus and equal in the scientific accomplishments of its proponents.

Jonathan Wells, Biologist

A "Science Lesson"

In a sermon expressing appreciation for the encouragement and support of "Father," the Reverend Moon, Unification Church theologian Jonathan Wells explains his early conviction that "I should devote my life to destroying Darwinism." This is the reason for his pursuit of a Ph.D. in biology at Berkeley *following* his advanced theological studies. Wells offers readers of the sermon a short but complete statement of purportedly scientific reasons for concluding that the standard biological account of life's history on Earth and of the mechanisms of animal development is false and deceptive. This statement is crucial: it is the central part of the sermon that provides a candid portrait of the author's thought and motivation.

It is a remarkable document, more so even than Wells's book, *Icons of Evolution*, whose purpose is to convince the public that science education today, especially in biology, indoctrinates children in a dishonest materialist myth: Darwinism. Like all sermons, this is a religious affirmation, meant for the viewers of Wells's Internet website, comprising presumably many members and friends of the Rev. Moon's Unification Church, as well as persons sympathetic to the activities of the CRSC and ID theorists. Nevertheless, it does include that indispensable section on science, making it also a *science lesson*. It teaches the main ideas of evolutionary science—as he wants the science understood. Wells's sermon is the conscious effort of a nominal specialist in developmental and evolutionary biology to reduce to the simplest, clearest terms the "failure" of those disciplines. He is telling his readers that a major discipline within modern evolutionary biology—embryology—is wrong, thus condemning both as fatally flawed. And this same intention informs everything else Wells has done, qua scientist and as author of *Icons*. We must therefore evaluate the sermon-lesson's scientific merit, starting with its key scientific assertions:

> According to the standard view, . . . development of an embryo is programmed by . . . its DNA. Change the genes . . . change the embryo, even to the point of making a new species. . . . Experiments similar to this [the "experiments" referred to in the movie *Jurassic Park*] have actually been performed. . . .
>
> In every case, if any development occurred at all it followed the pattern of the egg, not the . . . foreign DNA. . . . My [own] experiments focused on . . . reorganization of the egg cytoplasm after fertilization which causes the embryo to elongate into a tadpole; if I blocked the reorganization, the result was a ball of belly cells; if I induced a second reorganization after the first, I could produce a two-headed tadpole. . . . [T]his reorganization had nothing to do with the egg's DNA, and proceeded quite well even in its absence. . . .
>
> In a developing organism, the DNA contains templates for producing proteins—the building materials.
>
> To a very limited extent, it also contains information about the order in which those proteins should be produced—assembly instructions. But it does not contain the basic floor plan. The floor plan and many of the assembly instructions reside elsewhere (nobody yet knows where). Since development of the embryo is not programmed by the DNA, the Darwinian view of evolution as the differential survival of DNA mutations misses the point. . . . Darwin's theory may explain "microevolution" . . . such as minor differences among closely related species of salamanders. But it cannot account for "macroevolution." . . . Darwin's theory is incompatible . . . with . . . embryology

> . . . [and with] the fossil record. According to Darwinism, all creatures are descended from a common ancestor. Yet the oldest fossils show that almost all . . . major groups . . . appeared at about the same time, fully formed and recognizably similar to their modern counterparts. Darwin's theory predicts a "branching tree" . . . fossil record, yet that . . . is nowhere to be found. The fossils provide no evidence that all creatures are descended from a common ancestor. So the two major claims of Darwinism . . . are unsupported by evidence.[6]

Let us now examine the main ideas:

1. Wells: *According to the standard view, . . . development of an embryo is programmed by . . . its DNA. Change the genes . . . change the embryo, even to the point of making a new species.* That is emphatically *not* the standard view, although it may be that of Hollywood and *Jurassic Park*. It is grossly oversimplified. Yes, the ultimate information resource for an embryo, as for its species, is the gene pool, which is constituted of DNA; but it is not the *only* and *immediate* source of information for development. The actual development of an individual plant or animal results from biochemical and physical interactions between its genome (complete set of genes) and a huge storehouse of materials, preformed structures, and prepared signals for embryo building. That storehouse is the *egg* or *ovum*, usually the largest by orders of magnitude of all the body's cell types. Embryonic development consists of the interactions of that whole developing system with the physical environment in which its processes take place. Much of modern (post-1960) embryology has been devoted to working out, formally and in great chemical detail, the gene-cytoplasm-embryo-environment interactions. So modern biology does *not* say that one can change an embryo merely by changing its genes. But this is not what Wells's misleading gloss tells the naïve reader.

2. Wells: *Experiments similar to this have actually been performed. . . . In every case, if any development occurred at all it followed the pattern of the egg, not the . . . foreign DNA.* Yes, of course. But so what? As we have just shown, modern biology does not claim that *all* of embryonic development is programmed by an embryo's own genome; therefore, Wells's point does not damage evolutionary theory. That sort of experiment (which Wells describes badly) has indeed been done, and the typical results in many species have been known for more than a century. There is nothing new about it, and nothing informative about the role of DNA in development. Unfertilized eggs can be "activated" artificially; if any sort of development subsequently happens, it is called "artificial parthenogenesis" ("parthenogenesis" means, literally, "virgin birth"). Even some unfertilized, *enucleate* eggs (eggs deprived of their nuclei and therefore of the nuclear DNA genes) can be activated artificially. In some few species (such as ants and bees), parthenogenetic development of the unfertilized egg (which normally contains the maternal pronucleus, with genes—DNA—solely from the mother) is the natural route of embryonic

development. In artificial (as opposed to natural) parthenogenesis, a certain amount of aberrant visible development usually follows this activation of the egg's metabolic machinery. It amounts typically to only a few cycles of cell division ("cleavage") or, more rarely, to conversion of the egg, a single huge cell, by continued cleavage into a sphere of many small, relatively undifferentiated cells. "Activation" of the egg need not be done with DNA: in the case of frog eggs, merely pricking the surface membrane with a (preferably dirty) microneedle may suffice. For unfertilized sea urchin eggs, as Jacques Loeb showed early in the last century, simply raising the salt concentration of the seawater into which eggs alone have been placed will cause parthenogenesis.

That is the *unsurprising* part of the story. The surprising and important part, the part that preoccupied molecular developmental biology from the early 1960s until the mid-1970s, is what goes on at the molecular level in early development, both naturally and artificially activated, for example in frog or sea urchin eggs—*the process that enables even such usually incomplete development in the artificially activated case*. Does all that happen without benefit of genes, without DNA? No. All the earliest protein synthesis—without which nothing happens, no development, no change—hence all the crucial chemistry of the earliest developmental events in an embryo—is directed by "morphogenetic substances." These chemical agents induce and guide the morphological, or structural, development of the organism. They are stored earlier in the egg during its long manufacture (oogenesis) in the maternal body. The critical morphogenetic substances include, among others, a large population of messenger RNA molecules, the *maternal messages*, which have been *transcribed* (copied into RNA) *from the egg's own genes*, thus playing a vital role in the embryo's development. The presence and function of these messages, even in the earliest moments of development, has been known since about 1964. Their information comes originally from the maternal DNA that encodes them.

Therefore, early development certainly "follow[s] the pattern of the egg" (Wells's words) because that pattern was *already determined* in the egg's DNA long before activation or fertilization by sperm. All this was known and accepted in developmental biology by the early 1960s, so there is today no excuse for a biologist's not knowing it or not acknowledging it.[7] Every standard textbook of developmental biology tells the story in detail.[8] Christiane Nusslein-Volhard and E. Wieschaus (with Edward Lewis) received a Nobel Prize in 1995 for working out the genetic details and the specific body plan consequences of these processes in early insect development; but the facts were known to developmental biologists long before that.[9]

3. Wells: *[T]his reorganization* [in Wells's frog embryo experiments] *had nothing to do with the egg's DNA, and proceeded quite well even in its absence.*

As we have shown, this statement is not only grossly misleading but false.

4. Wells: *To a very limited extent, it* [the DNA] *also contains information about the order in which those proteins should be produced—assembly instructions. But it does not contain the basic floor plan* [embryo's body plan]. *The floor plan and many of the assembly instructions reside elsewhere (nobody yet knows where). Since development of the embryo is not programmed by the DNA, the Darwinian view of evolution as the differential survival of DNA mutations misses the point.* No, it does not miss the point, but Wells's statement does miss it, and by a wide margin. Again, the mechanism by which the "floor plan" of the species is recreated in each developing embryo is one of interaction among genes, gene products such as RNA messages and proteins (usually DNA-binding proteins) that act on the genes, and the physical environment in which development proceeds. For some species, the genetic mechanisms have been elaborated in wonderful detail since the 1960s. And, again, there is no modern development textbook that does *not* explore this subject. For many but not all species, the floor plan is laid down in the maternal body as an asymmetric (uneven) distribution of the morphogenetic substances in different areas of the egg. That distribution starts off the whole cascade of body plan construction events. The asymmetric distribution in sea urchin eggs of their maternal RNA sequences, copied from the egg's DNA, was first demonstrated twenty-five years ago.[10] And, in any case, little or none of this may apply to the fertilized eggs of mammals (including ours), where direct control of development by the ovum's DNA begins much earlier than it does among invertebrates such as sea urchins and vertebrates such as frogs (amphibians). Nevertheless, it is now quite clear that the same broad patterns of gene expression, and sets of homologous genes (DNA!), including the now-famous homeotic ("Hox") genes, *control* steps in the emergence of body plan in mammals and in fruit flies.[11]

5. Wells: *Darwin's theory may explain "microevolution" . . . such as minor differences among closely related species of salamanders. But it cannot account for "macroevolution."* Here we finally see where Wells is leading his readers. As should be obvious even to a nonbiologist reader, nothing in the preceding or following sentences of Wells's argument has any *evidentiary* connection with this bald conclusion, which depends on a particular and, in this context, idiosyncratic use of "macroevolution"—a use popular with creationists. In pointing to the "large-scale differences" for which "Darwin's theory," natural selection, cannot account, Wells is in effect arguing against the descent of these animal groups from a common ancestor, implying that there are no real genetic relationships among them. These large differences are taken as evidence *against* their genetic kinship, thus opening the way for separate, special creation of these groups by a supernatural designer.

Embryology is steadily enhancing our understanding of common descent. Since DNA plays a vital role in embryonic development, the reason for Wells's earlier dismissal of the value of DNA in development is evident: it allows him to discredit embryology, one of the linchpins of evolutionary theory, as a source of evidence for common descent. Wells's denial of macroevolution is therefore in effect the flat denial of common descent, as he has made explicit elsewhere:

> The diversity of the earliest stages of development, here illustrated [referring to a diagram of the "cleavage stages" of a fish, a frog, a chicken, and a mouse] strictly within the vertebrates, provides one of the strongest challenges to the neo-Darwinian conception of homology and macroevolution. Given the hierarchical, step-wise logic or "architecture" of animal development, early stages such as cleavage and gastrulation lay the groundwork for all that follows. Body plan structures in the adult, for example, trace their cellular lineage to these early stages. Thus, if macroevolution is going to occur, it must begin in early development. Yet it is precisely here, in early development, that organisms are least tolerant of mutations. Furthermore, the adult homologies shared by these vertebrates commence at remarkably different points (e.g., cleavage patterns). *How then did these different starting points evolve from a common ancestor?* [12](emphasis added)

("Least tolerant" implies, of course, what is indeed fact: rare mutations do occur that *can* give rise in early development—rarely, to be sure—to new fundamental body plans.) Wells and other ID theorists persist in using "macroevolution" to stand for a mysterious process, distinct from and unrelated to mechanisms of "microevolution" (by which they may or may not mean "speciation," depending on the sophistication of the audience). They depict macroevolution as a mysterious process through which large changes of body plan have occurred (or, for creationists, have never occurred!) in nature. The appearance of such a range of body types, they insist, must be due to some form of high-level intervention (God's design work). Phillip Johnson incorporated this idea of a mysterious process into ID creationism early on, even before there was a Wedge movement, in "Evolution as Dogma: The Establishment of Naturalism": "If classical Darwinism isn't the explanation for macroevolution . . . there is only speculation as to what sort of alternative mechanisms might have been responsible. In science, as in other fields, you can't beat something with nothing, and so the Darwinist paradigm remains in place. . . . If God created the first organism, then how do we know he didn't do the same thing to produce all those animal groups that appear so suddenly in the Cambrian rocks?"[13]

For the lay reader, this is a slightly misleading use of "macroevolution." When thoughtful contemporary biologists use it (many do but some prefer not to use it), macroevolution can stand for species formation; more incisively, it stands for embryonic processes by which many coordinated developmental events change *in concert* (hence "macro") to redirect the course of development, thus to change the body plan. An example of this is the mass movements during gastrulation of cells and tissues, which are themselves controlled by regulatory genes (especially genes concerned in the properties of the embryo's "morphogenetic fields"). Macroevolution as used in contemporary biology has no particular implications of gene-independent, non-Darwinian processes and certainly no implications of mystery. It is a term of art, for example, in the important emerging discipline of evolutionary developmental biology ("evo-devo"). A full account of the meanings of macroevolution as used in modern evolutionary and developmental biology is available.[14] Concisely put, the consensus among the experts (with certain important exceptions, such as the late S. J. Gould) is that the processes underlying macroevolutionary changes are the same as those that drive microevolutionary ones, continued over vast intervals of time, although there are arguments of detail against this "extrapolationism." But *none* of those arguments denies that a critical process in all evolutionary change is Darwinian natural selection and that evolution has occurred through descent with modification. Most important, however, is the increasing confidence of developmental geneticists, based solidly on experimental results like those with the body plan organizing Hox genes, that ordinary mutation and gene expression processes in embryonic development can produce and have produced dramatic—and heritable—changes in animal body plan.

6. Wells: *Darwin's theory is incompatible . . . with . . . embryology . . . [and with] the fossil record.* On the contrary, the neo-Darwinian account of life's history is the only one so far that *is* compatible with the mountains of evidence available from comparative anatomy, embryology, genetics, and paleontology. No honest reader of the scientific literature can fail to recognize that standard modern biology, for all its gaps (all real knowledge has gaps!), fits the observations to date very well. Every alternative account offered so far (such as special creation, internal striving, goal seeking, élan vital, intelligent design) either fails to account for the observations or is contradicted by them. Kansas State University geologist Keith Miller, who happens to be both an expert on fossils and a Christian, has shown in a short but pithy technical article how serious misrepresentation of the Precambrian/Cambrian fossil record and of microevolution characterizes many evangelical Christian declarations.[15] And, as serious students of phylogeny know, the mounting evidence of molecular genetics places the origins of the different animal groups far back in the Precambrian.

7. Wells: *According to Darwinism, all creatures are descended from a common ancestor. Yet the oldest fossils show that almost all . . . major groups . . . appeared at about the same time, fully formed and recognizably similar to their modern counterparts. Darwin's theory predicts a "branching tree" . . . fossil record, yet that . . . is nowhere to be found.*

We have touched in chapter 4 on this willfully eccentric account of the Cambrian radiation, propagated by all members of the Wedge leadership. The "oldest fossils" show no such thing as Wells declares. The oldest fossils are three-billion-year-old *microbial* cells and associations of such cells; that is, they are not yet animals at all. Fossils of animals, multicellular organisms of many types, appear in deposits far older than the Cambrian; those fossils are, so far as discovered, less disparate in body type and less diverse within type than Cambrian fossils—no surprise here. The fossil Precambrian fauna so far discovered may or may not include ancestors of the Cambrian, but it is simply not true that "all of the major groups appeared at about the same time." There were many Precambrian life forms. Fifty to a hundred and fifty million years or so before the mid-Cambrian period (i.e., 575–675 MYA, the "Proterozoic" period), during which there were already multicellular, soft-bodied, invertebrate animals, is not "about the same time"; and "almost all" is not the same as "all." Nor is it true that a "branching tree" pattern of the fossil record is "nowhere to be found." In fact, all internally consistent modern biological classifications are branched. The possible Cambrian "chordate" *Pikaia* was not a fish, a reptile, a bird, or a mammal, and it resembled none of them, although it might have been their ancestor. They are all "chordates"; there were no such things as fish, reptiles, birds, or mammals anywhere in the Cambrian world—all *those* chordates came much later. These later appearances are "branches" from the—or some—ancestral form.

Wells is promoting a basic misconception of the meaning of the *disparate* body types present in the mid-Cambrian—the phylotypes, most of the "phyla" we now name and then a few—that did nevertheless diversify enormously later in evolution.[16] Either he understands this and has purposefully obscured it, or he does not understand it and ought not to be teaching anyone about evolution.

8. Wells: *The fossils provide no evidence that all creatures are descended from a common ancestor. So the two major claims of Darwinism . . . are unsupported by evidence.* It is Wells's statement that is unsupported. To repeat: there were complex metazoan (multicellular) plants and animals on Earth long before the Cambrian. The data from molecular phylogeny, although there is lively argument about them and an outpouring of research literature,[17] point with increasing unanimity to the existence of genes (hence organisms) in the Precambrian and earlier periods that must have been *ancestral* to the Cambrian and all later versions. (For a reminder of how the ID promoters distort for their own purposes the facts of the Cambrian radiation, please see the discussion in chapter 4, including the notes.)

One of us taught and directed research in biology at universities for more than forty years and served as chairman of a large biology department and of a National Institutes of Health study section. In neither of those places—at any time after the early 1960s—would a summary statement like this one from Wells have passed muster even if submitted by a graduate student, let alone by a supposedly mature scholar. But the arguments he provides in *Icons of Evolution* are no more comprehensive and sophisticated than those of the science of his Unification Church sermon. This new work, vastly expanded, attacks the same straw men, albeit for the specific purpose of "reforming" K–12 science education. Launched into the nonsectarian and secular world, however, and unlike the sermon for Moon followers, *Icons* has potentially great political power. It is an accusation of the kind that the media and some politicians love, a match for the periodic revelations of commercial, clerical, or pedagogic transgressions: official science includes a worldwide conspiracy by the elite to hide the awful truth about Darwinism; the science is superficial, misleading, or wrong; fakery is taught to "the children" as truth. There is no fire in any of this, but there are, for most of Wells's readers, thick clouds of smoke.

The Icons

A good description of the aims and style of Wells's *Icons of Evolution—Science or Myth? (Why Much of What We Teach About Evolution Is Wrong)* introduces Eugenie C. Scott's review of it in *Science* magazine.[18] We quote from her review for comparison with the preceding long quotation from Wells's scientific avowals on his website:

> If someone were to charge that textbooks present atomic theory using evidence that is erroneous, misleading, and even fraudulent, and that we should therefore question whether matter is composed of atoms, eyebrows would be raised—at least at the accuser. If someone further claimed that distinguished physicists crassly participate in this fraud to keep the research dollars rolling in or to promote a materialist philosophical agenda, scientists would be angry at the attempt to besmirch the reputations of respected scholars. And if the same person proposed that citizens should encourage local school boards to insert anti-atomic theory disclaimers in science textbooks, discourage Congress from funding research in atomic theory, and lobby state legislatures to restrict its teaching, it is doubtful that such exhortations would receive much attention. Such would be the fate of Jonathan Wells' call to arms in *Icons of Evolution*, if biological evolution were not substituted for atomic theory in the above scenario. . . . Unlike atomic theory, evolution has obvious theological implications, and thus it has been the target of concerted opposition, even though the inference of common ancestry of living things is as basic to biology as atoms are to physics.

This analogy is important in any examination of the Wedge and its works, but for mysterious reasons, even intelligent observers fail to understand it. It applies to most ID productions, not just to Wells's *Icons*. In hope that our readers, especially the scientists among them, will see this, we expand Scott's analogy. Consider gravity. The phenomena of gravity are central to physics. For the macroscopic physical world that we know from immediate experience, gravitational phenomena are well enough understood, since Newton, to allow us to predict and control them in a satisfactory way. In that sense, we understand gravity—thanks to science. The underlying physics, on which of course most of engineering is built, is classical mechanics, to which Newton's laws of motion are fundamental and indispensable. Practical physics—including the exquisite precalculation and execution of complex orbital maneuvers by space vehicles moving in a post-Newtonian, relativistic universe—is solidly based on Newton's laws.

There are, however, known conditions under which those laws are not followed exactly and gravitational situations to which they are not applicable at all. In fact, it has been said, justly enough, that we do *not* understand gravity. One of the liveliest and most fundamental branches of theoretical physics is concerned with reconciling the best current picture of gravity from general relativity with phenomena in the atomic and subatomic realms that can be understood only through quantum mechanics. This is the "quantum gravity" problem. There is as yet no reconciliation. So we certainly do not understand gravity completely, which implies, at least for epistemological purists, that we don't really *understand* mechanics, classical or otherwise. It is probable that some new, deeper explanation of gravity, different from anything now used or taught, will replace or be superimposed on current theory, if not the everyday practices, of the physics of motion. Such transience is always true of good scientific theory: science, unlike religious or political doctrine, is never complete.

But suppose we now present this argument: "If that is so, how dare they teach about gravity in the schools, teach Newton's laws and mechanics, without calling attention to the unsolved problems? Whenever we teach students about motion, we should stress that what they are learning is just a theory that may be wrong. We should teach them about the problems of quantum gravity—*teach the conflict*! And, until we have that complete theory, all the way down, we must reassure schoolchildren of the worthiness of any private beliefs *they* may hold or be told about gravity, even if they make no sense as physics. If we do not, if we teach K–12 students physics in the ordinary way, we are inculcating a biased, seventeenth-century Newtonian view of reality. Worse, we are silencing legitimate alternative science. Children and their parents should protest. Concerned citizens should demand that government stop supporting physics until it admits the truth about gravity."

Of course, this argument is absurd. Learning mechanics is essential to learning any other physics. Learning *some* physics is essential even to the most limited cultural literacy, which it is the duty of all schools to provide. And even the simplest classical mechanics is a tough and dispiriting study for most K–12 students. So, students are exceedingly unlikely to learn mechanics together with a fistful of abstract and obscure warnings and qualifications. The problems for classical mechanics implicit in modern theoretical physics are important for progress at the frontier, but they are a pointless diversion from the learning of elementary science in school—and, for that matter, in college, except for those going on to a degree in science or engineering.

Yet Wells's arguments in *Icons* are exactly this kind, although his criticisms are driven by nothing remotely as serious as quantum gravity. In effect, Wells's complaints are the equivalent of casting heavy doubt on physics because we do not completely understand gravity.The "icons" of his title are (claimed) cases of textbook errors, oversimplifications, flawed illustrations, or even of frauds, taught as evidence for evolution. These cases, he says, are the failed "proofs" of Darwinism, and he suggests that *they* are the evidence on which evolutionary biology now stands. But they are not proofs. Even if all of those cases really were as flagrant as he asserts, his arguments would fail, and, as we shall see, the cases are most assuredly not what he asserts. He would almost certainly prefer that standard evolutionary biology not be taught at all, but, failing such a prohibition, he wants it to be taught only with emphatic warnings. He wants it taught, if taught it must be, with regular assurances built into the curriculum, pasted into the textbooks, that current evolutionary biology is, or may be, false, that the history of life on Earth, and the mechanisms of that history, are, or may be, very different—no need to specify in what way—from those that have come to be accepted in science worldwide over the last century and a half.

Wells's argument would be bizarre even if the icons were just as he sells them: as oversimplified or "faked" proofs of evolution offered in (some) textbooks. But they are neither. There are now a dozen detailed, point-by-point reviews of Wells's *Icons*, from scholars who know each subject and see it from multiple points of view in contemporary evolutionary science. These reviews are available to every reader,[19] and we hope that some of ours will examine them alongside *Icons* and the plaudits of Wells's CRSC promoters. The reviewers' consensus is that Wells's icons are nothing more than the making—for politico-religious purposes—of an enormous mountain from a scattering of molehills. That would be so even if the molehills were accurately presented. In general, however, they are *not* accurately presented. We will illustrate here with three examples of the possible ten. Serious responses to Wells's *Icons*, by specialists in each subdiscipline, should be consulted by anyone who is honestly concerned with the teaching of biology (and therefore of evolution) in the schools.[20] Wells, meantime, is busy as we write making more mountains—of words—via speeches, Wedge-friendly websites, and publications, attempting to rebut the arguments of his critics and to frighten Christian parents and teachers into political action against "Darwinism."

"Proofs" Said to Be False

Unfortunately, most scientists who know even a little about biology and evolution, and who are unaware of the reach and momentum of the Wedge in America, tend to smirk at such books as *Icons*, dismissing them as fodder for the scientifically benighted, dressed with what sounds like scientific language. After all, such offerings have a long history in creationism. We encounter this blindness to political reality among our scientific peers every day. It is a grave mistake. To people who know little or nothing about the subject of evolution—that is, almost everybody—the Wells arguments can seem both convincing and exciting; and they have the momentum of his religious fervor. *Icons* touches a raw nerve in the current war over the effectiveness of public education. "Here," it says, "is what your children are being taught in the public schools as proofs of evolution." "But," it insists, "they are not proofs at all, and some of them are outright fakery. Others are simply wrong. Demand a stop to the callous indoctrination of your children in this materialistic mythology!"

We ask the reader to consider: how many parents, already anxious about what their children are learning, how many journalists and lawmakers—weak in science but strong in politics, eyeing anxiously the support and votes of those parents—can resist joining in a call for an end to the "censorship" of such startling findings as these from Jonathan Wells?

And so to the "icons," which are supposed to be commonly taught examples or case histories of evolution about which Wells urges everyone to ask tough questions, especially if they appear in a textbook or are referred to by a teacher. We list them in order.[21]

1. Abiogenesis—life from nonlife—the synthesis of organic (biologically important) compounds, building blocks of life, from simple, nonliving chemicals, in the famous Miller-Urey experiment of 1953. That experiment doesn't work, says Wells, if it is done properly; therefore, it is wrong about the origin of life and thus no support to Darwinism. To teach about it is fraud.
2. The (Darwin's) "tree of life," that is, the evidence of a multiply branched descent with modification of contemporary and more recent species of animals from fewer, common ancestors in the past. Wells denies that there is any evidence for this, discounting in the process all of molecular phylogeny and insisting that there is no basis for assuming the existence of ancestors to fauna of the Cambrian "explosion."
3. "Homologies" of structure, for example, the close similarities of all the bones of vertebrate limbs, or the structures and amino acid compositions of proteins with related functions from widely separated taxa. These similarities, commonly held to be evidence of descent from common ancestors, are not that at all, says Wells. His treatment of it makes the (universal) use of homology in the evolutionary sciences nothing more than arguing in a circle.[22]

4. Haeckel's drawings of vertebrate embryos. These, Wells asserts, were faked. Moreover, the evolutionists know it and have known it for a long time. But until very recently they have said nothing about it *to protect these drawings as key "proof of Darwinism."* Haeckel's drawings are still reproduced in textbooks.
5. Archaeopteryx. This fossil form is commonly cited as proof of evolution, a "missing link" between the dinosaurs and modern birds. But it is not a missing link, Wells argues; it is not an ancestor of living birds, and it is therefore no proof of evolution.
6. The celebrated case of the peppered moth. This famous example of Darwinian selection operating to powerful effect in nature, over short time-intervals, is a fake, Wells says, because, among other reasons, the photographs of moths resting on tree trunks, used in many textbook accounts, were staged (the moths were glued on). So the phenomenon for which these moths are famous (industrial melanism, protective coloration) is no support for the Darwinist idea of natural selection.
7. Darwin's finches—another famous and often cited case of rapid natural selection in the wild. It isn't so, says Wells, for several reasons, most notably that observed population changes in beak morphology were not a result of macroevolution.
8. Fruit flies with four wings. Says Wells: the case of a four-winged fly, appearing spontaneously in a species of two-winged insects (flies are "Diptera": two wings), is not, as claimed in some books, evidence for a neo-Darwinian mechanism of evolution.
9. Fossil horses and directionality in evolution. Almost everything said and taught (illustrated in textbooks) on the evolutionary lineage of horses, a paradigm case, is wrong, says Wells, because evolutionists have rigid, materialist preconceptions about such phenomena as directionality and purpose.
10. Hominid evolution and humans. The gap between the anthropoid apes and humans is not really filled, says Wells, by the many hominid fossil sequences presented today as the human lineage, for, among other reasons, there was once a fraud in the business ("Piltdown Man"). The experts long failed to recognize this; besides, there are constant disagreements today among them about what fossil species, among the scores now known, are ancestral to which others and about what the fossil sequences actually mean. Paleoanthropology, in other words, is untrustworthy on questions of human antiquity.

The following comments address three of these "icons," but an informed, honest examination of any of the others produces the same result; and these three are among the most discussed. We have chosen them also because they stand for three kinds of systematically misleading exposition, at least one of which characterizes each of Wells's icons: (1) irrelevance of the given argument to evolution, or to "Darwinism"; (2)

misrepresentation of the importance of the material at issue to the state of evolutionary science; and (3) focus on trivialities in order to magnify the scientific significance of a minor point.

MILLER-UREY

In the nineteenth century, it was widely believed that most of the known compounds of carbon, other than the smallest and simplest, could be made only by living things. Examples of the simplest were oxides of carbon, such as CO_2 (carbon dioxide); the gas methane (one carbon atom with four hydrogen atoms bonded to it tetrahedrally, CH_4); the ions and molecular species derived from CO_2 in aqueous solution; the inorganic carbonates; and the like. But complex carbon compounds of moderate to high molecular weight were assumed to be the characteristic substances of life. Hence, they were called "organic" compounds. The branch of chemistry devoted to them was, and still is, "organic chemistry." Organic chemistry, together with physiology, spawned "physiological chemistry," which later became "biochemistry"; and the union of biochemistry with genetics and biophysics, the reach of that marriage becoming basic to all biology, achieved universal recognition as "molecular biology" in the 1960s.

Here is the point: organic compounds, even relatively simple ones such as the amino acids, which when polymerized in long chains make up proteins, were judged to be the unique markers of life. So the following question was in the air in the 1930s: where did these building block chemicals so characteristic of life come from in the first place? Was it life first, or the building-block organic compounds first? Because it seemed that they couldn't come spontaneously from *in*organic compounds, these organic molecules had to have arisen in some other way. There would have been little doubt about the answer offered by most life scientists even as late, say, as 1930: "Life comes first. These molecules are constituents or products of the living substance, which is called 'protoplasm.'"

By the late 1940s, geochemistry and geophysics had made sufficient progress to allow some agreement about what the primitive Earth, from three to four billion years ago, was like. First, it was a geologically violent place, bombarded by fiery rocks from space, microscopic to city size, shrouded in an atmosphere ablaze with electrical discharges (lightning), and wet by ceaseless rains. The first atmosphere was nothing like the present one in its chemistry: it had no or very little free oxygen. It almost certainly contained ammonia (a gaseous compound of nitrogen that dissolves in water) and, very likely, among possible very simple carbon compounds, perhaps of volcanic origin, the gas methane. The question of nonbiological synthesis ("abiogenesis") of organic compounds was under discussion and of particularly intense interest to physical and inorganic chemists. Among those was Nobel Prize–winner Harold Urey. Stanley

Miller, Urey's graduate student, undertook to test a hypothesis current in that and other laboratories: that high energy, as from ultraviolet radiation or electrical discharges, acting on inorganic molecules in water vapor and in aqueous solution and in equilibrium with a mixture of primitive atmosphere gases, might catalyze the synthesis de novo—from scratch—of some carbon ("organic") compounds of moderate molecular weight.

Miller did the experiment using a simple spark-discharge apparatus, a recirculating, sealed, glass system. The results were published in 1953—half a century ago. Miller's experimental arrangements, which, like his results, have been endlessly reproduced, yielded in easily detectable quantities several of the real building blocks of biological macromolecules, plus other carbon compounds and their polymers. A typical run done today, by a student, might yield acetic acid; urea; the amino acids alanine, glycine, and aspartic acid; and lactic acid.

The experiment is in fact still done sometimes—even in elementary courses. *It works*. Meanwhile, with the passage of that half-century, geochemistry and geophysics have matured. There is now much doubt that the primitive atmosphere could have been of the strongly reducing kind (i.e., hydrogen-rich and oxygen-poor) Miller and Urey assumed and used. Opinion is weighted today toward mildly reducing or oxidatively neutral atmospheres, with other gases (such as CO_2) in the mixture.[23] If the experiment is done with such an "atmosphere," it works differently, often with lower yields of amino acids than in the strongly reducing gas mixture. Does it fail? *No*. It just produces different organic compounds, among them some once thought to be products only of life. Some reaction products are more interesting than the original ones; most tend to be in lower yield than in the strongly reducing system. Others are less interesting. But what the justly celebrated work of Miller and Urey did was to show that *abiotic synthesis of the molecular building blocks of life is possible*. That was important science, and it was a first. It showed that the primitive Earth, for all its violence, could have acquired pools of organic molecules, among which were some precursor monomers of the carbohydrates, proteins, and nucleic acids of living cells. Wells has given a radically false impression of the condition of the science that has followed this work. Today, there is no longer any doubt about the possibility of spontaneous formation and abundance of organic molecules—including some of the building blocks of life—very early in our planet's history. Moreover, it is now known that they are remarkably abundant in outer space, and this knowledge has given new life to speculations that the precursors to life on Earth might have been seeded here by meteorites and cometary debris during the long intervals of Earth's heavy bombardment from space. Indeed, likely mechanisms for the synthesis of amino acids in comets and smaller, icy objects in space have now been demonstrated in the laboratory.[24]

Was the Miller-Urey experiment then, or is it currently, offered by responsible scientists and educators as *the* mechanism of the origin of life? *No*. Never. If anyone does offer it that way (and it might be difficult to find a teacher who does), he or she is wrong and should be educated on the subject. But such experiments—and there are others discussed today, of disparate design, that lead to the same results—discovered for the first time a number of quite plausible mechanisms for abiogenesis of organic chemical compounds on Earth or in space.

What does this have to do with Darwinism? Nothing, necessarily. Darwin himself had little to say about the origin of life. Evolutionary biology still has nothing definite to say about *the* origin of life on Earth. Neo-Darwinism—modern evolutionary biology—is about what happened after the appearance of the first cells. There is, of course, a tentative position emerging from evolutionary science and the activities in origin of life research, a quite separate, interdisciplinary field involving astronomy, chemistry, biology, and geology: the continuing hope of finding mechanisms, perhaps *the* mechanisms, for an origin(s) of life by natural processes. That is all. New data—and theory—appear in that literature every day, catalyzed in recent decades by continuous advances in nucleic acid chemistry and by advances in space science and planetary science. A recent example, just one of many, of the latter is the "Jigsaw Model" of the origin of life, proposed by John F. McGowan.[25] Origin of life research is a vibrant field of study, as full of the normal arguments of frontier science as it is of new discoveries in planetary chemistry—unimagined in the days of the Miller-Urey experiment. There is now available in English an excellent, scientifically sound, readable book on the whole subject and its excitements, written by philosopher-historian of science Iris Fry.[26] This volume takes up in detail the important strand of "RNA-world" studies. It even looks now, however, as though a practical chemistry for the self-replication of *DNA* is about to emerge, implying that primitive DNA might have been able to reproduce itself without benefit of the enzymes (proteins) that must now catalyze the process in living cells.[27] Wells's stagy hysteria over the Miller-Urey experiment fails to convey the state, purposes, and meaning of such science. It would all be amusing if it were not so silly.[28]

HAECKEL'S ARTWORK

We come next to those "faked" drawings by Ernst Heinrich Haeckel, a celebrated naturalist, embryologist, evolutionist, and admirer of Darwin in the last quarter of the nineteenth century. Haeckel, who coined such everyday biological terms as *ecology* and *phylum*, was an immensely productive writer and illustrator, as well as a renowned naturalist.[29] Whatever he believed, he believed strongly, and for that he proselytized. A conservative, religious youth, anxious always to please his aged and traditionalist parents, he metamorphosed in adulthood into a scientific leader of radical inclinations—in politics as well as in science. He pushed against the boundaries of comparative morphology (then the most exciting branch of "natural history"), of embryology, and finally of evolution and philosophy of science. Needless to say, in a time when speculation in such fields was much livelier and more welcome than it has been for the past three quarters of a century (there is always more active speculation in developing disciplines than in mature ones), Haeckel was much honored. Not unexpectedly, he was also reviled by his many enemies, including scientific and other conservatives. His radical political views were a source of consternation to fellow scientists who otherwise respected his work.[30]

In embryology, his great love, he popularized the bold proposal (expanding an idea that was, however, already in the air) that became known as "the biogenetic law": ontogeny recapitulates phylogeny. This means that during development, the embryo retraces the steps of the evolution of its species, so that the early stages of vertebrate embryos (for example) are all very much alike, while modifications that uniquely identify the species (wings, or limb structure, or a beak, or respiratory arrangements, skin morphology, etc.) appear late in development, on top, as it were, of the earlier phylogeny. There was a point to this: Haeckel hoped to bypass the painstaking and frustrating search for fossils. By careful study of developing embryos, of *developmental* morphology, he maintained, we would eventually discover the evolutionary lineages of animals, including *Homo sapiens*. That was his purpose: to trace through developing embryos the evolutionary histories of all living things, including humans.

There is a kernel of truth in the idea of recapitulation, but it is wrong in detail. It was shown to be wrong as soon as embryos began to be studied for their own sake and in minute detail with the excellent new compound microscopes of the last quarter of the nineteenth century. Such study began also to include what was soon to become indispensable: experimental manipulation. This new kind of embryology, appearing in Germany late in the nineteenth century and growing fast in explanatory power in the early twentieth, not least in America, was named by Wilhelm Roux, its founder, *Entwicklungsmechanik*—developmental mechanics.

Haeckel was, like many naturalists of his day, an energetic illustrator. (One had to be!) So he illustrated his books and papers copiously—and flamboyantly. Adequate photographs, especially photomicrographs, were not to be had. His representations of exotic marine animals (of his own collection and naming), of phylogenetic trees of the animals, are beautiful: in style they recall the putti and acanthus leaves of the baroque. And they are inaccurate, as was pointed out even in Haeckel's lifetime, for example, by Ludwig Rutimeyer, paleontologist at the University of Basel.[31] His illustrations of vertebrate embryos—prepared, note this, 133 years ago—are like that: beautiful and inaccurate, shaded toward what he wanted to see. Moreover, the biogenetic law was abandoned by the middle of the twentieth century.

In the inspired progress of embryology that followed the introduction of experimental methods (including a Nobel Prize for one of its famous practitioners, Hans Spemann, in 1935), it became clear that the early stages of vertebrate development are *not* all the same; that the unique departures characterizing the classes (fishes, reptiles, birds, mammals) do not happen only near the end of development. So Haeckel's theory is long defunct. But the important point is that this entire shift of opinion came from biologists: it was the self-correction of science, not the insights of theologians and philosophers, or, least of all, the creationists, that detected and corrected the error. Drawings like Haeckel's became irrelevant, technologically and substantively, to the advance of evolutionary science.

There is no particular reason why Haeckel's old drawings should still show up in a textbook. They are not needed. They prove nothing. But they are easy to reprint, the point they are supposed to make is roughly correct, they cost nothing, and the authors and publishers of textbooks are often not as scrupulous as those of the primary scientific literature *must* be. The same is true of all such school textbooks in all fields, including religion.

Now, we may ask two questions: (1) Are Haeckel's overly imaginative figures ("fudged," as biologist Massimo Pigliucci says, not "faked," as Wells says), which a *very* few school textbook publishers still foolishly reprint from time to time because they're pretty, false through and through, misleading about Darwinism? (2) Are these, and Haeckel's notions of the appearance of vertebrate embryos, offered today as "proofs" of evolution? *No*. We have excellent photographs, to which students can obtain easy access. Many or most college students of introductory biology actually see the embryos in the laboratory, either alive or fixed and prepared for examination, at least the embryos of amphibians (tadpoles), birds (chick), and mammals (often a preserved early pig embryo). Many dissect these or study their microscopic structure in stained thin-sections in the introductory biology courses. High school students see photographs, or they visit a museum to see the real embryos preserved in fixatives.

And guess what? Vertebrate embryos, for most of the longest period of middevelopment, *do* look remarkably alike, pretty much, but not exactly, as Haeckel figured them in some of his drawings. These remarkable resemblances—fish, salamander, tortoise, chick, hog, calf, rabbit, human—were seen and drawn in figures long before Haeckel's time.[32] They were, in fact, the observational core of the great work of Karl Ernst von Baer, the nineteenth-century embryologist who, almost single-handedly, turned the subject into an empirical science.[33] There can hardly be a college biology student today, or one in an advanced high school course, who has *not* seen at least a few photographs of the derivatives of pharyngeal pouches and arches of vertebrate embryos and learned how the structures seen in due course, in fish embryos, as gill slits take part later in formation of entirely different structures—for example, the Eustachian tubes—in the mammal. That's how evolution works: by scavenging what is already there. That is the reason for the remarkable similarities among the embryos of all vertebrates.

For a hundred years, most embryologists have been aware of Haeckel's fudgings. A few years ago, a temporary arousal of interest in them prompted a news note in *Science*, in whose title the word "fraud" appeared, and Wells pounced on it. The explanation for that use in *Science*, and the reasons why annoyance with Haeckel's old drawings is justified but also, in fact, insignificant—for modern embryology and *certainly* for evolutionary biology—are there to be gleaned from Elizabeth Pennisi's note.[34]

The similarities and identities of structure among vertebrate embryos of different classes, considering how much the adults finally differ, are very great indeed. Only a sophisticated and eager student, with no other information than the specimens before him under the microscope, could distinguish a microscopic section of a fetal pig at nine millimeters from that of a chicken embryo at three and a half days of incubation (these pig and chick embryos are at comparable stages of development). In modern drawings or photomicrographs of such embryos, the two dozen or so labels naming the main internal structures are identical, as are most details of the structures themselves. Thus, Wells's solemn avowal of Haeckel's drawings as fraudulent "proofs of evolution" is a tempest in a teapot. Fortunately, there is now a lengthy and authoritative justification for this judgment: a searching review by Michael K. Richardson and Gerhard Keuck of all the arguments (including those recently appropriated by Wells), beginning with those from Haeckel's contemporaries. The essay's abstract puts the matter of the embryo drawings in a nutshell:

> Haeckel's much-criticized embryo drawings are important as phylogenetic hypotheses, teaching aids, and evidence for evolution. While some criticisms of the drawings are legitimate, others are more tendentious. . . . Despite his flaws, Haeckel can be seen as the father of a sequence-based phylogenetic embryology.[35]

Finally, after all his complaints, and all the analyses by scientists that these complaints have prompted, it turns out that among the double handful of textbooks Wells reviews in *Icons*, only *two* actually contain Haeckel's drawings, although, according to Eugenie Scott's *Science* review (cited earlier), "all of them present, in varying degrees of detail, the scientifically accepted inference that comparative embryology reflects common ancestry."

Finally, we consider those famous peppered moths. The discussion of this "icon," too, is a rhapsody on trivialities. Now, as always since 1859, religious opponents of evolutionary biology (as opposed to scientific ones, some few of whom exist), plus a very few postmodern literati who write on science, have done their work not by offering new scientific evidence, but by belittling the content, the authors, the publishers, and the teachers of existing science. In the Wells story on *Biston betularia*, the peppered moth, as presented in *Icons*, we have an archetype of the new "creation science." Old-time biblical literalists didn't pay close, analytical attention to the scientific literature; they merely cherry-picked through the surfaces and made their own slanted summaries of any and all disputes they could find. The new way is to study the scientific literature and the background of it rather more closely, so that chapter and verse of the disputes may be cited as necessary (but also, of course, selectively). As in the old creation science, the study remains a search for exploitable quotations and arguments within the science. Those having been found (inevitably), one can insinuate by elaboration and appropriate rhetoric the worthlessness of the whole enterprise—in this case, of evolutionary biology.[36]

The visible parts (wings and body) of the peppered moth normally resemble coarsely ground black pepper on a light gray background—that is, a randomly mottled, light neutrality. It was assumed that there is adaptive value in this: mottled gray is indistinct or invisible against the background on which this species of moth commonly alights—the bark, also usually gray and mottled, of tree trunks, branches, and twigs in the native (English) habitat. This is due in part to the natural color of the bark and in part to an abundance of associated lichens. Why did scientists assume that the moth's coloration is *adaptive* (in this case, protective)? First, the color is determined and inherited by a known genetic scheme. Second, the primary predators of moths are birds, for whom the moths are presumably a delicacy. Birds have notoriously sharp eyesight. So the hypothesis was that the peppered moth's color is *camouflage*, making it invisible against the trees.

Until the middle of the nineteenth century, *B. betularia* in Britain, at least those observed by humans, were peppered gray. Rarely, however, a darker (melanic) form of the same moth was seen. Some of those were and still are nearly black. The basis of this originally aberrant melanism of wings and body is a replacement of the wild type pigmentation gene by one of several mutant versions (alleles). The dark-colored moth stands out sharply against light gray, mottled backgrounds. Moths of both kinds, the pale gray type and the melanic (pigmented) type, have been placed on tree trunks and photographed in order to show the effectiveness or futility of their color as camouflage, light against light, dark against dark, and vice versa. Wells labels such photographs fakes.

The melanic moths began to appear in quantity in the 1840s and 1850s, especially in the vicinity of Manchester. Fifty years later, the vast majority of moths near England's industrial centers, often 98 percent or more, were the melanic form; the original, peppered variety were the rare, and disappearing, exceptions. This is, of course, a *very* short interval for so sweeping a population-level genetic change in the expressed characteristics (phenotype) of a species.

By the end of the nineteenth century, the essentials of the modern argument had been proposed and debated. Now for the interesting part: as explained, for one example among many, by biologist (and industrial melanism expert) Bruce Grant, J. W. Tutt had suggested that the original speckled gray was protection against predators when the moths rested on trees, but that in the regions of heavy industry, the lichens had been decimated by air pollution, and factory smoke and soot had blackened most tree-trunk bark and branches. There, the melanic moths had suddenly a great advantage. They should be much less visible to their predators than the original peppered phenotype. Natural selection, Tutt suggested, had brought about this shift in population characteristics, and with remarkable speed.

Although no fieldwork or controlled experimentation was done for some fifty years thereafter, population genetic theory was applied in the 1920s to test the plausibility of selection as the mechanism of this rapid, population-wide, hereditary shift in pigmentation. There were denials as well as support for this prototypically Darwinian hypothesis. Then in the 1950s, entirely separate researches demonstrated that population dominance of genetically alternative light or dark forms of the mouse *Peromyscus* is determined *by the color of the background* on which the population lives and that the selection agents are their predatory enemies: owls. Contrast between coat color and background has much to do with the mice surviving or not surviving. This newer finding was, in short, a nice example of natural selection for camouflage.

In England, biologist H. B. D. Kettlewell undertook experimental studies on *Biston betularia*, testing the hypothesis that the switch from peppered to dark forms had been a response to the new industrial environment that blackened the trees in the very short interval (from the point of view of geological time and evolution) of about a hundred years. The specific hypothesis was that these moths are most vulnerable to their predators—birds—when at rest on the trunks and limbs of the trees that are their habitat, and that in the heavily industrial (coal-burning) regions, with the lichens killed and the trunks and limbs of trees blackened, the peppered coloring was no longer protective, while melanism *is* protective, hence selectively positive. Kettlewell did predation experiments with collected, marked, and released moths. The results were a vindication of the hypothesis that natural selection is effective in changing the population gene pool. *And that is the modern definition of evolution*. This instance of industrial melanism in the peppered moth became a favorite example (but by no means the only one) of Darwinian natural selection in action. There arose in consequence a biological industry of confirming and extending Kettlewell's results, and—typically for the competitive game of science—another industry of denying or, more commonly, offering reservations about them.

Some of the reservations turned out to have substance. The genetics are not so simple as was once thought. Some attempts at repetition of Kettlewell's experimental designs did not succeed. Bird vision is much better understood now than it was then, and there are necessary qualifications of the original picture. Evolutionary biology has grown lustily in the last fifty years, and reviewers of the industrial melanism literature have not been hesitant to focus on mistakes made by their predecessors. (No better rebuttal than *that* is needed for the crude creationist claim that science conspires to conceal the failure of Darwinism.) Things are not so simple as they were in the 1950s. And it is this healthy internal skepticism that Wells has mined from the science literature, and inflated, in his reporting of the "staged" pictures of moths resting on tree trunks.

Is, then, the whole peppered moth story of Darwinian natural selection also "staged," also a fake, providing no support for the idea of natural selection, as Wells wants his readers to believe? *No!* Bruce Grant has published quite recently, in the mainline journal *Evolution*, a comprehensive survey of the responses, from *within* evolutionary biology, to Kettlewell's original work and of the state of knowledge in the broad field of biological crypsis (camouflage) and industrial melanism. This treatment of the subject is in part an analytical summary of the science, including newer experimental approaches, and in part a metareview of books and earlier reviews. It illustrates the eagerness with which scientists seize on perceived weaknesses in the most popular theories and results to get at the truth (and to display their own excellence). We quote two passages from Grant's essay, long enough to convey its flavor and his convictions. Note that *all* the authors named are working evolutionists, "Darwinists."

> Sargent et al. subvert the traditional explanation for industrial melanism by presenting an equivocal analysis of the evidence. Coyne laments that our "prize horse" of examples is in bad shape in his review of a recent book, *Melanism: Evolution in Action*, by Michael E. N. Majerus (1998). Although Majerus assesses the bulk of the literature and controversy that has accumulated since Kettlewell's (1973) *The Evolution of Melanism*, he could hardly agree less with Coyne or with Sargent et al. that the basic story is built on a house of cards. Majerus argues convincingly that industrial melanism in the peppered moth remains among the most widely cited examples of evolution by natural selection for two reasons: First, the basic story is easy to understand. Second, *the evidence in support of the basic story is overwhelming*. . . . That these changes [rise and decline of melanism with increasing and decreasing air pollution, respectively] have occurred in parallel fashion in two directions, on two widely separated continents, in concert with changes in industrial practices suggests that the phenomenon

> was named well. The interpretation that visual predation is a likely driving force is supported by experiment and is parsimonious given what has been so well established about crypsis in other insects. Majerus allows that the basic story is more complicated than general accounts reveal, *but it is also true that none of the complications so far identified have challenged the role assigned to selective predation as the primary explanation for industrial melanism in peppered moths*. Opinions differ about the relative importance of migration and other forms of selection. It's essential to define the problems, to question assumptions, to challenge dogma. This is the norm in all active fields of research. Majerus has succeeded admirably in communicating this excitement to the reader. I would add this: Even if all the experiments relating to melanism in peppered moths were jettisoned, we would still possess the most massive data set on record documenting what Sewall Wright (1978) called "*the clearest case in which a conspicuous evolutionary process has been actually observed*." Certainly there are other examples of natural selection. Our field would be in mighty bad shape if there weren't. Industrial melanism remains one of the best documented and easiest to understand.[37] (emphasis added)

That is why the color of *Biston betularia*, the peppered moth, continues—properly—to illuminate elementary discussion of natural selection, among the several other evolutionary mechanisms known to operate, in a few elementary textbooks. It is a very good choice. The raising of trivial objections continues, however, in some quarters. These are not objections to the fact that the population-wide changes of *Biston*'s camouflage are a splendid display of evolution in action. They are caviling about technical details of mechanism and about who identified which one. Bruce Grant has dealt effectively with this in a recent review in *Science*.[38]

The rest of Wells's book is no better as science than are these examples of its arguments. The problem is that it takes as long (and as much space) to rebut them, and to document the rebuttals from primary sources, as it must have taken Wells to write his book. Moreover, to make an intellectually honest rebuttal in a short speech, or worse, in a "debate" before audiences largely innocent of the science and often hostile to it (many of the audiences Wells addresses) is a hopeless undertaking. The only chance is to use rhetorical tricks, just as the ID promoters and other creationists do. But that is not work that serious scientists are likely to take up, or are generally capable of performing. And no lay audience is likely to remain awake and alert through a substantive, scientifically complete dissection of all Wells's claims and charges.

The following extract is from a letter sent by Professor Grant to *The Pratt Tribune* of Pratt, Kansas, following a visit by Jonathan Wells and a subsequent flow of anti-evolution invective from the paper's readers:

> In your paper, Ms. Katrina Rider "asserts" the peppered moth story is a hoax. She conveys the impression that dead moths were glued to trees as part of a conspiracy of deception. She seems unaware that moths were glued to trees in an experiment to assess the effect of the density (numbers) of moths on the foraging practices of birds. Taken out of the context of the purpose of the experiment, the procedure does sound ludicrous.

But, should we blame Ms. Rider for her outrage upon learning that moths were glued to trees? No. Instead, I blame Dr. Jonathan Wells, who wrote the article she cites as her source of information. While he has done no work on industrial melanism, he has written [an] opinion about the work. To one outside the field, he passes as a scholar, complete with Ph.D. Unfortunately, Dr. Wells is intellectually dishonest. When I first encountered his attempts at journalism, I thought he might be a woefully deficient scholar because his critiques about peppered moth research were full of errors, but soon it became clear that he was intentionally distorting the literature in my field. He lavishly dresses his essays in quotations from experts (including some from me) which are generally taken out of context, and he systematically omits relevant details to make our conclusions seem ill founded, flawed, or fraudulent. Why does he do this? Is his goal to correct science through constructive criticism, or does he a have a different agenda? He never mentions creationism in any form. To be sure, he sticks to the scientific literature, but he misrepresents it. *Perhaps it might be kinder to suggest that Wells is simply incompetent, but I think his errors are by intelligent design.*[39] (emphasis added) (Signed:) Bruce Grant, Professor of Biology, College of William and Mary

Making Room for God

Thus, by virtue of the Wedge and its "scientific" voices, impressionistic and political—rather than scientific—discourse on evolution reigns supreme in the public arena. As Frederick Crews observes,

> Political suspicion on the left [Crews refers here to the political left's deep antagonism to evolutionary psychology and to Darwinism generally]; fear of chaos on the right. Who will stand up for evolutionary biology and insist that it be taught without censorship or dilution? And who will register its challenge to human vanity without flinching? The answer seems to be obvious at first: people who employ Darwinian theory in their professional work. But even in this group we will see that frankness is less common than waffling and confusion. The problem, once again, is how to make room for God.[40]

The intended *Icons* audience, then, because they believe that they must "make room for God" in the classroom (and only for this reason), take Wells's assertions as knock-down arguments for "reforming" school science. The charges of conspiracy are inflammatory—*by design*—and evidence-free. To fan the flames a bit more, in an interview about *Icons* Wells portrays himself as a martyr for honest science:

> [Interviewer]: Do you think publishing *Icons of Evolution* will jeopardize your career as a professional biologist in America?
>
> Wells: Yes. Darwinists have . . . near-totalitarian control over American biology. Someone hoping for a career as a . . . biologist . . . must toe the orthodox line . . . to publish in prestigious journals and obtain research grants. Without . . . [these], an up-and-coming biologist cannot compete in the job market. I knew this when I started writing *Icons* . . . but . . . it was better to speak out against a great evil (i.e., misrepresenting the evidence to defend an anti-religious . . . view) than to keep quiet . . . to further my career.

In the very next breath, he admits that he is doing no scientific research:

> [Interviewer]: Are you currently doing research?
> Wells: My current research is limited to reviewing . . . scientific literature. I am not now doing laboratory research. I have begun several . . . experimental projects, but they are on hold until the current controversy is resolved.

Yet he fully intends to keep writing books about "the controversy":

> [Interviewer]: Should we expect any other controversial books from you in the future?
> Wells: Yes. I'm . . . working on . . . *The End of the Genetic Paradigm*, about the doctrine that genes are the secret of life. The . . . media tell us . . . genetic defects are . . . the root of human disease, . . . genetic programs control embryo development, and . . . genetic dispositions guide . . . [human] behavior. But the data fall far short of justifying these claims . . . [T]hey are deductions from neo-Darwinian theory rather than inferences from biological evidence. As you can imagine, the implications are far-reaching.[41]

His ambitious publication plans notwithstanding, the science so far contributed by Jonathan Wells is trifling when it is not plain wrong. But charges of conspiracy are, unfortunately, effective politics and powerful public relations.

Michael Behe, the Icons, and the War on K–12 Science

The political impact of Wedge activity is not due to informed and critical reading of Wells's arguments by parents, teachers, administrators, and legislators, although many of them have probably read the book and some of his defenses of it. As the sampling of the *Icons* should indicate, reading the book is not enough: judgment of the validity of its arguments requires knowledge of the science being *mis*represented. What makes the misrepresentation effective, therefore, is not just its reaching a large general audience but dogged, systematic *repetition* in politically significant venues by ostensibly qualified scientists, namely, the CRSC core group. That repetition gives Wells's *Icons* the appearance of a genuine exposé, of urgent truths about wrongdoing in science education. An example must suffice to make this large point. This one uses the Haeckel's drawings icon.

We have shown already that the commotion made by Wells and the CRSC membership about those drawings of vertebrate embryos is irrelevant to contemporary embryology and evolutionary science. Professional biologists, in particular, are aware of the facts, given that it is almost impossible to get an advanced degree in biology, even today in the era of emphasis on molecular genetics and computer simulations, without having at some point studied real vertebrate embryos, either in photographs, or as living, preserved, dissected, or fixed and sectioned samples, and noted their remarkable similarities. Having seen and studied them by modern methods, not with Haeckel's primitive techniques, the serious student cannot doubt that these are systematic, comprehensive similarities of internal structure and external appearance. The similarities are, in fact, "homologies." Haeckel is not a part of contemporary scientific argument. He and his drawings are as irrelevant to it as are the theories of Democritus to nuclear physics.

But here, now, is Michael Behe, ID spokesman and professor of biochemistry at Lehigh University, addressing a hearing of Pennsylvania's Senate Education Committee in June 2001. This hearing was about certain proposed revisions to the state standards for K–12 science, revisions that were proposed and were being pushed by creationists. They were carefully crafted to appear innocuous: they claimed merely to encourage teaching the students the skills of judging scientific theories—"critical thinking" about science. In fact, however, their purpose was to require the teaching of arguments against evolution. We quote a small but representative section of Behe's testimony, in his comments praising the inculcation of a "questioning attitude."

> But can this questioning attitude be applied even to Darwinian evolution by natural selection, for which, we are often told by Darwinists, no contrary evidence exists? As a professional biologist, I think it would be pretty easy, as recent newspaper stories demonstrate. . . . The story concerned some drawings made by a 19th century German scientist that showed embryos of many vertebrate species to be "virtually identical" in appearance in their earliest stages of growth, as Darwinian evolution would expect [Behe is wrong; it "expects" no such thing]. For over a hundred years the drawings have been presented in high school biology texts as strong evidence for Darwinian evolution, and they are still in current textbooks. But the drawings are false—the embryos don't look like that. . . .
>
> Pennsylvania biology students could be asked to discuss the following questions:
>
> - If inaccurate drawings showing nearly identical embryos were cited as textbook evidence to support Darwinian evolution, does the fact that the embryos look considerably different count as evidence against Darwinian evolution?
> - If not, what might possibly count as evidence against Darwinian evolution by natural selection?
> - If for over a century scientists quietly accepted misleading drawings in high school textbooks that agreed with the dominant theory, might there be other difficulties that have been neglected?
> - How do we know that it was natural selection that drove enormous changes in the distant past?[42]

Behe's remarks, like the rest of his testimony, had one purpose: to convince legislators and the managers of the state education system, under the guise of encouraging "critical thinking" by schoolchildren, that there is something terribly fishy about ("Darwinian") evolution. But it would be astonishing, given his education and his job, if he did not *know* that the Haeckel's drawings fuss is a trivial matter that, whatever its details, affects contemporary evolutionary science not at all. There is in the fuss nothing that can "count as evidence against Darwinian evolution." Behe's testimony, in short, is intelligently designed to implant a falsehood—that the rare reprinting of Haeckel's (scientifically) ancient drawings in a few textbooks reveals deep error in basic evolutionary biology. He is informing them, citing his qualifications as an expert, that the fudging of some illustrations a century ago is a current cause for deep distrust, even in elementary school, of the modern science. In a letter to Pennsylvania state senator James J. Rhoades after the hearing, Roger D. K. Thomas (John Williamson Nevin Professor of Geosciences, Franklin and Marshall College), who also testified, took special note of Behe's departure at the hearing from his supposedly more even-handed treatment of evolution in *Darwin's Black Box*: "Dr. Michael Behe has a different view of the matter [from mine], as you know. In his testimony yesterday, he took a much more traditional position against evolution than his book, Darwin's Black Box, had led me to expect. . . . Yesterday, he chose to raise questions . . . about the reality of evolution in general."[43] In light of the credentials Behe cites, his testimony suggests that he understood perfectly well that what he said to the committee was grossly misleading.

William A. Dembski (et al.): Catch My Next Book!

The Treadmill

Among serious environmental scientists who attempt to distinguish, for the public, real problems from Chicken Little problems ("the sky is falling!"), there is a familiar jest. It is about a volume on the state of the environment, published annually, that announces the end of the world, but includes among its pages an order blank for the next year's volume. The verbal output of the leading ID figures qualifies as a version of this dismal joke. Their production of speeches, papers, news releases, World Wide Web arguments, and books might be named "the [ID] critic's treadmill." Here is how it works.

Wedge books are aimed first and foremost at that portion of their audience who are, unfortunately, uninformed about the way science works and about the current state of evolutionary science, that is, those who will actually believe that "Darwinism" has failed. So to their least sophisticated audience, Wedge members simply announce, as they have done from the start, the wonderful news that Darwinism is dead. All that's left to do, they say, is mopping up and getting Darwinism out of the curriculum—or, should the latter fail, as is at first likely, getting ID solidly *into* it, now. Since this element of the audience is already convinced that Darwinism is both false and evil, the tidings are glad; they buy each new ID version, knowing what to expect in the next speech or volume: more good news. To audiences at the next level, educated laymen, that is, readers who actually try to understand the argument, the ID promoters speak first of other things, especially of their scientific breakthroughs and the feeble opposition of entrenched Darwinists. But in appropriate places they acknowledge that the work of toppling Darwinism is not *quite* finished. Still, they indicate that the prospects are excellent, because Darwinism, or naturalism, or materialism, is in disorderly retreat, its troops deserting.

However, such glib reassurances fail with fully informed audiences, because the audience knows something of the science (and the logic, and the mathematics). In this audience are scientists and philosophers of the unfortunately small band who take the trouble to write reviews of books by others than competitors in their own disciplines. Overwhelmingly, the independent reviewers—and book authors—in this category have found and reported grave deficiencies in ID and in every one of its core arguments, especially Phillip Johnson's repetitious indictments of naturalism, Wells's *Grand Guignol* of scientific conspiracy, Behe's "irreducible complexity," and Dembski's "design inference." There are now many such reviews and books, some of which we cite—almost as many now as there are predictable blurbs and rave reviews, in their own outlets, from CRSC members and media mercenaries.

The Wedge's unremitting campaign for respectability requires the public to believe that Wedge scientists and scholars and those who applaud their work are disinterested investigators, seeking only the truth, and the equals or betters, in stature and in the quality of their results, of the "naturalists" (i.e., all other scientists and scholars). On that basis only can they argue for and hope finally to win, for their version of creationism, authority in the science classroom. If Wedge scientists and theorists are *not* the equals (in science) of their opposition, or if opposition arguments against those Wedge advocacies are plentiful, legitimate, and strong, if they come from recognized scientists and philosophers and are the consensus in the relevant disciplines, then Wedge-style ID cannot be passed off as representative science, much less as a reputable alternative to standard science, entitled to equal time. It is, rather, in those circumstances marginal; at best it is a denial of standard science. It is not theory but the denial of theory. As such it does not belong in, and should not waste the time of, the K–12 science classroom, even if some day some part of the ID argument turned out, against all the present indications, to be true.

How then do the ID theorists respond to the strong professional criticism? They continue to publish and confer and advise all the more frenetically, although never yet in the appropriate place: the scientific journals. Nearly all of Dembski's and Behe's fusillades are fired from their own websites, or from other sites committed to religious apologetics, or via books issued (with a few exceptions) by religiously committed presses. Dembski boasts of the advantage of going straight to the popular audience rather than running the gauntlet of legitimate scientific peer review: "I've just gotten kind of blasé about submitting things to journals where you often wait two years to get things into print. . . . And I find I can actually get the turnaround faster by writing a book and getting the ideas expressed there. My books sell well. I get a royalty. And the material gets read more."[44] Clearly, then, avoiding the scientific venues is a deliberate strategy. To be sure, Wedge authors take note of some of their critics; but typically they do not respond to them adequately. Instead, they concentrate torrents of words on the peripherals of major critiques, or add *new* arguments to their old ones without addressing fully the identified problems of the original. Then they refer to an obscure, or even better, a still-forthcoming book or article by one of them, in which the full answer is supposedly already given or is about to appear (examples follow). That forthcoming book or article does appear—these authors repeat themselves in print at an astonishing pace. But it offers the same arguments, perhaps superficially altered; and this happens on a regular basis for each Wedge author.[45]

Therefore, they have a neat, portable, and convincing (for their selected audiences) defense against professional criticism: they can plead that they are, after all, engaged in *research*, and that things change in the course of research. They can accuse a hostile reviewer of failing to attend to their latest book, or article, or speech, as Dembski recently did to physicist Matt Young, when he chided Young for citing his "semi-popular" work such as *Intelligent Design: The Bridge Between Science and Theology* (InterVarsity Press, 1999): "[A]s a physicist claiming expertise in information theory [Young] has no excuse for not engaging my technical work. . . . Charitable readers with the requisite technical background . . . see the merit of my previous information-theoretic work. Uncharitable readers like Young . . . have been eager to attribute confusion on my part. . . . But let's make a deal. . . . [E]ngage my technical work . . . by reading and citing *The Design Inference* and . . . my newest book *No Free Lunch*. Having engaged that material, give me your best shot."[46]

The book about whose use Dembski complains is a very recent one. Much of it is devoted to an extended summary of Dembski's arguments, especially those of his "technical work." But it was to that public face of the broad argument that Young had responded, with appropriate simplicity and clarity. Thus, any attempt on the part of the critic to keep up means climbing onto the ID treadmill: response and counter-response. Many a potential critic of an ID production must decide, after a glance at the larger body of serious literature in his own field to be read, that life is too short and the gain for the pain of this ID treadmill is much too small.

Whenever a grave fault is found in an argument of Behe, or Dembski, or Johnson, or Wells, it can be dismissed as out of date with respect to other ID "discoveries," or overridden with heavily hedged versions of the original. The newly hedged versions can then be sold to the least sophisticated part of the audience, with, as Dembski correctly observes, royalties expected.

Among all the advocates of ID, William Dembski creates the steepest treadmill, more so even than Behe. Dembski must now do this, because he is the current intellectual leader of the Wedge. His writings are the target of scrutiny by many scholars and researchers in the fields of technical (as opposed to religious) activity he writes about. Thus, he publishes even more impetuously than do the others at CRSC. No scrupulous reviewer of Dembski's work is safe from the rebuttal that he has not kept up with Dembski's most recent or most "technical" thought. Of course, that ought *not* to matter. Dembski is among those who announced long ago that Darwinism was already at least in deep trouble—or dead (although to other audiences he asserts now, as quoted in our epigraph, his hopes for friendly collaboration with its proponents). This constant tinkering with and inflation of what were supposedly knock-down anti-Darwinian arguments in the original form would not be necessary if they really had been knock-down arguments.

Dembski's Argumentative Style

However, the method of technical argumentation employed by Dembski cannot be described, as can that of Wells, for example, as rabble-rousing, or as trifling. In mathematical inquiry—as opposed to Christian apologetics (which is his most fundamental and enduring enterprise[47])—Dembski's style can appear straightforward, using available methods of scholarship in his fields. He has been very well educated and is competent in their use. In Dembski's strictly mathematical and statistical offerings, he does not, as do other ID theorists in their published pleadings, complain of conspiracies or censorship by cabals of materialist philosophers, scientists, and editors (except for his colleagues at Baylor, about whom he has complained quite a lot; he also engages in conspiracy mongering in his more popularly oriented writing, such as "Intelligent Design Coming Clean"[48]). When Dembski's disinterested-logical-scientific style of argument is on display, as in his original and main technical and not explicitly ID opus, *The Design Inference* (published not by a house with religious commitments,[49] but by a university press),[50] it appears erudite almost to a fault. *The Design Inference* brought recognition (not necessarily approval) to Dembski not only from the Wedge's nonmathematical, nonevolutionist admirers but also from peers in the relevant fields. This recognition helps now to identify Dembski as the current intelligence of intelligent design. His most recent (only as we write!) opus, *No Free*

Lunch, reinforces that identification. He is not content, as are most disputants who favor special creation ("intelligent design" *is* a case of special creation),[51] to defend his beliefs by rooting around for weak spots, however trivial, in the biological literature and then announcing his findings. In the part of his work devoted to mathematical evidence of design, Dembski has a quite different project: to prove by formal reasoning that life is necessarily a product of active, purposive intelligence—of an agent.

Dembski, like other ID people, sometimes (but not always!) denies that there is the implication of agency in his argument—usually when he is addressing an audience with possible scientific competence. But design without agency is trivial in the context of biological objects and the history of life on Earth: the presence of design, as ordinarily defined, is already perfectly obvious. Any complex mechanism that functions to bring about predictable change in itself or its environment has a "design" by any normal or comprehensible use of that word. The question is not whether something does or does not strike us as designed, but rather who or what did the designing, and how.

We do not mean to suggest a modesty in Dembski that his ID colleagues lack. Dembski's typical claim, at least up to the publication of *The Design Inference*, and unless he is pushed hard by expert criticism, is to have *proven* the argument from design, or at the worst to be only a hair's breadth away from rigorous proof. The implication is, of course, that he alone has met the immemorial challenge to logic, mathematics, natural science, metaphysics, and moral philosophy, the challenge that had eluded them all until just now: to establish the truth of life's willed designing by an incomprehensibly intelligent agent outside nature. That is why one nonmathematical admirer has called him "God's mathematician" (see chapter 9). Dembski's claim is to have rehabilitated, with the aid of probability theory, information theory, statistics, and his own insights, British (especially Paley's) natural theology, thus showing where David Hume, all the subsequent great skeptics, and all theoretical biologists have gone astray in dismissing the argument from design. That is not the voice of a modest younger scholar.

David Berlinski, a mathematician associated with the Wedge and beloved of neo-conservatives for his wordy and splenetic antagonism to evolutionary biology, atomic physics, cosmology, and other modern science that annoys him, provides a strong blurb as front matter (not merely on the dust jacket) to *The Design Inference*. In it he commends Dembski's style of discourse as, among other virtues, "modest."[52] To that, physicist-engineer Mark Perakh replies, in a searching review of Dembski's writings, "Dembski's style reveals his feelings of self-importance, which is obvious not only from his penchant for introducing pompously named 'laws' but also from his categorically claimed conclusions and such estimates of his own results as calling some of them 'crucial insight,' 'profoundly important for science,' or 'having a huge advantage' over existing concepts."[53]

Intelligent Design: The Bridge Between Science and Theology represents, perhaps better than all his other works to date, the unique mixture of evangelical Christian fervor, abstract philosophic-mathematical argument, and brash self-confidence that characterizes Dembski's writings and delights creationists. It is, in fact, a collection, with slight modifications, of materials already published elsewhere, particularly in Christian theological journals and in *The Design Inference*. It is a careful work, not of new substance but of popularization of the "scientific" arguments, not the theological ones (which do not change). It is suitable for its intended audience, that is, people interested in theology, but it does *not* fail to present all the technical positions. Dembski advises his readers that

> I shall show that detecting design within the universe follows a well-defined methodology [Dembski's own methodology]. Moreover, when applied to the irreducibly complex biochemical systems of Michael Behe, this methodology convincingly demonstrates design.
>
> The implications of intelligent design are radical in the true sense of this much overused word. *The question posed by intelligent design is not how we should do science and theology in light of the triumph of Enlightenment rationalism and scientific naturalism. The question rather is how we should do science and theology in light of the impending collapse of Enlightenment rationalism and scientific naturalism.* These ideologies are on the way out. They are on the way out not because they are false (although they are that) or because they have been bested by postmodernism (they haven't) but because they are bankrupt. They have run out of steam. They lack the resources for making sense of an information age whose primary entity is information and whose only coherent account of information is design.[54] (emphasis added)

The "coherent" account, of course, is to be understood by his readers as Dembski's own, according to which rationalists and naturalists—including almost all the world's scientists—have not yet recognized that their brief candle is guttering out. (For those readers who may be uneasy about the philosophical status today of the argument from design, and of the natural theology whence it sprang, as well as its relationship to the other traditional arguments for the existence of God, any recent dictionary of philosophy will suffice.[55] There, references pro and con will be found, with proper examination of both in light of contemporary philosophy of science.)[56]

Technical philosophic detail is not needed, however, for our comment on Dembski's contributions. For our purposes, the argument from design, ancient or modern, is just this: human life, all its component organisms, and their component parts, are too complex and too clearly the products of an antecedent *idea* (their complexity, in other words, is not random). They manifest a purpose. They are too much like things that we *know* to have been designed to have arisen without external guidance of the process. Therefore, they cannot have arisen in nature spontaneously, that is, "by chance."

To the Designer by Elimination

The foundation on which Dembski's technical argument against Darwinism rests is that Darwinism, hence most of modern biology—in his account of it—insists that diversity and the adaptedness of living things arose "by chance." He clearly believes that if, by logic, mathematics, and introspection, he can show that chance cannot explain diversity and adaptedness, then he has taken that great, single step needed to prove that design, meaning a designer, an agent, is the first cause of life (and everything else). This is a classical argument by elimination. It is in fact the old "God-of-the-Gaps" argument. Dembski denies that agency is a necessary element of his argument, or even of "intelligent" design; but elsewhere in his works agency is very much in the forefront of his argument, as it must be if the works have a theistic purpose. Dembski cannot abandon agency—that would lose him most of his audience. (Dembski encourages people to believe he takes seriously the possibility of the designer as an "extraterrestrial." But in1992 he himself discounted that possibility with his "Incredible Talking Pulsar," describing the pulsar as "the mouthpiece of Yahweh" and the intelligence communicating through it as "not a super-human intelligence . . . realized in some finite rational material agent" but "a supernatural intelligence . . . who is both intelligent and transcendent." See note 20 in chapter 9.) For a serious reviewer, however, there is no keeping up with these switches, from book to book and chapter to chapter, on the question of agency. This point is of particular concern to Dembski's critics. Zoologist Wesley Elsberry, for example, in his review of *The Design Inference*, has documented it carefully.[57]

It is appropriate nevertheless to start even the shortest discussion of Dembski's arguments with the best-known dismissal of the original argument from design, whose rehabilitation and elevation to the top of all intellectual life is Dembski's announced project. Of course, there is nothing wrong with reinstating the argument from design, if it can be done. The fact that the great eighteenth-century philosopher David Hume weakened it to irrelevance in his *Dialogues Concerning Natural Religion*, or that most philosophers of science thereafter have agreed with him, is not a reason to avoid trying. The need is to introduce truly new arguments in favor of design (or better teleological arguments for the existence of God than Paley's), or sound empirical evidence that Hume could not have known and that is *not* undone by his reasoning.

In *An Inquiry Concerning Human Understanding*, Hume recognized that for every claim of a miraculous (or supernatural, or hugely improbable) event, there must be some evidence on which the claim is based. Such claims, though extraordinary, are nevertheless factual claims; they therefore require supporting factual evidence. We justifiably believe, after all, only on the basis of *some* sort of evidence, even if that is no more than the testimony of witnesses to a purported miracle (Hume recognizes the problematic nature of such testimony). As Hume also points out, our "only guide" concerning evidence for factual claims is experience, that is, empirical observation. And the credibility of miracle claims stands or falls with empirical evidence.

The evidence germane to any claimed miracle will include that in favor of or against it. Hume's test is simple: if, after an inquirer sums up the evidence on each side, the evidence in favor of the supposed miracle is stronger than all the evidence against it, then it is rational to believe the claim. But if the evidence against the claim is stronger, then a miraculous explanation is irrational. Hume's argument continues then to show that, notwithstanding the human inclination toward acceptance of miraculous claims, the evidence *for* a miracle *cannot* be stronger. But that part of the argument need not concern us.[58] What should concern us is that modern versions of the argument from design are in effect claims of miracles, and there has not been a successful creationist counter to Hume's basic argument;[59] nor has there as yet been a convincing counter-argument from Dembski, who, among all the CRSC troops, has had to try hardest and most often to counter it. Had there been a convincing counter-argument, every philosophy department in the world would be emphasizing it in the introductory course!

In theology and in some versions of ID-creationism, there are recurrent weak replies to Hume. A favorite is that life's design need not be miraculous, or words to that effect: the designer could work or have worked through discoverable natural processes. That is indeed a possibility; but the pathway from designer to such processes, the actual mechanism of the designer's activity, has yet to be discovered and so remains, so far as we now know, miraculous. Everything else said by advocates of ID, however, makes it plain that supernatural agency and supernatural processes, specifically, *are* being proposed. What they assert is the intelligent conception, anticipation, and invention of the mechanisms of life and its actual assembly from the raw—inorganic—materials of a prebiotic Earth or of some place not on Earth. Dembski asserts unambiguously that supernatural agency is the only acceptable answer to the question of the source of the world's creation and order:

> This is the mystery confronting the scientist. Why is the world ordered and whence cometh this order?
>
> There are but two options: Either the world derives its order from a source outside itself (à la creation) or it possesses whatever order it has intrinsically, that is, without the order being imparted from outside. So long as the order is coming from outside, we are dealing with a world that is a creation. On the other hand, if the order belongs to the world intrinsically, we are dealing with nature. The question *Whence cometh the order of the world?* is one of the most important questions we can ever ask. . . .
>
> Throughout Scripture the fundamental divide separating humans is between those who can discern God's action in the world and those who are blind to it. . . . This severing of the world from God is the essence of idolatry and is in the

> end always what keeps us from knowing God. Severing the world from God, or alternatively viewing the world as nature, is the essence of humanity's fall.[60]

So in Dembski's view, modern science is idolatry. Moreover, his description of creation as a divine speech act qualifies fully as a miracle:

> God's act of creating the world is thus the prime instance of intelligent agency.
> Let us therefore turn to the creation of the world as treated in Scripture. The first thing that strikes us is the mode of creation. God speaks and things happen. . . .
> The language that proceeds from God's mouth in the act of creation is not some linguistic convention. Rather, as John's Gospel informs us, it is the divine *Logos*. . . . For the divine *Logos* to be active in creation, God must *speak* the divine *Logos*.[61]

This is, in short, creation by divine fiat. Given everything else we have seen and know about nature so far, such divine creation of the mechanisms of life would be a miracle, whether or not so named. However, the evidence for the scientific view of the origin of both the prebiotic Earth and its biotic processes—the sum total of the hard facts and consensus interpretations of geology and paleontology, evolution observed in living species (to which even CRSC spokesmen accede, albeit as "microevolution" only), the supporting evidence from applied biology and medicine, and from cosmology—is *enormously* more weighty than any purported evidence for the miracle. No mathematical refinement to the design arguments has changed that fact.

When the large output of words on paper from Behe, Wells, et al. is not aimed at finding little faults in the literature of modern biology, or at dodging expert criticism, it simply declares that strong, empirical evidence *does* exist for supernatural design in living things. Indeed, in both sources just cited, Dembski asserts that "the crucial breakthrough of the intelligent design movement has been to show that this great theological truth—that God acts in the world by dispersing information—also has scientific content." So far, however, they have not made the content of that "breakthrough" available to neutral scientific inspection. Despite such boasts of success, therefore, Dembski is forced to find an alternative to hard empirical evidence. That can be only through abstract means—logical, mathematical, or statistical—of circumventing the need for empirical evidence. If Dembski is going to prove that intelligent design exists, or happens, or did happen—and that design does not result from the operation of natural laws or by "chance"—then an abstract means is his only alternative.

Dembski's Solution: Adding Design to Chance and Necessity

In *The Design Inference*, Dembski elaborated in words and abundant symbols his happy solution to this problem. Two key chapters of the succeeding Dembski opus, *Intelligent Design*, are a careful restatement thereof, stripped of the more tedious (and excessive) mathematical formalisms in *The Design Inference* but nevertheless complete, and addressed to intelligent lay readers. Gert Korthof, a biologist and one of many expert critical reviewers of Dembski's argument, places this solution in comparison with the well-known aphorism on scientific explanation originated by the articulate Nobel laureate geneticist, Jacques Monod. Korthof writes, "In the natural sciences there are only 2 types of explanations: natural law and chance. Jacques Monod explained this in his famous and aptly titled *Chance and Necessity*. Dembski adds a third type of explanation: Design."[62] And so he does. The core argument of ID is now, together with Behe's claims for irreducible complexity, Dembski's *explanatory filter*, a hopeful expansion of Monod's categories that adds design, as an independent and fundamental category of authentic explanation, to necessity and chance. This is the "well-defined," foolproof methodology to which Dembski refers. Dembski offers no empirical evidence except (1) Behe's "irreducible complexity" argument, which is not empirical and is, as we have seen, rejected by nearly all scientists in the relevant fields: genetics, cell biology, and biochemistry; (2) the "fine-tuning" of universal constants, a popularized statement of the strong anthropic principle, the meaning and interpretation of which is at best opaque, by their own admission, to most cosmologists (and for a sharp dismissal, see the statement of Nobelist Steven Weinberg)[63]; and (3) certain data on amino acid sequence and function in enzymes, data he is not qualified by training or experience (in molecular biology), or by the evidence of which he writes, to judge.

Dembski's design inference is on its face a rather simple algorithm, which is to say that it is a fixed procedure for answering a question or solving a problem. Followed faithfully, Dembski says, the design inference will always yield the same correct result. Stipulate first that explanations for all the objects and events in the world are of three *and only three* (not just two) kinds. That which Monod called "necessity," Dembski calls "regularity," meaning, approximately, "natural law." These phenomena are said to occur with high probability from a given set of initial conditions. For example, the motions of the planets are explainable through Newton's laws. (Note that the relations among explainability, probability, and natural law are not in the least obvious, but Dembski implies that they are.)

Other phenomena are not predictable consequences of initial conditions and therefore cannot be explained by "regularity." Presumably, such events occur with low probability and can therefore be assigned to chance. But certain phenomena have not only a low probability of occurrence but also recognizable patterns that preclude mere chance as a plausible explanation. (No one would "explain" Mt. Rushmore's sculptures as a consequence of weathering.) So, if we can recognize such a case—

extremely low probability of occurrence but with recognizable, meaningful patterns—then we have eliminated not only regularity but chance as well. If so, we are left with the third category of cause or explanation. For things that are highly improbable but not accidental (or random, or spontaneous), Dembski argues, the cause or explanation must be—design. Under this obligatory triadic schema, it is the only thing left. So the explanatory filter is an algorithmic method of logical inference, by which it is claimed we can reliably discover design, anywhere in the physical world, without false positives, by a process of elimination.

If the algorithm is to be employed practically in detecting intelligent design, this last is critically important. But if we can indeed employ such a process of elimination as Dembski's conceptual filtering system for living things, with no more physical data than we already have, then we can claim, without troubling to find direct physical evidence of it, that life must have a design*er*. We have discarded every other form of explanation we (Dembski) can think of. What is left at the end need not be proved (and Dembski does not prove it): it is just all that's left. Dembski admits, of course, that we cannot do this elimination *quite* so simply. We need very strong reasons and qualifications, beyond the probability value but inherent in the properties of a low-probability object or event, for eliminating "chance."

Dembski's claimed discovery, densely elaborated in ordinary and mathematical language and growing ever more complicated in succeeding books, is a collection of rules and reasons for that most critical step: the elimination of chance. Here it is, still in bare outline: first, the phenomenon in question must be complex, that is, it must occur with low probability. Then, that complexity must display a pattern identifiable independently of the phenomenon in question; it must show *specified* complexity. Of course, "complexity" implies high "information" content, although in a very loose way. So it is complex specified information (CSI) that is Dembski's infallible indicator of design. Show that a very rare thing or event has CSI, and you have eliminated chance as well as regularity. That is the claim of Dembski's proposed design inference: you have proven that the thing needing to be explained is *inexplicable*—beyond the certainty that it was designed. It should be obvious that elaborating this as a foolproof method for detecting design in nature entails logic, information theory, and statistics. And those, when he is not writing theology, are the tools for Dembski's constantly growing appliance, the design inference.

Why is it "constantly growing"? Because what must have looked to him and his admirers, a few years ago, like a knock-down case for design (therefore against "chance"; therefore, presumably and hopefully, against "Darwinism," which creationists misrepresent as chance) has turned out *not* to be at all knock-down. As criticism mounts, Dembski has had to respond. Unlike some of his ID peers, he has responded in detail to some critics, although not necessarily directly to their objections, and rarely satisfactorily. In the accumulation of these responses, however, the explanatory filter has acquired many hedges and qualifications—verbal and symbolic. The argument is now either hedged or sabotages itself at each nodal point of the algorithm: it has lost its original simplicity and brio. It is in fact difficult to follow now (because of ambiguities and inconsistencies), even for experts who have taken the trouble to follow it.

For most of the readership toward which, for example, *Intelligent Design*, rather than *The Design Inference*, was aimed, that doesn't matter. They are likely happy with what sounds like a good bottom-line result and do not argue about the adequacy of the statistics, mathematics, thermodynamics, information theory, or biology. They are delighted that a fellow true believer can speak for them in those arcane languages. But for those who can inquire into the adequacies, the argument has grown weaker with each qualification and in each new publication. Detailed critiques show this; here we can represent them only by abbreviated example, but we supply references to the literature. However, two things can be said about the structure of Dembski's argument concerning features that are not unique to it or indeed to the CRSC's work. These concern (1) probabilities (especially vanishingly small probabilities) and their obverse in this particular context, their information; and (2) the description of Darwinism and the ID-encouraged public impression of it: that it is a theory in which *chance* is the explanation (and thus ultimately the meaning) of life.

IMPOSSIBLE PROBABILITIES

First, then, there is the fascination with elementary probabilities in complicated, especially biological, processes and the meanings of such numbers. This becomes a set of arguments about inconceivably small (or very large, approaching certainty) probabilities. It is a domain in which the proper application of mathematical statistics to particular situations whose details are not known or knowable is highly uncertain, even for experts. But fascination with extreme probabilities in biology accelerated greatly forty years ago, when the new genetics illuminated processes underlying the long-known, microscopically visible behavior of chromosomes (hence, also of the genes they bear) in cellular and organismal reproduction. Students in a college biology course (who presumably had learned a little mathematics) might have been asked to calculate the number of different chromosomal *kinds* of fertilized eggs (zygotes) that could be produced by a sexually reproducing pair of a species (like *Homo sapiens*) with twenty-three pairs of chromosomes (forty-six total), excluding all known complications such as random chromosome breakage and reunion. The result, without the (real) complications, is a number larger than seventy quadrillion ($>7 \times 10^{16}$). In a useful elementary probability textbook of the period, exactly this calculation was demonstrated, illustrating the enormous expansion of genetic variability provided by sexual reproduction.[64] But the same sort of calculation was being used, and continues to be used, to demonstrate the ridiculously small probability—effectively, the impossibility—of getting any particular one of those equally possible zygotes (fertilized eggs)!

But ridiculously improbable things happen all the time, as Dembski, to his credit, admits. At least one of those zygotes would happen every time! Very low probability does *not* mean impossibility. An identifiable boundary or threshold between improbable and impossible is a time-honored speculative claim, especially, but not uniquely, of creationists against evolution. In fancy form, Dembski, too, uses it. Misuse of the basic rules of counting and exploitation of the public's innumeracy, consciously or from not knowing what is actually being counted, has been a continuing effort of creationists, especially of the few who are mathematically inclined. Mathematical deductivists, who believe that scientific arguments must be constructed as formal, logically valid mathematical arguments, and who may have some axe to grind against biology or cosmology or materialism generally, love to "prove"—by displaying huge negative exponentials—that standard evolutionary biology (Darwinism) is false *because* the mathematical probability of evolution is so low that it is effectively impossible. Trotting out absurdly small probabilities, such as for the spontaneous (one-step, not intelligently designed) appearance of even a small protein of specific amino acid sequence, remains the commonest form of argument that something other than "chance" must be the explanation and that therefore Darwinism is a snare and a delusion.[65] This, when developed with mathematical notation, has now been named (with delightful irony) the Basic Argument from Improbability—BAI. But such an argument is specious: no protein ever assembled itself in one step from all its amino acids, and no reputable scientist has ever supposed that it could; that is not how proteins are synthesized. Nevertheless, by showing how wildly improbable that would be (it is!), one can imply for the mathematically and biologically unsophisticated that proteins can never have evolved at all.

Physicist-engineer Mark Perakh, in his review of Michael Behe's *Darwin's Black Box*, describes this situation economically:

> Behe seems to assume that an event whose probability is 1/N, where N is a very large number, would practically never happen. This is absurd. If the probability of an event is 1/N it usually means that there are N equally probable events, of which some event must necessarily happen. If event A, whose probability is very low (1/N), does not happen, it simply means that some other event B, whose probability is equally low, has happened instead. According to Behe, though, we have to conclude that, if the probability of an event is 1/N, none of the N possible events would occur (because they all have the same extremely low probability). The absurdity of such a conclusion requires no proof.[66]

And the distinguished biochemist Russell Doolittle has commented on the matter as follows: "[T]he next time you hear creationists railing about the 'impossibility' of making a particular protein, whether hemoglobin or ribonuclease or cytochrome c, you can smile wryly and know that they are nowhere near a consideration of the real issues."[67]

William Dembski is not guilty of the silly form of this argument, but he does use it in more sophisticated forms, as his mathematically literate critics have shown. His newest book, *No Free Lunch,* is an attempt at a comprehensive reply to all the expert criticisms to date, including those on probability. The book carries small probability–mongering to extremes. For example, long sections of chapter 5 (Dembski's attempt to dispose once and for all of expert rebuttals of Behe's "irreducible complexity") are effectively a user's manual of applied BAI. Here, by these methods, Dembski also tries at last to present as evidence for ID the innocuous but exploited (by the Wedge) enzyme structure-function findings of Douglas Axe. This effort is not made, however, by molecular biologist Axe, who wrote the real article. His publications in scientific journals are silent on ID. The effort is in the opinions and arguments of Dr. Dembski (who is not a biologist, molecular or otherwise). Dembski provides only a vague statement on the possible ID significance of Axe's results, and nothing more. His attempt is hedged with such safe, logic-sabotaging locutions as "*preliminary indications are that* proteins permit a perturbation tolerance factor of no more than 10 percent (thermodynamic considerations *seem to preclude* proper folding for more than this percentage of random substitutions)"[68] (emphasis added).[69] So the "evidence" Dembski has promised repeatedly, the calculations his critics have demanded—the clear, quantitative traces of intelligent designing—are a mere "preliminary indication" in yet another verbally gifted exercise with infinitesimal probabilities.

But he continues to insist that his logic has yielded a foolproof method for finding sound empirical (inductive) evidence for design. But proof of design through deduction is not a promising or, in the view of some, even a possible mission—deductive arguments are not empirical evidence. Preliminaries and vague possibilities, probabilities huge or infinitesimal for systems whose detailed operations are unknown (e.g., all the steps, over billions of years, by which a particular protein came to its present composition), are not adequate evidence for design, and they are certainly no justification for announcing that they *are* evidence. Meantime, weary critics are beginning to determine what, if anything, is really new in *No Free Lunch*. They are not impressed. Wesley Elsberry, a close analyst of the book, reported after a first reading:

> Dembski has, so far, not analyzed potential counterexamples [to irreducible complexity and intelligent design in complex biological systems]. I proposed at Haverford College last June that Dembski "do the calculation" for the Krebs cit-

> ric acid cycle and the impedance-matching apparatus of the mammalian middle ear. Dembski has not done so.
>
> In other places, Dembski fails to take up the arguments of critics, as in Dembski's mischaracterization of a program written by Richard Dawkins. Two out of three of the steps that Dembski says characterize the program are, in fact, Dembski's own invention, appearing nowhere in Dawkins's work.[70]

Still, some mathematicians today and in the past (including some Darwin contemporaries) have used and continue to use plain and fancy forms of BAI. It is, among other esoteric pleasures, a means of trying to show that mathematical brainpower triumphs over 150 years of plodding field, laboratory, and theoretical work in biology. A short but elegant dismissal of this enterprise (i.e., low probability of some biological phenomenon; ergo, Darwinism false) from mathematician Jason Rosenhouse responds to the new outburst of BAI since the founding of the Wedge. BAI is in fact Rosenhouse's coinage. For the mathematically literate, his long letter (in fact an essay) in a recent issue of *The Mathematical Intelligencer* should, and perhaps may, discourage this category of foolish arguments against evolution.[71] He has enlightening things to say also about IC and certain creationist "improvements," old and new, on the science of thermodynamics. Of the latter, Dembski provides a sterling example in the announcement of his new scientific law: the "Law of Conservation of Information." So far, thermodynamicists and information theorists are deeply unimpressed.

DARWINISM AS CHANCE?

What then, of "Darwinism" portrayed (falsely) by ID as the doctrine that *all* the physical phenomena of life are due to *chance*? That, too, is a time-honored tactic of creationism. To put that attribution in its lowest common denominator (LCD) form: according to the creationist rendition of Darwinism, animals, or living cells, or—now, in the new day of ID—metabolic pathways, or even single protein molecules, in all their mind-boggling complexity (in fact, their improbability according to BAI), *have* assembled themselves purely spontaneously, by "chance," at random. We are supposed to laugh in disbelief at this, as indeed we should. Dembski, like the rest of the CRSC, does nothing in most of his expositions to disavow the claim that Darwinism is really as described in this LCD. By choice of words and style of writing, especially in his more popular and apologetical offerings, he *encourages* it, although he certainly knows better. But he always leaves sufficient space for his own escape in case of challenge.

Dembski knows that the most common (but not the only) Darwinian process is natural selection (which he disparages by calling it "the Darwinian mechanism")—the culling for greatest reproductive success, from among all the reproducing variants in a population, where the number of variants is always very large. The selectable "variants" are the genomes—the gene-sets or *genotypes*—of individual organisms, because those genomes are the most important (but not the only) contributors to *phenotype*—the size, shape, and function of the individual organism in the given environment. And it is the phenotype—the individual bird, tree, or human—that lives to reproduce or fails to do so. The range of "variants" available at any moment is a matter of chance, since it is due to random mutations of and recombinations among genes in preceding generations and in the one about to reproduce. Selection, however—the choice of individuals who will survive to reproduce their particular variations, which is, in effect, the choice of which variants have an opportunity to become the norm in succeeding generations—*is the very opposite of chance*.

Selection, a metaphor Darwin himself used to mean the environmental preservation of an individual genotype, is *determined* by environment, by environmental opportunities and constraints in each generation, and therefore by how, and how often, the environment changes, either by its own laws or because the population relocates.[72] These variants become part of an environmental setting that constrains, or channels, the development of the populations in which they occur. The selecting environmental pressures are in effect "regularities," that is, temporary, local laws of nature, which have a discernible, determinative effect on which phenotypes reproduce and pass their genes to their offspring. Originally deleterious mutants and other variants can, and sometimes do, become favorable when the environment changes even slightly. Regardless of which genotypes are preserved by their environment or why, the fact remains that immediately subsequent to the variations, the determinative factors of an organism's environment begin to operate as *nonrandom* elements in the process of natural selection.

All these phenomena are massively documented and visible to any investigator. Immigrants to a population, some of whose properties differ from those of the indigenous members, are yet another source of variation; and the environment does change, constantly and erratically. Mostly it changes imperceptibly over short periods, centuries. Sometimes, though, it changes catastrophically, within minutes or hours. Always, over long periods, the changes are enormous. Earth is a very active planet: geologically, it is alive. Earth itself has been evolving—physically, "geologically"—for more than four billion years. So it is dishonest to lead readers and audiences to think that once "chance" is eliminated by deductive argument as a mechanism of evolutionary change, Darwinism is finished.

By reading original ID documents such as those we cite, including Dembski's energetic but highly selective defenses against critics, our readers will find that except in technical replies to a few of those critics, the advocates of ID and of all "scientific creationism" actively encourage the scientifically naïve to confuse chance and randomness with Darwinism. By stipulation, the argument of the "explanatory filter" means (until Dembski hedges it, as eventually he does) that anything which *cannot* be explained by the frequent, regular operation of known natural laws, or simply as an accident, *must* be both very rare and designed. That is not so, even accepting the idiosyncratic stipulation of only three alternatives (whose defined probabilities are never fully justified). Moreover, even if it were so, and we were to call the successfully filtered object or event "designed" according to Dembski's inference, there would be no necessary intelligence in that design, if "intelligence" means what it usually means. There would be no implication of *will* or *conscious purpose*—unless one were to make the additional, unjustified stipulation that environmental change itself is intelligence! "Design" is perfectly comprehensible as a natural concept, and it is interpreted this way all the time in other natural contexts: the naturally occurring design/pattern left by waves on a beach, or the design/configuration of atoms in a molecule of water, or the design/symmetry of snowflakes, and so on. We see "design" in the genetic patterning of a population by the environmental preservation of particular genotypes. There is nothing extraordinary about design in this sense; it is a perfectly ordinary way to understand it. Design, therefore, does not logically entail a designer, and we use the word in this noncontroversial, nondesigner sense all the time.

Nor is there any reason why an object cannot be due to chance *plus* regularity: genetic variation more or less by chance; selection by demonstrably "regular"—lawful—processes. Dembski's heroic labors in disputing and obscuring this flaw in his argument, in response to critics who have identified it, have been so far in vain. We cite some published representatives of this and related complaints from experts, together with samples from Dembski's responses. These are samples only; we could have made similar reference to any one of a dozen other objections.

DNA DESIGNED?

Biologist Gert Korthof, who has reviewed *Intelligent Design* and is aware of other Dembski and CRSC offerings that celebrate the supposed "discovery" and "proof" of ID, focuses exclusively but at length on DNA and real (functional) genes, especially on the problem of quantifying and assigning meaning to the *information content* of genes in DNA (most of the body's DNA is not working genes, but rather pseudogenes or "junk" DNA).[73] He is concerned with the extent to which functional DNA *has* been shown by Dembski, or *can* be shown using Dembski's method, to be intelligently designed. This is as much as Korthof, who seems to maintain neutrality on the merits of neo-Darwinism generally, analyzes in his review. His firm conclusion is that Dembski has not shown that DNA is intelligently designed. That, Korthof explains, means that the whole design inference technique fails in its purpose of detecting design:

> But it also follows from the [Dembski's] definitions that a piece of 'junk DNA' (non-coding DNA) of 1000 bases *has about the same information content as a gene of 1000 bases.* This is because both sequences fall into the category of mathematical random sequences. So clearly Dembski needs an extra criterion to detect meaningful DNA and he proposed 'specification.' However a definition of 'specified DNA' is absent in his book. That means that Dembski is not (yet) in a position to make meaningful claims about Complex Specified Information in DNA, let alone claim that a specific piece of DNA is 'intelligently designed.'[74] (emphasis added)

In this review and in those from the physicists, there is detailed argument that so far as Dembski's algorithm goes, his basic information-content argument, which is central to use of the algorithm in demonstrating design, is both immaterial and wrong. And this is after giving the design inference every possible benefit of the doubt in its various ad hoc definitions (e.g., of a "universal probability bound," the cut-off point beyond which the improbable becomes the impossible) and its blind spots (e.g., to the fact that the 3-D structure of a protein *cannot* be calculated from its gene's DNA sequence; thus, the DNA information, however calculated, cannot in any case be the "design" of the protein).

Illustrating one objection to the claims of the design inference, Korthof speaks of the Fibonacci sequence. Leonardo Fibonacci (de Pisa, A.D. 1170–1250) discovered it in seeking the solution to an algebraic problem he had set himself. Each number of a Fibonacci sequence is the sum of the two previous numbers, as in 1,1,2,3,5,8, so that, in general,

$$a_{n+1} = a_n + a_{n-1} \text{ for } n > 2.$$

Such series are sometimes observed in nature. Their recognition as such is rather a surprise. One example is the spiral arrangement of leaves on the stems of some higher plants. Korthof shows that the application of Dembski's design filter to a case of Fibonacci numbers found describing a biological structure, following all the (Dembski's) rules, positively identifies that arrangement ("event") as having CSI. If so, it must be explained by and be a product of intelligent design, as Dembski wants it defined. But clearly it is not that. The Fibonacci-number arrangement of leaves is not complex in any ordinary sense, although it would be under Dembski's idiosyncratic definition of complexity. It has a quite simple, entirely *natural* explanation, as do many other, similar instances of such apparently complex (and often beautiful) arrangements in biology.[75] Thus, it yields a false positive to the design inference.

Dembski's eventual response to this is typical. It is the same as the response he gives to all other examples of false positive results obtained from his explanatory filter, the design inference. He dismisses the argument on grounds that the ID is not in the design itself, but in some generative processes antecedent to the actual emergence of the CSI—in the present case, of the leaf positioning. As critics produce example after example of CSI in events that are clearly *not* "designed" in the ID sense, he claims that the universe as a whole was designed at the very beginning of time ("fine-tuned" is the term now in use by creationists).

Now, any diagnostic procedure that gives false positives is at least not foolproof. But the one indispensable and unchangeable requirement for the whole of Dembski's argument is that it gives *no* false positives. A finding of ID by Dembski's explanatory filter, in some biological object, may well then be false even though the object clearly manifests "design" in the ordinary sense of recognizable organization for some function. An ostensibly intelligently designed system may well be a system that seems "designed" only in the ordinary, trivial sense, but not by anything we or Dr. Dembski would consent to calling "intelligent." So the filter is *not* that long-promised, infallible pointer to the defeat of Hume and the rehabilitation of Paley (and Thomas Aquinas).

Dembski's first response to this and Korthof's other polite objections was a case of the treadmill. Responding via e-mail to Korthof's objection that natural selection *can* produce the phenomena Dembski claims are the work of a designer, he simply assures Korthof that it cannot, asserting that his *previous* publications *do* present effective arguments against natural selection. He then hedges by conceding that his arguments can be made more effective, but characteristically he concludes by informing Korthof that he is writing yet *another* book that will do this.[76]

We offer no comment on this response beyond the expectation that readers will find it inadequate. In due course, Dembski took up this argument again and, as noted, dismissed it this time by attributing ID not to the case in point (the Fibonacci pattern itself) but to some remote process of its origin. He uses various methods of defending against objections like Korthof's, and he trots them out at length in *No Free Lunch* (e.g., pp. 12–14). But at no point does he concede, in his boastful campaign for the Wedge, that the design inference as given needs, well, quite a lot of work before it can be taken seriously.

HOW NOT TO DETECT DESIGN

Branden Fitelson, Christopher Stephens, and Elliott Sober, philosophers at the University of Wisconsin, chose the title "How Not to Detect Design" for their 1999 review of Dembski's *The Design Inference*. Theirs is a professional analysis of Dembski's technical argument as it appeared in a university press monograph. It focuses on the logic—symbolic and otherwise—and to a lesser extent the statistics of the explanatory filter.

The conclusion is not a happy one for Dembski. So serious are these authors about their conclusion that they place it at the beginning of their review, rather than more conventionally at the end. It is the second of the two introductory paragraphs to the essay:

> Dembski's book is an attempt to clarify these ground rules [of the design inference]. He proposes a procedure for detecting design and discusses how it applies to a number of mundane and nontheological examples, which more or less resemble Paley's watch. Although the book takes no stand on whether creationism is more or less plausible than evolutionary theory, Dembski's epistemology can be evaluated without knowing how he thinks it bears on this highly charged topic. In what follows, we will show that Dembski's account of design is deeply flawed. Sometimes he is too hard on hypotheses of intelligent design; at other times he is too lenient. Neither creationists nor evolutionists nor people who are trying to detect design in nontheological contexts should adopt Dembski's framework.[77]

There are eleven subarguments in Dembski's broad claim that his design inference is both universal in application and *reliable*, producing no false positives or negatives. Fitelson et al. touch on them all and find trouble with each one. For example, they find repeated internal contradictions to the stated foundational principles of the design filter. They cite sequential page numbers on which a necessary connection between *design* and *agency* is either affirmed and then denied, or denied and then affirmed. They find that the stipulated parsimonious ordering of "regularity," "chance," and "design" (high/moderate/low probability) is arbitrary and indefensible. They explore the idiosyncratic *specification* conditions (by which chance is eliminated, leaving only design as the explanation), and they find that those conditions do not, and cannot, identify the "specificity" that Dembski must have in order to eliminate regularity and chance. They argue that the critical probability threshold (or "probability bound") on which Dembski's handling of chance depends is not justified.

These are a few of the reservations set forth by Fitelson et al. (For qualified readers who wish seriously to judge the quality of Dembski's work, there is really no alternative to consulting *The Design Inference* with at least this critique alongside it.) Because of its high visibility among philosophers of science, Dembski felt obliged to reply to this essay promptly and directly, not by referring to a forthcoming book or an obscure lecture. And so he did, although on his own website. This time, the answer was five pages long (in response to the fifteen dense pages of Fitelson et al.). And, although some three of those five pages are padded with bravado, two others do speak to the charges of flawed argument.

But they do not put those charges to rest. If anything, they highlight what is becoming a recognizable Behe-Wells-Dembski method of attempting to dispose of expert criticism. There is likely to be, first, a powerful hubris, confident assertions that the critics just don't get it. For example, in his reply Dembski writes:

> Not everyone agrees. Elliott Sober [one of the Fitelson et al. authors], for instance, holds that specified complexity is exactly the wrong instrument for detecting design. . . . In this piece I want to consider the main criticisms of specified complexity as a reliable empirical marker of intelligence, show how they [the criticisms] fail, and argue that not only does specified complexity pinpoint how we detect design, but it is also our sole means of detecting design. Consequently, specified complexity is not just one of several ways of reinstating design in the natural sciences—it is the only way.[78]

A careful reader will notice that nothing has as yet been said in these sentences that even hints at a substantive rejoinder to Fitelson. But Dembski here identifies specified complexity, first, as the (only) reliable empirical marker of *intelligence* and second, as a detector of *design*. We must conclude that he means it all; that is, reliable detection of design by an intelligence, *not* just design alone. Here, Dembski does not hedge: he claims that the filter detects intelligent design, whether Sober and his colleagues agree or not. Beyond that, the reader learns only that Sober disagrees with Dembski and therefore that Sober has failed to get it. Next, Dembski provides an omnibus explanation for all objections to his much-disputed way of using probability in connection with information and complexity:

> Now, statistical decision theorists have their own internal disputes about the proper definition of probability and the proper logic for drawing probabilistic inferences. It was therefore unavoidable that specified complexity should come in for certain technical criticisms simply because the field of statistical decision theory is itself so factionalized (cf. Bayesian vs. frequentist approaches to probability).

The meaning is not hard to see: Dembski's own practice of statistical decision theory rises above the quarrels within that discipline; *his* methods are self-evidently correct and there need be no doubts about his application of statistical decision theory. (Yet if that application is questionable, his entire argument disintegrates.) He is satisfied, so his readers should be satisfied, too. Implication: his claim that intelligent design is now established, based on his (idiosyncratic) views of the meaning and use of probability and information, is perfectly sound. Still, leaving no hole unplugged, Dembski acknowledges later that there may well be further quibbles from the experts in decision theory: "Such diversity of formulations is fully to be expected given the diversity of approaches to statistical decision theory. *Consequently, the concept of specified complexity is likely to undergo considerable fine-tuning and reformulation in coming years*" (emphasis added).

Fair enough! But if "considerable reformulation" is imminent, why does Dembski—and why especially do his admiring colleagues and sponsors of the ID movement—tell the world (including members of the Congress of the United States) that their recent discoveries of "specified complexity" and "irreducible complexity" have *already* overturned, or thrown the gravest doubt upon, the body of scientific theory and practice (neo-Darwinism) on which modern biology depends? Why does the Wedge insist that ID belongs in the school science classroom *now*? Among serious scholars, even in ordinary universities, let alone excellent ones, such a gap between the status of one's research and the content of one's announcements about it could be cause for dismissal, or at least for deep suspicions about one's honesty.

The most characteristic, perhaps the most diagnostically useful, paragraphs in Dembski's response to Fitelson et al. address the criticism (made also by others among Dembski's expert reviewers) that "design," if actually identified by Dembski's procedures, cannot demonstrably mean *only* design as the work of an intelligent agent. But Dembski is not at a loss:

> [I]t's not clear why this should be regarded as a defect of the concept. It might equally well be regarded as a virtue for enabling us neatly to separate whether something is designed from how well it was produced. Once specified complexity tells us that something is designed, not only can we inquire into its production, but we can rule out certain ways it could not have been produced (i.e., it could not have been produced solely by chance and necessity). A design inference does not avoid the problem of how a designing intelligence might have produced an object. It simply makes it a separate question.

A neat trick, this. The critics' argument is that a finding of "design" by use of the filter, a finding of specified complexity, might show in principle (not necessarily in practice) that something has features that point to design. The critics, for purposes of argument, concede that much. But then they argue that such a finding cannot rule out something *other* than intelligence as its cause, such as a *combination* of chance and necessity (which produces, for example, the uncontroversial cases of design mentioned earlier). That, Dembski is now saying, isn't *really* a problem. What he has done, he suggests, is to separate the question of design from the question of how the designing was done and to what effect it was done. We suspect that Fitelson et al., concerned about the confusion of epistemic with deductive arguments in *The Design Inference*, were not satisfied by these responses to their analysis. Among professionally qualified reviewers, such dissatisfaction is common.

There is much more to be said about Dembski's slippery methods of replying to critics. But summaries are not the best possible conveyance. The best (assuming that the relevant original, too, has been read with care) is for the serious inquirer to compare critique and response, point for point, provided that he or she has some knowledge of the subject matter. Readers who care to make such a comparison, and are willing to follow the detail, may examine as an easy first example the dispute between Dembski and one of his articulate critics, philosopher Robert Pennock.[79] This dispute was published as a lengthy online exchange at the Metanexus website. Pennock's fifteen-page response to Dembski's earlier essay, which was itself a protracted response to Pennock's book-length critical study of ID, *Tower of Babel,* provides a vignette of Dembski's defensive technique as well as substantive new criticism of the design inference. In "The Wizards of ID,"[80] Pennock cites and discusses in full Dembski's essay "Who's Got the Magic?" A typical paragraph from Pennock's article represents not only his own complaints about Dembski's astonishingly inconsistent arguing before the public but the complaints of many other Dembski critics:

> In his response to me, Dembski cites variations of deism to show that God could have created without miraculous violations of natural laws. This response is puzzling, in that I had myself discussed the deist option in my book as one way that a person could accept evolution and scientific methodology while still retaining belief in God as Creator. I gave this as one of several counter-examples to IDCs' [Intelligent Design Creationists'] rejection of such a possibility. Johnson explicitly dismisses deist views throughout his writings. Indeed, to try to set up the (false) dichotomy that he needs to legitimate purely negative argument, he goes much further and dismisses any form of theistic evolution. *Dembski's response is all the more puzzling, since he adopts the same position.*[81] (emphasis added)

PHYSICS

Because the scientific claims of such Wedge figures as Chien, Behe, and Wells are sufficiently circumscribed to be tested by observations, it is possible in a reasonably short discussion to provide a good sense of their scientific merit. That is not possible with Dembski's oeuvre. He insists that he has given full empirical legitimacy (for example) to Behe's notion of irreducible complexity; but nobody has been able to confirm that he has succeeded. The increasingly convoluted arguments *for* such empirical legitimacy (including those in *No Free Lunch*) remain entirely unconvincing to biologists and mathematicians who actually study them. Beyond that are the wide scope and range of Dembski's references, the critical parts of which are statistical-mathematical and physical. Evaluation of his core claims is not something that can be done briefly by calling attention to plain facts or to some simple logical error. It must be done at length, comparing what is new in Dembski's claims with what they are supposed to correct or improve in existing science. Fortunately, several physicists and mathematicians have done so. The congruence of their independent conclusions about what is wrong with Dembski's uses of physics (including the logic of physics) means that, despite his enormous range of reference, example, and anecdote, physical scientists are no more impressed with the science of the design inference and ID than are the biologists, mathematicians, and philosophers.

Physicist Perakh has written a sixty-page review of Dembski's ideas, amusingly titled "A Consistent Inconsistency" (cited previously). This is a patient commentary on Dembski's output up to *No Free Lunch* and the one most explicitly centered on physics, engineering, and cryptology (the author's fields of expertise). It opens with quotations of elaborate encomiums from six of Dembski's fellow ID creationists. But following their ecstatic praise, for contrast, Perakh provides references to and brief discussion of twelve strongly negative analyses, some of the most dismissive from logicians, mathematicians, and physicists like himself. Among the negative reviewers, at least two are committed theists. (There are a good many more theists—practicing Christians as it happens—who have in other contexts found serious fault with CRSC-style ID. See later discussion.) All these critiques focus on Dembski's idiosyncratic thermodynamics and information theory, that is, on his central justifications for the explanatory filter, without which the broader arguments are useless.

Before he proceeds to the substance of his physical arguments, however, Perakh reflects on Dembski's "mathematism," which he defines as follows:

> If a mathematical formula is derived in physics, or some technical science, or engineering, it compresses into easily comprehensible form certain essential relations between various data, which otherwise would be much harder to review and manipulate. This immensely facilitates some useful procedure. If, though, mathematical symbolism is used for the sake of symbolism itself, it does not advance the understanding of a subject, at best simply saving space and time in the discussion of a subject, and at worst making the matter more obscure because of esoteric symbolism which requires a lengthy deciphering.
>
> Actually Dembski's book "The Design Inference" contains little of genuine mathematics, but is full of "mathematism," this term denoting the use of mathematical symbolism as embellishment.

Perakh then reinforces this grievance with one of many possible examples: a long and tedious sampling from pages 48 and 49 of Dembski's *The Design Inference*.

The critique proper begins with what the author sees as a general violation of physical logic inherent in Dembski's one-dimensional algorithm for inference to design. In the scheme, at the first node of the explanatory filter, a yes/no decision must be rendered: yes = law, no = chance. But the logic of physical reality is simply not like that. Perakh offers examples that violate Dembski's stipulations. The first concerns the impacts, in one (selected) square meter of tennis court on one side of the net, of tennis balls flung by a tennis-practice machine on the other side. Assuming a large number of throws and depending on the total number of them, the probability of impact of some of the balls on the observed square meter will be determined to be low or, rather more likely, *very* low. Should a number of the balls land in that spot, a naïve observer might then attribute the "event," in toto, to chance. The physical reality, however, is that chance determines only the exit speed of the ball from the pitching machine. But whatever the speed, all remaining essential phenomena and their quantities—the number and the timing of impacts on the chosen square meter—however low the calculated probability, are fully determined by the "regularity," that is, the law(s), of classical mechanics, which set(s) the trajectory of each ball.

This case is therefore one of law and chance acting together and inseparably in an "event" whose cause is to be investigated. Perakh then describes the Galton board, a machine in which many balls fall under the influence of gravity through baffles and into bins placed at increasing distances from the point of egress. The drop of each ball is governed by chance, but the final distribution of balls in bins—the pattern—is eventually determined when a sufficient number of balls have already fallen into the bins. This distribution, which is a function of position to the left and right of center when a large number have fallen, is infallibly the familiar Gaussian (the "normal") distribution. The balls will always be distributed in the same way. Perakh is showing here that "the situation is in a sense opposite to the case of the tennis balls: while for the tennis balls chance operated through law, now the law (Gaussian distribution) operates through chance."[82] In both cases, the tennis balls and the Galton board, neither law nor chance separately can be identified as the unique cause of such events, of which there are uncountably many in the real world. We are back to the fact that natural law and chance can act together, inseparably. The situation worsens when we move to the last node of the filter, where we are supposed to choose again, yes/no, between chance and design. Here, we confront the internal contradictions of Dembski's proposals on agency and on design without a designer. In short, even on brute physical logic, the design inference fails to work. This, however, is only a part of Perakh's paper. Most of it is devoted to arguments from the author's special subjects: probability, information, and complexity. Like the other experts who have written, Perakh finds those uses unsound.

The physics-qualified commentators, though, whatever their own specialties, focus on the least dispensable of Dembski's claims, one that he and his CRSC colleagues now propagate widely among their lay audiences: that he has discovered a new law of physics (or of information theory), the Law of Conservation of Information. This "law" is the formal equivalent of the second law of thermodynamics—which asserts that the entropy of a closed system (such as the universe) can only remain constant or increase. In his treatment of information, Dembski adopts a part of Claude Shannon's famous entropy theory of communication, one of the basic ideas of modern information theory, for estimating information content. Thus, Dembski's proposed new law is a law of thermodynamics. But it is a "law" only if it is true. The Dembski conservation argument says that complex specified information cannot be created by natural processes,[83] because by his newly discovered law, information is conserved in natural processes.[84] (Dembski's law would mean that all the information content of all the DNA of every organism that has ever lived or ever will live was either fixed at the moment of creation, never to change except by the action of what/whoever initially fixed the amount, or that what/whoever initially fixed the amount must periodically intervene to change the amount.) If that is indeed a law, *and if* the appearance of CSI is an infallible indicator of ID (and obviously, in this context, of supernatural design), then indeed no naturalistic explanation of life is possible. Dembski's "law" is therefore a very bold announcement. It is also false. To see why, one needs to understand how his arguments fail to represent the true significance of the rule that initially shaped them—Shannon's rule for the entropy of a message, its relationship with the laws of statistical thermodynamics, and through those laws to general thermodynamics. They also fail to account for the empirical fact that the complexity and the information in living things *has clearly increased* over time.

To start, one must understand the underlying fault in Dembski's use of the Shannon formalization, which is concerned with message transmissions between a source and a receiver via a physical channel. That Dembski's appropriation of it is wrong is shown in detail, accessibly and independently, by physicists Victor Stenger and Matt Young. Stenger's commentary, a chapter from his recent book, is more extended. This is Stenger's summary:

> Dembski's definition of information, $-\log_2 p$, is of the same form as the Shannon uncertainty in the special case of equal probabilities [for all possible configurations of a system, or message]. . . . [W]e can immediately see that this definition is not conventional and will equal R [*the actual and needed amount of information carried*] . . . only for equal probabilities and when the transmission is perfect. . . . While Dembski refers to Shannon, he does not derive the expression for information he uses from Shannon's expression, nor justify it by any other method. His examples, however, indicate that he does not limit himself to equal probabilities within an ensemble of symbols or "events." Neither does he average over the ensemble [as the Shannon equation requires]. In fact, his so-called "information" is really just another way of writing the probability p of an event in logarithmic form.[85]

This is a mistake with consequences: in Shannon's formulation, the entropy of a message can and does decrease as information increases—as information flows in from outside the system (e.g., from the sender of a message). The entropy of the message after a transmission event is smaller than it was before; the information content has increased by an amount (in bits, in the Shannon formulation) exactly equal to the decline of entropy after transmission is complete. Dembski's measure of information is really just a measure of the entropy, not the information, of the event. Thus, there is no adequate justification for asserting that an increase of information due to natural processes is impossible. A decrease of entropy—that is, an *increase* of information—would be impossible *if the system were closed* (or, in chemical thermodynamics, "isolated"). But biological organisms and their parts are not closed systems. They are typical open systems, maintained by constant inward flow of negative entropy ("negentropy"), which is the same as information. In open systems entropy *can* decline, and information and complexity can increase. Fundamentally, this is what the active field of self-organization phenomena is about. So the thermodynamic underpinning (the second law, applying to *closed* systems) doesn't in any case apply in any simple way, and there is no justification for a "law" such as Dembski's Law of Conservation of Information. Misunderstanding or misuse of the Second Law of Thermodynamics has been a creationist hallmark for nearly a century.

But no confusion of thermodynamic and information symbology needs to be invoked in a basic judgment of Dembski's claims; that is, it is not just the symbols that are improperly employed. Stenger examines the possibility that a law of conservation of information does exist, even if Dembski's formalism for information is incorrectly employed. He then offers a simple thought-experiment, the result of which indicates that a natural perturbation of the model system can increase the system's real information content—however symbolized.

Physicist Matt Young, starting with Stenger's insights on the difference between Dembski's "information" and Shannon's "uncertainty," provides additional simple models in thought-experiments. They not only illustrate the basic laws but also demonstrate an inevitable increase of information content, under simple conditions, by purely natural processes—*including* natural selection.[86]

The point of Dembski's "Law of Conservation of Information," of course, is to reinforce an argument that is not convincing on the basis of his explanatory filter alone, namely, that something *other* than a natural process is needed to achieve the design so clearly evident in living things. This is design in the sense that the complex machinery of biological systems is usually (not always!) well adapted to what it must accomplish in existing environments, internal and external. Earlier, this concept might have been expressed as an efficient "homeostasis" exhibited by functioning physiological systems, and thus by the living organisms bearing them. Still, the assertion that *information* is conserved is heartening for theists and especially for creation science. Dembski's Law of Conservation of Information is foreshadowed in his early apologetical writings, on purely theological grounds, where he identifies it as his "Law of Priority in Creation" and attributes to it a biblical origin:

> I would like to see this law elevated to a status comparable with the laws of thermodynamics. The law is not new with me. It is found in Scripture:
>
> "Jesus has been found worthy of greater honor than Moses, just as the builder of a house has greater honor than the house itself." [Hebrews 3:3, NIV]

> The creator is always *strictly greater* than the creature. It is not possible for the creature to equal the creator, much less surpass the creator. The Law of Priority in Creation is a conservation law. It states in the clearest possible terms that you can't get something for nothing. There are no free lunches. Bootstrapping has never worked.[87]

Dembski's "law" seems to be that long-sought creationist discovery by virtue of which naturalism, including Darwinism, collapses. Such a law would also conveniently mandate that whatever cosmologists have discovered or may yet discover scientifically about biological origins must be classified as an act of supernatural intervention, at least at the beginning of the world. After all, information does increase with time, at least in certain parts of the universe. If information cannot be created by natural processes, it was then and must still be created by—something else. But despite the ultimately supernatural sanction of its earlier, biblical version, there is no scientific basis for Dembski's new law.

Yet a shred of possibility remains. More sophisticated creationists have argued that no evolutionary process such as natural selection can create information in DNA and the genome, because exchanging nucleotide sequences in a stretch of DNA (via mutation) does not, in principle, change its Shannon information content. The four nucleotide monomers of DNA remain there, and the amount of information remains the same, no matter how they may be switched around. Richard Dawkins, by profession a zoologist and evolutionist but with much experience in information theory, has produced a highly readable argument *refuting* the claim that natural processes cannot increase the information content of the genome. His is an intuitive presentation of the general concept of information content, of the sources of binary units or bits in the measure of information content, and of the standard contemporary measures of complexity. It can be useful for anyone whose hopes for a revival of dualism in science have been raised unjustifiably by current versions of Dembski's law and similar arguments. The information content of DNA, as of whole organisms, populations, and entire ecosystems, *has* increased enormously in the course of evolution, by known mechanisms.[88] Dawkins is not alone, of course, in having made this clear: in an economical but devastating probe of Dembski's arguments, mathematician–computer scientist Jeffrey Shallit, in his review of *No Free Lunch*, lists important contributions to the subject (all ignored by Dembski) dating back to 1961.[89]

A Worthy Scientific Alternative?

The book you are reading is primarily about the Wedge and only secondarily about the ID "science" that Wedge public relations activities promote. For the other lead scientists of the movement—Chien, Behe, Wells, and the like—it is possible to recapitulate succinctly the scientific arguments, because they are at least in part empirical and subject to testing with facts. But Dembski's case is not so simple. The problem is not merely that his arguments are abstract. Except for a surfeit of homely analogies intermixed with some erudite historical vignettes, they are continuously qualified and modified by propositions that cannot be conveyed except at their own length. The full range of those arguments cannot be represented, much less countered point by point, under a mere subhead of a chapter like this one. The required book-length refutation, an unrewarding task that would not convince true believers and would be redundant for working scholars, would be about something *other* than the Wedge itself. It would be valuable, nevertheless. The closest approach to such a refutation so far is the extended critique of *No Free Lunch* by Richard Wein. But we think that presenting the full story of the Wedge movement and its activities is now more important for science and for the future of public education in science.

The idea that the ID proponents hope to implant in the public consciousness, especially in behalf of the campaign to topple evolutionary biology from its place in biological education, is that the claims of CRSC-ID science *equal* the scientific claims of standard science, and therefore merit at least equal time in public school science classrooms. But the ID claims are *not* equal, and it is easy to demonstrate that even for Dembski's proliferating product. His large but sadly misguided "scientific" output can be dealt with fairly, but briefly and in plain language, *only* by showing that it is dismissed by experts whose concordant opinions, and whose professional and scientific qualifications, cannot be doubted or dismissed. Of course, he *could* be right and all of them wrong. But such a thing has happened only rarely in the past millennium. It is not only rare but, in Dembski's case, most *improbable*. Attending to the judgments of scientists who spend their professional lives working and producing in the fields concerning which ID proponents make their pronouncements makes more sense. So until Wedge scientists present peer-reviewed genuine data from their original research, there can be no justification for the Wedge's interference in school science curricula or in public education generally.

We don't teach schoolchildren controversial physical marginalia in the physics curriculum. Nor do we undertake the equivalent, for that matter, in elementary Sunday school religion classes: we don't teach the old Christian heresies or the stories of false messiahs to children. We should not teach marginalia along with elementary biology. Here, as elsewhere, the teacher's obligation is to present good, mainstream science. Conflicts over marginalia should indeed be taught, but only at the appropriate educational level and by instructors who are qualified to judge and present fairly both sides of a conflict. That excludes all education from preschool to high school and most undergraduate college education in biology. So far, ID, including Dembski's theoretical arguments for it, remains no more than a part, albeit a relatively sophisticated part, of "creation science" marginalia.

Dembski's project of feverish writing, hedging, and restating is no longer anything like science or even mathematical speculation. His is perhaps the saddest of the ID *personal* stories, for Dembski began his crusade a decade or so ago with intelligence and scholarly promise. His first seriously technical book, in fact, his doctoral dissertation, *The Design Inference,* though brash and error-ridden, was nevertheless provocative. What has happened since then is exemplified by two communications—one from the University of Chicago philosopher William Wimsatt, once an academic advisor to Dembski, and the other from ID critic Richard Wein. Professor Wimsatt, internationally known in philosophy of science, contributed a dust jacket blurb for *The Design Inference,* published in 1998. He is, moreover, one of five references listed by Dembski on his official curriculum vitae, posted on his website.[90] Recently, Wimsatt encountered on the Internet one of the glad tidings advertisements for Dembski's newest book, *No Free Lunch.* His response to it was a letter posted to Ian Pitchford's evolutionary psychology discussion group:

> I could not in conscience fail to respond to the ad for Bill Dembski's new book, "No Free Lunch", and to the general tenor of the political push generated either within or by others using the so-called "intelligent design theory". This is not a theory, but a denial of one, and a denial whose character is widely misrepresented, at least in the press.
>
> "Shows" and "refutes" are success verbs appropriate to mathematics and logic, though they are used widely (and often inappropriately) as if they had the same force in the empirical sciences, and, with less demanding standards still, vernacularly to describe adversarial processes in the courtroom.
>
> (It is perhaps this linguistic confusion which allows a lawyer from the University of California at Berkeley [Phillip Johnson] who either cannot tell the difference between methods of advocacy and methods of truth seeking or cynically tries to get others to confuse them, to claim to be competent to act as an "expert witness" on the quality of scientific argumentation on topics of which he knows nothing.)
>
> Unfortunately "popular" presentations of "Intelligent Design" have tended to give the impression that it rested solely on mathematical demonstrations. Anyone who could have succeeded in showing that natural selection is incapable of generating biological structures according to standards from mathematics or logic would have constructed a mathematical proof that would have dwarfed Godel's famous Undecideability theorem in importance. As one who read Dembski's original manuscript for his first book, found much to like in it, and had appreciative remarks on the dust jacket of the first printing, I can say categorically that Dembski surely has shown no such thing, and I call upon him as a mathematician to deny and clarify the implications of this advertising copy.
>
> The key issue for intelligent design is to apply a mathematical apparatus (which is nice, but has deep problems in its application) to probability estimates that come from elsewhere. The Neo-but-still-pseudo-scientific creationists take probability estimates that are problematic at best, and commonly just irresponsible or unfounded (and in any case, not accepted by any reputable natural sci-

entists I know who are speaking about their own subject matter), and run them through Dembski's apparatus.

So does the fact that the argument has a mathematical component validate it? No. The answer is found in the computer programmer's lament: "Garbage in, garbage out". A deductive argument with faulty premises shows nothing at all.

William C. Wimsatt.
Professor of Philosophy.
Committees on Evolutionary Biology
and Conceptual Foundations of Science[91]

No Free Lunch is, as indicated, the current culmination of Dembski's *and Behe's* arguments against modern evolutionary science. Those are the only arguments within the ID movement that ever had any prospect of affecting the real science and, more broadly, education and culture. Dembski touts his *specified* complexity and Behe's *irreducible* complexity, both exemplified in the bacterial flagellum, as jointly sufficient to topple Darwinism. Yet his failure is manifest even to scientists who are Christians, and who would therefore be understandably sympathetic to Dembski's concerns about naturalism. The most recent criticism from this quarter is from Howard Van Till, a physicist and astronomer (emeritus) at the evangelical Calvin College. In his review of *No Free Lunch*, "*E. Coli* at the *No Free Lunchroom*: Bacterial Flagella and Dembski's Case for Intelligent Design," Van Till makes a summary evaluation in his abstract: "[A] critical examination of Dembski's case reveals that, 1) it is built on unorthodox and inconsistently applied definitions of both 'complex' and 'specified,' 2) it employs a concept of the flagellum's assembly that is radically out of touch with contemporary genetics and developmental biology, and 3) it fails to demonstrate that the flagellum is either 'complex' or 'specified' in the manner required to make his case. If the bacterial flagellum is supposed to demonstrate ID, then ID is a failure."[92]

Dembski's and Behe's arguments have failed to make any impact at all because they are not only wrong but have remained so despite careful, detailed assessments by expert critics from the relevant scientific and mathematical disciplines; and they are rapidly bringing disrepute to their authors. It is beyond the scope of this chapter and this book to take up even the "new" arguments in *No Free Lunch*, which appropriate to ID certain mathematical theorems governing "evolutionary" algorithms, as employed mainly in computer science. But the necessary analyses have been published, one of them by Richard Wein after a long period of testing and discussion within a private group of mathematicians, physicists, and biologists. This work is now available to the public online. It is a remarkable essay, seventy-two single-spaced pages of close argument on every important argument made or attempted in Dembski's book—and necessarily on all the preceding ID offerings from Dembski *and* Behe. No reader seriously interested in ID theory, whether as proselyte or critic, who can absorb a little logic and mathematics, should miss the opportunity to study Wein's essay. But for our immediate purpose, a single introductory paragraph is enough:

> Some readers may dislike the frankly contemptuous tone that I have adopted toward Dembski's work. Critics of Intelligent Design pseudoscience are faced with a dilemma. If they discuss it in polite, academic terms, the Intelligent Design propagandists use this as evidence that their arguments are receiving serious attention from scholars, suggesting this implies there must be some merit in their arguments. If critics simply ignore Intelligent Design arguments, the propagandists imply this is because critics cannot answer them. My solution to this dilemma is to thoroughly refute the arguments, while making it clear that I do so without according those arguments any respect at all.[93]

Most recently, Mark Perakh has added his own critique of *No Free Lunch*,[94] as has mathematician Jason Rosenhouse in his elegantly concise review of the book published in the most appropriate scientific journal, *Evolution*. That review includes the following comment on Dembski's long-promised, long-awaited calculation of complexity in a real biological object: "The flagellum is irreducibly complex you see [this is Rosenhouse's *irony*], implying that it can be treated as a 'discrete combinatorial object.' . . . The text soon becomes a dazzling congeries of binomial coefficients, perturbation probabilities, and sundry mathematical notation, all in the service of a computation that may as well have been written in Klingon for all the connection it has to reality."[95]

Thus, even if we were to grant for argument's sake that ID *is* science of a sort, near the end of his identification of "creation science" *as* science (albeit very bad science), philosopher Elliott Sober points out that

> The long-term track record of "scientific creationism" has been poor. Phrenology eventually was discarded; although it showed some promise initially, it failed to progress in the long run. Creationism has fared no better; indeed, it has done much worse. It was in its heyday with Paley [1802], but since then, the idea has moved to the fringe of serious thought and beyond.[96]

One ought now to interpret "scientific creationism" as explicitly including "Intelligent Design."

Some readers may be made impatient not only by the lofty abstraction of Dembski's versions of ID *but also* by the less lofty but equally abstract rejections of them from mainstream philosophers, mathematicians, and scientists. Such impatience is understandable; therefore, a simple but fundamental point about *all* arguments for creation by intelligent design may be helpful. The proponent of intelligent agency has a three-way choice: to specify that the world-designing agent has a specific range of capabilities, that it has an infinite range of capabilities, or that the true range of its capabilities is unknowable.

To date, no theist or ID theorist has proposed an explicit (hence delimited) range of the designer's powers, a proposal that would offer hope of testing. Some have indicated, and many seem to believe, that this range is infinite. Others seem convinced that the range of the designer's powers is unknowable. The latter are what we have to deal with so far in "theistic science." But contrary to the claims of ID theorists, both choices are useless for serious scientific inquiry. If the agent's range of powers is infinite, anything in the world we observe can be explained by those powers; so there is no possible disproof of agency—hence no point in running tests. If, on the other hand, the range is unknown or unknowable, then no test of agency is possible, since we cannot identify a specific performance for which to test. In short, no inquiry can ever disprove a remote, deistic hypothesis of ID. But no demonstration of ID is possible, either, until the range of the designer's powers is specified and demonstrated empirically. So far, that has not been done.

6

Everything *Except* Science I

Each day . . . I am interacting with a remarkable group of young scientists, historians and philosophers connected to our Center *for the Renewal of Science and* Culture. *Some who literally have risked their careers to pursue the truth (as I wrote you last year) are now finding financial support and academic positions, getting major publishing houses to print their books and appearing at key international conferences, in debates and on television. Many of you are helping Discovery to help them, and that help is absolutely crucial.*

Bruce Chapman, President,
Discovery Institute *Journal*, 1999

Michael Behe, William Dembski and I have gained sufficient recognition and public support through our books and lectures that book-length refutations are beginning to appear.

Phillip Johnson, *The Wedge of Truth*

What the Wedge and ID creationism lack in scientific accomplishment they compensate for in the ambitious schedule of public relations and self-congratulation that have been a constant feature of the program from the start. The Wedge Document lists specific goals by means of which the progress of the CRSC in pursuing its strategy can be estimated. Wedge members maintain a dizzying schedule of activities—new ones show up on the Internet constantly, making it difficult to keep track of them. Not all are of equal significance; some are relatively minor, whereas others, such as the conferences, are highly visible and have significant impact. Not all activities are organized by the CRSC fellows themselves. Some have been arranged by others—their allies and even some of their opponents. When CRSC members participate, however, in events organized by others—whether allies or opponents—the Wedge's goals are advanced all the more effectively. Phillip Johnson's goal, an important place for ID at the table of public discussion, is met, and the CRSC registers as a source that must be consulted. The Wedge publicizes and exploits these activities—just as much as those it organizes itself—to enhance its public image as a player in a supposedly serious, high-level evolution/creation "debate." In this and the following chapter, we survey these image-enhancement activities, following the sequence in which these undertakings are proposed in the Wedge Document. We identify at least one activity, in the Wedge Document's given order, for every major goal set forth there.

The Wedge Document provides detail about goals in the "Five Year Strategic Plan Summary," the statement of its "Goals" (which extend over the next twenty years), and in the "Five Year Objectives." All activities there identified and followed to the present demonstrate that the primary Wedge strategy—relentless and efficient political and public relations—is a well-funded, systematic program that, despite some setbacks, has advanced considerably toward the goals set in the original document. Because the activities discussed here would be impossible without funding, and because the primary sources of this funding reveal much about the character and purposes of the Wedge, we examine this element first.

Money

Included in the Wedge Document's list of planned "Activities" is "Fund Raising and Development." Money has flowed into the CRSC's coffers at an increasing rate since its establishment in 1996. In the March/April 1997 *Reports of the National Center for Science Education* (NCSE), Eugenie Scott, in an article about the CRSC entitled "Anti-Evolutionists Form, Fund Think Tank: Old-Earth Moderates Poised to Spread Design Theory," begins with a reference to the CRSC's generous startup funding: "A press release dated August 10, 1996, announced that two private foundations have granted the Seattle-based Discovery Institute nearly a million dollars to establish the Center for the Renewal of Science and Culture. The Center will sponsor conferences, disseminate research and support postdoctoral students." The CRSC's website shows that this support includes generous stipends for recipients, with the CRSC now granting "full-year research fellowships between $40,000 and $50,000" and "short-term research fellowships between $2,500 and $15,000."[1]

Funding for the CRSC is channeled through the Discovery Institute, the parent organization. DI, which was granted 501(c)3 status in 1996 (www.guidestar.org), announced the formation of the CRSC in its August 1996 *Journal*, reflecting clearly the tone and strategy later formalized in the Wedge Document:

> For over a century, Western science has been influenced by the idea that God is either dead or irrelevant. Two foundations recently awarded Discovery Institute nearly a million dollars in grants to examine and confront this materialistic bias in science, law, and the humanities. The grants will be used to establish the Center for the Renewal of Science and Culture at Discovery, which will award research fellowships to scholars, hold conferences, and disseminate research findings among opinionmakers and the general public. . . . Crucial, start-up funding has come from Fieldstead & Company, and the Stewardship Foundation which also awarded a grant.[2]

In its spring 1998 *Journal*, detailing its 1997 activities, DI also adds to its list of startup funding sources the "Maclellan Foundation, and the SG Foundation."[3] DI announces in the *Journal* that "more than $270,000 in research grants were distributed" to CRSC fellows from 1996 through 1997. According to president Bruce Chapman, DI's budget increased from $120,000 to "over $1.2 million" between 1991 and 1997.[4] The budget thus rose tenfold, an increase due mostly to the grants for the startup of the CRSC in 1996. According to the Guidestar financial information from DI's IRS Form 990, DI's income from contributions for FY 1997 totaled $1,413,751.[5] While there is no way to estimate what percentage of DI's total 1997 income was earmarked for the CRSC, it is clear that 45% was received from grants, and 58% of its 1997 expenses went to "Research & Programs."[6] If DI received almost a million dollars in startup funding for 1996/1997, then the bulk of DI's income from contributions was earmarked for the CRSC, boosting it to preeminence among other DI programs. When compared to DI's income from contributions for 1995, prior to the establishment of the CRSC in 1996, new funding more than doubled between 1995 and 1997, the period during which the CRSC was established.[7]

1995	1996	1997
$628,997	$800,117	$1,413,751

1998–2000

DI's income from contributions for 1998, according to its Form 990 (www.guidestar.org), was $1,417,103, only marginally higher than for 1997. In 1999, however, their financial ship came in, thanks again to their previous benefactors: Fieldstead & Company, the Maclellan Foundation, and the Stewardship Foundation. DI announced its good fortune in its 1999 *Journal*:

> [T]hree enlarged grants to the Center for the Renewal of Science and Culture (CRSC) have enabled it to expand the number of fellowships it is supporting for scholarly work on the theory of "intelligent design." . . .
>
> Crucial decisions in the fall of 1998 at the Fieldstead & Co. . . . increased its grant to Discovery to $300,000 per year for the next five years. The Maclellan Foundation . . . also increased its grant to $400,000 for 1999, while the . . . Stewardship Foundation . . . voted to increase its CRSC grant to $200,000 per year for the next five years. Special grants are likely to bring the overall CRSC budget to over $1 million for 1999.[8]

As the number of CRSC fellows has expanded, so, of course, have DI's expenditures for fellowships. According to DI's IRS 990 forms for 1997, 1998, 1999, and 2000 (the only years yet available on Guidestar), the total cost of "Fellowships/Research" for each respective year (some of which could have been for fellows in other programs) was $261,374, $284,152, $329,265, and $404,295—an increase of 55% between the CRSC's first and fourth years of operation. According to Larry Witham in the December 29, 1999, *Washington Times*, DI had a total operating budget of $2 million for 1999.[9] Witham reports that "Stephen Meyer [CRSC director] said CRSC got $750,000 [38%] of that, and he hopes its budget will grow to $1 million in 2000."[10] (The CRSC's 1999 allotment from DI is roughly three times the National Center for Science Education's 1998 FY revenues of $258,957.) Meyer's hope could have come close to fruition, since DI began the 2000 fiscal year with a fund balance of $2,304,275 and ended with a fund balance of $1,375,154, indicating DI's expenditure of almost $1 million that year.[11] With $1.5 million from Fieldstead & Co. and another million from the Stewardship Foundation assured over the next five years (beginning in 1999), along with $400,000 in 1999 (and possibly more later) from the Maclellan Foundation, CRSC has acquired a minimum of $2.9 million in guaranteed, direct funding through 2003—an admirable record of financial success for such an organization, accomplished during its own specified time frame in the Wedge Document: "We believe that, with adequate support, we can accomplish many of the objectives of Phases I and II in the next five years (1999–2003), and begin Phase III."

Following the Phases of the Wedge

Despite the absence of the scientific productivity that was to mark the successful completion of Phase I, "Scientific Research, Writing & Publicity," the Wedge from the beginning has been tireless in Phase II and III activities, "Publicity & Opinion-making" and "Cultural Confrontation & Renewal," specified in the Wedge Document. Taken together, the activities discussed here and in chapter 7 show the intensity of the Wedge's campaign. The sheer volume of their undertakings is the best evidence of the Wedge's seriousness but also prevents our recording it fully here. We therefore present a summary of these undertakings, such being the sole means of conveying the solid reality of the Wedge strategy. To those who have braved the tedium and followed the Wedge's activities closely, it is quite clear that the central, nonnegotiable aim of the Wedge is to be so often in the public eye and to insinuate itself so persistently into the cultural and academic mainstream that their opposition will tire and admit defeat.

Our chronicle of these activities for each major category of "Wedge Projects" therefore provides a picture of the systematic nature of the Wedge's advance, which continues despite setbacks in Washington State, in Kansas, and in the demise of the Michael Polanyi Center at Baylor University, which we discuss later. Such setbacks do not sway the Wedge from its goals. Apart from Phases II and III, the Wedge has been productive—in a formal sense—in only one area of the original Phase I: "Writing" (we discuss "Publicity" under Phase II). The writing really belongs under Phases II and III, since it reflects no genuine scientific research, but it does generate abundant publicity.

Phase I: Book Publications

Wedge members are now close to meeting one of the Wedge Document's Five Year Objectives: "Thirty published books on design and its cultural implications." Twenty-three books have been written to date; Phillip Johnson has written six of them himself. Several criteria identify books that fit into this category: (1) the book promotes intelligent design and related issues. This criterion covers not only the books Wedge members claim are scientific but also their usual philosophical or quasi-philosophical productions. (2) The authors are either CRSC fellows, as most are, or close ID supporters. The Wedge Document does not require that the books must be written by CRSC fellows, and several written by Wedge members have co-authors who are not fellows but are obviously sympathetic to Wedge goals. (3) The book's date of publication is within the time frame covering the existence of the Wedge. Even though Phillip Johnson says the Wedge movement began in 1992 with the SMU symposium ("Darwinism: Science or Philosophy?"), one can argue, as we do, that it began with Johnson's decision to make anti-Darwinism his life's mission; hence, the Wedge's book schedule actually began in 1991with Johnson's first book. (4) The book is used in such a way as to advance the Wedge's goals. These books are heavily promoted; most are sold on the CRSC and Access Research Network websites. The following are the books that advance the Wedge strategy.

1. Beckwith, Francis J. *Law, Darwinism, and Public Education: The Establishment Clause and the Challenge of Intelligent Design*. Rowman and Littlefield, 2003.
2. Behe, Michael. *Darwin's Black Box: The Biochemical Challenge to Evolution*. The Free Press, 1996.
3. Behe, Michael, William A. Dembski, and Stephen C. Meyer. *Science and Evidence for Design in the Universe: Proceedings of the Wethersfield Institute*. Vol. 9. Ignatius Press, 2000.
4. Buell, Jon, and Virginia Hearn, eds. *Darwinism: Science or Philosophy?* Foundation for Thought and Ethics (FTE), 1994.
5. Craig, William Lane, and J. P. Moreland, eds. *Naturalism: A Critical Analysis*. Routledge Studies in Twentieth Century Philosophy, Routledge, 2000.
6. Dembski, William A., ed. *Mere Creation: Science, Faith and Intelligent Design*. InterVarsity Press, 1998.
7. Dembski, William A. *The Design Inference: Eliminating Chance Through Small Probabilities*. Cambridge University Press, 1998.
8. Dembski, William A. *Intelligent Design: The Bridge Between Science and Theology*. InterVarsity Press, 1999.
9. Dembski, William A., and James Kushiner, eds. *Signs of Intelligence: Understanding Intelligent Design*. Brazos Press (Baker Books), 2001.
10. Dembski, William A. *No Free Lunch: Why Specified Complexity Cannot Be Purchased Without Intelligence*. Rowman and Littlefield, 2001.
11. Denton, Michael. *Nature's Destiny: How the Laws of Biology Reveal Purpose in the Universe*. The Free Press, 1998.[12]
12. DeWolf, David K., Stephen C. Meyer, and Mark E. DeForrest. *Intelligent Design in Public School Science Curricula: A Legal Guidebook*. Foundation for Thought and Ethics, 1999.
13. Johnson, Phillip E. *Darwin on Trial*. Regnery Gateway, 1991.
14. Johnson, Phillip E. *Reason in the Balance: The Case Against Naturalism in Science, Law and Education*. InterVarsity Press, 1995.
15. Johnson, Phillip E. *Defeating Darwinism by Opening Minds*. InterVarsity Press, 1997.
16. Johnson, Phillip E. *Objections Sustained: Subversive Essays on Evolution, Law and Culture*. InterVarsity Press, 1998.
17. Johnson, Phillip E. *The Wedge of Truth: Splitting the Foundations of Naturalism*. InterVarsity Press, 2000.
18. Johnson, Phillip E. *The Right Questions: Truth, Meaning and Public Debate*. InterVarsity Press, 2002.
19. Moreland, J. P. *The Creation Hypothesis: Scientific Evidence for an Intelligent Designer*. InterVarsity Press, 1994.

20. Moreland J. P., and John Mark Reynolds. *Three Views on Creation and Evolution*. Zondervan Publishing House, 1999.
21. Newman, Robert C., John L. Wiester, Janet Moneymaker, and Jonathan Moneymaker. *What's Darwin Got to Do with It: A Friendly Conversation About Evolution*. InterVarsity Press, 2000.
22. Wells, Jonathan. *Icons of Evolution: Science or Myth?* Regnery Publishing, Inc., 2000.
23. Wiker, Benjamin. *Moral Darwinism: How We Became Hedonists*. InterVarsity Press, 2002.

Other books written prior to the beginning of the Wedge have been absorbed into the movement; these include those Dembski identifies as marking the beginning of the ID movement.[13] All were at least coauthored by people who became CRSC fellows subsequent to their publication:

1. Denton, Michael. *Evolution: A Theory in Crisis*. 2nd edition. Adler and Adler, 1997 [first published in 1985].
2. Kenyon, Dean H., and Percival Davis. *Of Pandas and People: The Central Question of Biological Origins*. Haughton Publishing Co., 1996 [first published in 1989].
3. Thaxton, Charles, Walter Bradley, and Roger Olsen. *The Mystery of Life's Origin: Reassessing Current Theories*. Philosophical Library, 1984.

These books put the Wedge more than two thirds of the way toward the goal of thirty books on ID, and more are forthcoming. Dembski received a $100,000 grant from the John Templeton Foundation to write *Being as Communion: The Metaphysics of Information*.[14] In his May 2002 curriculum vitae, Dembski announced other works in progress, the titles of which indicate they will count toward the book publication goal: *The Design Revolution: Making a New Science and Worldview* (InterVarsity Press) and *Uncommon Dissent: Intellectuals Who Find Darwinism Unconvincing* (Intercollegiate Studies Institute).[15] The article "Intelligent Design in Public School Science Curricula: A Legal Guidebook" (apparently a version of the book by the same title) by David K. DeWolf, Stephen Meyer, and Mark DeForrest mentions an upcoming book at Michigan State University Press, *Intelligent Design, Darwinism, and the Philosophy of Public Education*, edited by CRSC fellow John Angus Campbell.[16] Paul Nelson's biographical sketch on the CRSC website notes that his "monograph" *On Common Descent* will be published in the University of Chicago's "Evolutionary Monographs" series.[17] If these books are published, the Wedge will have surpassed the Wedge Document's goal of thirty books on ID.

There are also books related peripherally to the Wedge, such as CRSC fellow J. Budziszewski's *Written on the Heart* (InterVarsity Press,

1997), a volume on natural law to which Phillip Johnson devotes a chapter in *Objections Sustained.* Two recent books bear endorsements from Wedge members: (1) Neil Broom, *How Blind Is the Watchmaker? Nature's Design and the Limits of Naturalistic Science* (InterVarsity Press, 2001), and (2) Cornelius Hunter, *Darwin's God: Evolution and the Problem of Evil* (Brazos Press, 2001).[18] Another book, plausibly seen as a statement of the Wedge's theological underpinnings, is *Unapologetic Apologetics: Meeting the Challenges of Theological Studies*, edited by Dembski and Jay Wesley Richards, with a foreword by Phillip Johnson (InterVarsity Press, 2001). The book consists largely of papers written by Dembski, Richards, and others as students at Princeton Theological Seminary. Part 5, "Science," contains chapters bearing the same titles as previously published work by Dembski: "What Every Theologian Should Know About Creation, Evolution & Design" (*Princeton Theological Review*, March 1996) and "Reinstating Design Within Science" (*Intelligent Design: The Bridge Between Science and Theology*, InterVarsity Press, 1999). Richards also has a chapter entitled "Naturalism in Theology and Biblical Studies." These chapters pinpoint Dembski and Richards's understanding of Christian apologetics as necessarily including ID apologetics.

Phase II: Publicity and Opinion-Making

The Wedge is making steady progress toward these Phase II goals, the most immediate of which are the Wedge Document's "Five Year Goals": (1) "To see intelligent design theory as an accepted alternative in the sciences and scientific research being done from the perspective of design theory," (2) "To see the beginning of the influence of design theory in spheres other than natural science," and (3) "To see major new debates in education, life issues, legal and personal responsibility pushed to the front of the national agenda." What follows represents the most significant, though by no means all, of their activities.

BOOK PUBLICITY

Wedge members' books are available at online retail outlets such as Barnes and Noble and Amazon.com. Books, videotapes, audiotapes, and their journal, *Origins and Design*, are marketed aggressively on Access Research Network. DI has an online bookstore through which it sells Wedge books and videotapes.[19] Book publicity is constant, exemplified by Phillip Johnson's promoting his work at the February 2000 National Religious Broadcasters convention in Anaheim, California.[20] According to the February 5–6, 2000, NBR *Convention News*, the convention drew more than 5,000 people, making it a profitable site for book promotion.

Most recently, DI has dedicated separate websites solely to marketing Jonathan Wells's *Icons of Evolution* and Dembski's *No Free Lunch.*

ARN promotes both books aggressively.[21] And the Wedge publicity is supplemented by supporters such as Hank "Bible Answer Man" Hanegraaff, who promoted Phillip Johnson's *The Wedge of Truth* enthusiastically on his December 19, 2000, radio interview of Johnson. In his sales pitch for the book, which he began reading the day before the interview "while I was watching the St. Louis Rams play the Tampa Bay Buccaneers" and finished the next morning before Johnson arrived, Hanegraaff asserted that by buying and reading the book, "you can be part of an army that is learning to think strategically about a war that *will* be won, and you can be part of the winning team as we make a difference for time and for eternity."[22]

OPINION-MAKER CONFERENCES

Each of the numerous conferences held or attended by the Wedge actually fits into this category, but the CRSC has attended at least one conference explicitly labeled "Opinion-Maker Conference," as recounted in its November/December 1997 "Year End Update." This event was clearly a networking opportunity for Wedge members:

> Opinion-Maker Conference: At the invitation of Ed Atsinger, President of Salem Communications, Inc., Steve Meyer and Phillip Johnson recently addressed a national conference of radio talk-show hosts. The talk-show hosts were extremely enthusiastic in response to Steve and Phil, and their presentation of the case for Intelligent Design. Afterward, Howard Freedman, National Program Director of Salem Communications Inc., and many of the talk-show hosts invited Steve, Phil, and other scientists to appear on their programs to discuss the evidence for design.[23]

Another opportunity for networking was the "Dinner in Dallas," also announced in the Update: "Ninety business, civic and academic leaders, and representatives from Christian non-profits attended a CRSC dinner in Dallas, TX, featuring Mike Behe, Phillip Johnson and Steve Meyer."

APOLOGETICS SEMINARS

The Wedge Document states that the CRSC seeks "to build up a popular base of support among our natural constituency, namely, Christians. We will do this primarily through apologetics seminars." William Dembski's Fieldstead & Company–supported seminar, "Design, Self-Organization, and the Integrity of Creation," fits into this category. The course description shows that the June 19–July 28, 2000, "Summer Seminar" at Calvin College was designed to attract Christian participants who supported ID: "The aim of this seminar is to see whether a rapprochement between design and self-organization is possible that pays proper due both to the divine wisdom in creation and to the integrity of the world as an act of creation. . . . [S]cholars with expertise in the following disciplines are especially encouraged to apply: complex systems theory, information/design theory, history and philosophy of science, philosophy of religion, philosophical theology, and any special sciences dealing with complex systems." Each applicant was required to submit "a one-page description of his/her vocation as Christian scholar and teacher." In furtherance of the Phase II goal of using these seminars "to encourage and equip believers with new scientific evidences that support the faith, as well as to 'popularize' our ideas in the broader culture," participants in Dembski's seminar were "expected to present their projects at a conference the following spring and publish them in appropriate venues," further facilitating the Wedge's publication goals.[24]

The spring follow-up conference was held at Calvin College in May 2001, and the Wedge and its supporters were well represented in the speaking schedule: Alvin Plantinga and Del Ratzsch (well known ID backers), and CRSC fellows William Dembski, Stephen Meyer, Bruce Gordon, Jonathan Wells, Pattle P. T. Pun, Scott Minnich, Paul Nelson, and Jed Macosko. Graduate students were offered scholarships to finance their attendance—possibly a way of recruiting the "future talent" that Phase II says the Wedge seeks to "cultivate and convince."[25]

Dembski has conducted other ID-based apologetics courses or seminars. He taught an ID course at California's Trinity International University over six days in January, February, and March 2000. As part of Trinity's spring 2000 theme, "Defending the Christian Worldview," Dembski's course description included his promotion of ID as the cause of Darwinism's "impending collapse" and his depiction of "Darwinian evolution" as "a failed scientific research program."[26] However, the Wedge has taken steps toward a permanent influence on Christian apologetics, which will also help advance the Wedge Document's Five Year Objective of ensuring that "seminaries increasingly recognize & repudiate naturalistic presuppositions." The Wedge clearly sees theological seminaries as a necessary target, as explained by Dembski and Jay Richards in the introduction to *Unapologetic Apologetics*: "If we take seriously that Christianity embodies humanity's chief truth—that God was in Christ Jesus reconciling the world to himself—then the most important school of all is the seminary. The seminaries teach our ministers who in turn teach their congregations about Jesus Christ. Whether they do so faithfully and truthfully depends on the training they receive at seminary."[27] Seminaries that recognize evolution and other aspects of what Dembski and Richards call "post-Enlightenment liberalism" are indulging in an "accommodationism," which is "so caught up in gaining the respect of the secular academic world that it loses its integrity as a Christian witness."[28] They warn that because of this accommodation, "the mainline and liberal seminaries can be a dangerous place for a student's Christian faith."[29]

Their solution to this danger is student activism on behalf of evangelical theology: "The enthusiasm of youth . . . is wonderfully capable of upsetting the status quo. What we are urging . . . is an intentional activism by evangelical students directed at the mainline seminaries both to renew and reclaim these institutions. . . . Evangelical students need to take up the mantle of public apologist."[30]

In 1995, while a student at Princeton Theological Seminary (PTS), several years after the formation of the Wedge, Dembski helped to found the Charles Hodge Society (named after a nineteenth-century Princeton theologian who rejected Darwin's theory of evolution) and to revive the *Princeton Theological Review*, a journal Hodge had founded.[31] Unhappy with the absence of apologetics from the curriculum, Dembski and Richards also helped found the Princeton Apologetics Seminar in 1995 in response to "the failure of the mainline denominations to take Christian apologetics seriously."[32] In his foreword to *Unapologetic Apologetics*, Johnson calls the PTS apologetics seminar "the mother of all seminary peer groups, an apologetics seminar where the tough issues are debated even in front of outside critics."[33] More important, Johnson points out that behind those seminars stands the Wedge:

> Behind this student movement is a more general intellectual movement that will bear fruit in the coming century. It is a bit thin on the ground for now, but so was the Christian faith in the first century. . . . Methodological naturalism is a branch on the materialist tree that will lose its power to intimidate when the tree is known to be hanging in midair. The Spirit moves when and where it chooses, and those who are moving with it are never afraid to perturb established branches and twigs that have lost sight of their own roots. That is the point of the intelligent design (or "mere creation") movement, to which Dembski and Richards have contributed much. . . . We come from creation by God, not from unguided nature, and people who wish to be rational must recognize that fact. Show me a mainstream seminary that is unafraid to say that without equivocation, and dare the wrath of the scientific and academic establishments for doing so, and I will show you an institution that deserves your enthusiastic support.[34]

TEACHER TRAINING PROGRAMS

Influencing teachers and students through ID classroom resources is perhaps the most subtle and dangerous of all the CRSC's goals, for it is aimed explicitly at future generations and, therefore, at the future of science. Rob Boston of Americans United for Separation of Church and State attended a conference entitled "Reclaiming America for Christ," held by D. James Kennedy, head of the fundamentalist Coral Ridge Ministries, at which Phillip Johnson was a featured speaker. Johnson's remarks show that the religious impetus behind the ID movement surfaces energetically when he is before a safely receptive audience:

> During Coral Ridge Ministries' Feb. 26–27 [1999] "Reclaiming America for Christ" conference, several speakers discussed strategies for injecting fundamentalism into the [public] schools. . . .
>
> [I]tching to get his religious perspective into public schools is Phillip Johnson. . . . Asserting that Darwinism is "based on awful science, just terrible," Johnson said the theory has "divided the people of God" . . .
>
> *The objective, he said, is to convince people that Darwinism is inherently atheistic, thus shifting the debate from creationism vs. evolution to the existence of God vs. the non-existence of God. From there people are introduced to "the truth" of the Bible and then "the question of sin" and finally "introduced to Jesus."*[35] (emphasis added)

The targeting of children and young people generally was part of the very early strategizing of the Wedge. The March/April 1995 issue of the American Scientific Affiliation's *Newsletter* reports efforts to sell *Of Pandas and People* to educators at a conference in Beaumont, Texas. The FTE, publisher of *Pandas*, flew Wedge member Robert Kaita of Princeton University to the conference, where Kaita testified that ID theory is "something that seems to be eminently reasonable."[36] He also assisted in efforts to convince a textbook board in Alabama to adopt *Pandas*, for which he was a reviewer (see "Acknowledgements" in the 1993 edition). Failing to convince boards in both Alabama and Idaho to adopt the book, Jon A. Buell, president and founder of the FTE (and editor of the Wedge book *Darwinism: Science or Philosophy?*) subsequently "took the approach of marketing directly to biology teachers, who are generally easy to contact, available for meeting, and receptive to new resources."[37] There is even some indication that targeting elementary school students was one of the original aims of the Michael Polanyi Center, established by Dembski in October 1999 at Baylor University (see chapter 7). The Polanyi Center's "Purpose of Center" webpage in January 2000 noted that, in addition to the undergraduate and graduate courses of study that the center hoped to offer, "the significance of the MPC's research and educational efforts are communicated more broadly through articles and books aimed at a popular audience, and through workshops for lay audiences and *grade school students*"[38] (emphasis added).

In 1996, at the first formal Wedge meeting, the "Mere Creation" conference at Biola University, Phillip Johnson announced that "some of us are preparing teaching materials to help home-schoolers, private schools, and even adventurous public school teachers to teach the kids what the textbook writers and curriculum planners don't want them to know. Of course the Darwinists and their lawyers will resist this ferociously."[39] The prospect of legal challenges, however, has not slowed the Wedge's preparations to push forward into the public schools. They are advancing on a number of fronts.

More Teacher Resources: Foundation for Thought and Ethics. The Wedge also wants to put its "learning" materials directly into teachers' hands. Through the FTE, with Jon Buell, the Wedge has launched direct marketing efforts to teachers in an attempt to persuade them, on their own initiative, to adopt *Of Pandas and People*, to which FTE holds the copyright. These efforts actually predate the formal founding of the Wedge; they go back to 1990, when FTE wrote to supporters asking for assistance in contacting the high school biology teachers in their area about the availability of *Pandas*. Citing their success in stimulating interest among teachers around the country, FTE indicated to recipients of the letter that "we are finding that the best approach to the local school system is through the biology teacher," pointing out that "biology teachers are generally easy to contact, available for a meeting on short notice, and receptive." Readers were urged to join the "quiet army," which would solicit teacher support for *Pandas*.

A similar letter went out in March 1995, attempting to stimulate reader activism by appealing to outrage over an unsuccessful attempt to have *Pandas* adopted in the Dallas area: "Some conservative Christian members of the Plano School Board and zealous coworkers drafted a resolution to have the Plano district buy copies for all biology students." The attempt having failed because a "board member who had recruited an angry mob blocked the move," FTE cited a successful case in which a "Darwinist" biology teacher in Atlanta had adopted the book after a class presentation based on it given by the father of a student who had challenged the teacher. This letter specifically referenced ID and asked for prayers—and money.[40] (Later, we discuss FTE's recent Internet initiative.)

New Mexico Book Distribution. The Wedge was recently assisted by a more direct, aggressive approach to putting its books in teachers' hands. This effort was undertaken by New Mexico ID supporters, assisted by John Omdahl, a professor of biochemistry and molecular biology at the University of New Mexico School of Medicine. Omdahl is featured in "Darwinism under Attack" in the December 21, 2001, *Chronicle of Higher Education*. The article highlights his practice of devoting part of the final lecture in his biochemistry and molecular biology class (where he avoids the word "evolution") to his reasons for favoring ID rather than evolution, a practice with which most of his colleagues are "very uncomfortable."[41] Omdahl is one of "100 scientists" who signed the Wedge's 2001 document, "A Scientific Dissent on Darwinism," produced in response to a PBS television series on evolution (mentioned later in this chapter). With John W. Oller, Jr., he co-authored "Origin of the Human Language Capacity: In Whose Image?" in J. P. Moreland's *The Creation Hypothesis* (1994). His support of ID precedes even the formation of the CRSC, dating back ten years, when he co-signed the Ad Hoc Origins Committee letter defending Phillip Johnson after Gould's scathing review of *Darwin on Trial* (see chapter 1).

In his effort to promote ID, Omdahl wrote a letter in March 2002 to science department chairs in seventy-seven New Mexico middle and high schools. Each letter accompanied a copy of Michael Behe's *Darwin's Black Box*. Although the letter was not on university letterhead, Omdahl used his professional title and affiliation as the associate dean of research at the UNM School of Medicine. In this letter, Omdahl informed teachers of Michael Behe's "new" explanation of aspects of Darwinian evolution that could not be explained by appealing to chance, assuring them that students would find the book relevant to their biology studies. He asked teachers to circulate the book through their science departments and then to have it placed in their school libraries (where, of course, children would have access to it). Omdahl also added a postscript: apparently anticipating that the book might be received with skepticism, he directed teachers to Behe's responses to his critics at his CRSC website.[42]

Omdahl's letter was the only accompaniment to the book, and his use of his university affiliation created the impression that the book donations were done under university auspices. But one teacher, Steve Brugge, the science department chair at Albuquerque's Eisenhower Middle School, had received additional information about the involvement of a conservative Christian organization, so Brugge wrote to Omdahl about the discrepancy:

> I am the science department chair at Eisenhower Middle School. . . . I received yesterday a copy of *Darwin's Black Box*. Enclosed also was a letter from you extolling the value of Behe's ideas.
>
> I have a number of concerns:
>
> Your letter would make it appear that the book is from UNM's School of Medicine. The book, in fact, is from the New Mexico Family Council. There is nothing in your letter acknowledging this fact. I have no way of knowing if this is an omission by design or simply sloppiness. It should, however, be corrected; people should know the real source of this unsolicited gift.
>
> Public school libraries buy books based on many criteria. Unsolicited books are not placed in a library's collection at the whim of any outside group. Our school librarian, in consultation with faculty, adds [to] and subtracts [from] our collection. . . .
>
> I personally cannot see its value in a middle-school library. This is not because of its content. Its reading level is simply not age appropriate. I have never seen it listed on any middle-school reading list; I would be delighted if you can point me to such a list—the New Mexico Family Council hiding behind the lab coat of UNM simply does not count.[43]

In his reply to Brugge, Omdahl disavowed any connection between UNM and the book project, explaining that it resulted from Behe's having visited UNM and Omdahl's sharing a similar scientific background (biochemistry) with Behe. He also disavowed Brugge's statement about the book's being sent by the New Mexico Family Council. He thus implied that he was acting independently, though he consistently used the plural pronouns "we" and "our" in reference to the book's distribution and requested the return of the book sent to Eisenhower, explaining that "we" were running low on copies.[44]

However, as Brugge's letter correctly points out, Omdahl was *not* a solitary actor in this effort. New Mexicans for Science and Reason posted on its website a March 27 e-mail "alert" distributed by the New Mexico Family Council (NMFC), which is "one of 40 Family Policy Councils throughout the country which works closely with [James Dobson's] Focus on the Family." In this alert, Phil Robinson claims credit on behalf of NMFC for sending the books, stating, exactly as Omdahl did, NMFC's desire that the books be circulated through the science departments and then placed in school libraries. Robinson's alert ended with the request that readers pray for the schools receiving the books and for a favorable response to the book by its recipients.[45] Robinson's claim that NMFC was responsible for the book's distribution was independently confirmed when the young-earth Creation Science Fellowship of New Mexico posted notice on its website that "Phil Robinson sent out Michael Behe's book 'Darwin's Black Box' to every middle and high school science department from Socorro to Los Alamos."[46]

Brugge responded on April 7, 2002, to Omdahl's denial that NMFC had sent the book, citing the NMFC e-mail alert. This time, he asked Omdahl to identify the "we" in his earlier message and requested information about funding:

> Perhaps you need to contact Phil Robinson and clarify his crystal-clear claim of being the source behind the book. And, in fact, if the NMFC is not behind the book, that still begs the question of its funding. I would be very interested in knowing who the "we" are in [your] note. Seventy-seven copies of Behe's book represent a considerable investment. My guess is that this was not done in your capacity as a professor at UNM, and this still makes me wonder why you used your UNM title on the letter sent to science chairs. All of this seems a bit fishy to me.[47]

And Brugge declined Omdahl's request to return the book: "I shall, thank you, keep the book." (The mailing would indeed have cost a considerable sum: the Touchstone paperback edition sells at Amazon.com for $9.99. Adding five dollars for postage and tax, the mailing of books to seventy-seven schools probably cost roughly $1155.) A month after Brugge's second message, Omdahl still had not responded with the requested information.

Responding to a letter from John W. Geismann, UNM Faculty Senate president, Philip Eaton, UNM Interim Vice President for Health Sciences, reported that the university provided no financial support for the project. He also said that Omdahl had thought he was avoiding the linkage of UNM to the mailing by not using university letterhead, though he had used his affiliation. Responding to a direct inquiry from Geismann, Omdahl admitted that it would have been better had he not used his affiliation, but denied any improprieties in his involvement.[48] Omdahl's denials, of course, evade the question of who *did* provide the money for the books. New Mexicans for Science and Reason succinctly summarized this effort: "So, a religious group has distributed a creationist text to hundreds of public school teachers, in such a way as to make it appear that UNM supports the action. Something smells funny in Denmark!"[49]

The World Wide Wedge: "FTE Online." As another way of engineering direct contact with teachers, one of the most important aspects of the Wedge's education strategy is the use of the Internet—which bypasses boards of education. The FTE has now decided to use the Internet to reach teachers (as well as parents who might contact teachers), through its new website, www.fteonline.com. Although the site is registered to FTE, William Dembski is both the administrative and billing contact, as well as FTE's "Academic Editor."[50] FTE's mission is "to restore the freedom to know to young people in the classroom, especially in matters of worldview, morality, and conscience" by offering "an enriching series of high school textbooks now used in both public and private schools." Teachers can "click on one of the textbook links below to learn more about the tools that can equip you to truly educate your students." Clicking will take them to *Pandas*, of course, with endorsements from Behe and Wells, as well as to other, non-ID texts. But the site is heavily weighted toward ID, with the major Wedge books available for online purchase.[51]

The CRSC Web Curriculum: "Legally Permissible!" The FTE effort sinks to insignificance in comparison to the "science education" program uploaded to the CRSC site in early 2000. Pursuant to its plan to influence not only K–12 but college science instruction, the CRSC posted its "Science Education Resources," which it called "the next logical step after print resources."[52] The plans called for a "modular web curriculum," which "will dominate the future of education, making education both more local-personal as well as more international," enabling American educators "to select what fits their local environment" and international educators and students "to take part in global conversations among peers" via Internet discussion lists and bulletin boards.[53] Materials provided on the site were clearly designed for public school use: "We supply curricular materials that are particularly designed to supplement standard science textbooks used in public schools."[54] When the site first appeared, CRSC allowed unrestricted public access to a set of "learning" (i.e., *teaching*) objectives and a lesson plan on the "Cambrian Explosion."[55] However, after a short period (until approximately mid-March 2000), CRSC restricted access to this "pilot curriculum"—no doubt because of its dubious constitutionality and its obvious scientific flaws. The rest of the site was accessible, although the page entitled "How to Use this Website" bore a legal disclaimer: "We are not responsible for the inappropriate use of our web materials."[56] The objectives, lesson plan, and other classroom resources were now accessible only with a user name and a password, for which there was no way to apply from the site itself. This meant that teachers wishing to obtain the materials had to contact people affiliated with the CRSC, allowing the latter to screen and approve them. Since summer 2001, the science education link has been removed, making the site completely inaccessible to the general public, although early in 2002 DI posted a notice that the site was "under development," which indicates plans to resurrect it.

The restricted-access teaching materials, entitled "Evolution: A Case Study Approach" and billed as "our first installment of supplementary curriculum for public high school biology courses," offered a case study at the time of its restriction: "Biology's Big Bang: Rival Accounts of the Cambrian Explosion of Life." This case study's "Cambrian Explosion Objectives" included learning "current evidence for the Cambrian explosion" based on "recent Chinese fossils," with a reference to "P. Chen" (apparently a misspelling of "Chien") so as to present him as an authority. As already shown, Paul Chien has done no scientific work on the Cambrian fossils. (The CRSC's *own science education bibliography* on this website listed not a single work by Chien.)[57] Included in Objective 4, "current competing explanations of the Cambrian explosion," was "design theory (P. Chen [Chien] and S. [Stephen] Meyer)"—along with punctuated equilibrium (N. Eldredge and S. J. Gould)," implying here that Eldredge and Gould support or at least do not reject "design theory, "which is not simply an error but a deliberate falsehood.

Another page designed to imply the scientific legitimacy of ID was "Cambrian Explosion: Featured Scientists," which contained a table (with most of the information still "forthcoming") listing names of "scientists/ scholars." Listed alongside recognized evolutionary biologists such as Gould and Richard Dawkins, and philosophers of science like Michael Ruse, were ID proponents Paul Chien, Michael Behe, and Stephen Meyer. The table was designed to inform readers of each person's "viewpoint & identity."[58] This restricted "Cambrian Explosion Lesson Plan" required students to "Role Play the Lives of Scientists, Science Educators, and Philosophers of Science." Ignoring pedagogical soundness in favor of wily self-promotion, the plan was to pair students off as opposing characters—ID proponents against pro-evolution scientists and philosophers: "[Phillip] Johnson/[William] Provine," "[Stephen] Meyer/[Eugenie] Scott," and "[Michael] Behe/[Michael] Ruse." These lesson plans were correlated with the Glencoe textbook, *Biology: The Dynamics of Life*, widely used in public schools.

In addition, a page entitled "Cambrian Explosion" contained a clear reference to Dembski's idea of "specified complexity" as one of the "key features of the Cambrian explosion": "The sudden emergence of the various animals of the Cambrian explosion represents an unsurpassed *discontinuous* increase in the specified complexity (or information content) of the biological world" (emphasis in original).[59] Nowhere, however, was Dembski, who has no credentials of any kind in biological or evolutionary science, referenced or identified as the author of this concept, and of course, there was no reference to its rejection by all the professional engineers, statisticians, and mathematicians who have evaluated it.[60]

The Wedge's creationist agenda and awareness of the potential legal ramifications of their ID educational resources are plain enough. The contents of the case study and its restriction to selected users are typical evidence of both. Further evidence of the thoroughly creationist ideology of the science education website was located on "Related Websites," under "Other Progressive Supplementary Science Curricula," from which there was a link to the creationist Access Research Network (discussed later).[61] On a page entitled "Our Curriculum is Scientifically Accurate" could be found the typical conspiracy mongering in which ID creationists engage, the purpose being to sow among innocents the seeds of distrust of science:

> The evolutionary establishment so far has dismissed Behe's design hypothesis as premature, arguing that molecular machines and the origin of life represent unsolved problems that may yield to further research. This response raises some troubling questions, however. If naturalistic mechanisms cannot yet explain the origin of the cell and its machinery, why aren't students told this? And if such mechanisms fail to explain relatively simple one-celled organisms, how do we know they can explain the origin of vastly more complex multicellular organisms? Are scientists telling the public everything they know about this subject?[62]

Teachers were assured that "Our Curriculum is Legally Permissible in Public Schools" and were urged "Don't let legal intimidation squash classroom innovation."[63] On a page entitled "Web Curriculum Lowers Political Hurdles," CRSC asserted that its "Web curriculum can be appropriated without textbook adoption wars"—meaning, of course, that teachers who wanted to use it could easily do an end run around curriculum and textbook adoption procedures.[64] The CRSC asserted on this page that "teachers will flock to our site as they see political hurdles minimized." These statements were clearly intended as an incentive to teachers with creationist sympathies who might be concerned about facing legal action. As added assurance, the CRSC promised those who took advantage of its resources that an "e-mail listserv for registered web curriculum users can include legal advice."[65]

The Science Education Resources site was constructed by Michael Newton Keas, an associate professor of natural science at Oklahoma Baptist University (OBU), who is being nicely remunerated by the Wedge for his services. The CRSC, where he is now a "Senior Fellow," provided the fellowship enabling him to devote his time to constructing this site. Keas reported in his September 1999 "Self Profile" that he received from CRSC "travel, laptop computer, and research grant for June 1999 and load reduction for spring 2000, all for the purpose of research and curriculum development related to the Science Education Resources website of the CRSC," which is "designed to supplement high school and college science courses (including US 311/312 at OBU)" and "is my primary responsibility as a Fellow of the CRSC." His "Progress Report for 2000–2001," updated in September 2001, indicated that he had been granted funds for a semester sabbatical from OBU, facilitating progress toward his goal to "develop a Science Education website and CD-ROM package for Discovery Institute."[66]

ID College Course Prototype. "US 312" is Keas's OBU course, "Unified Studies: Introduction to Biology." It is apparently a prototype for CRSC's college curriculum: OBU is listed on the Access Research Network website as one of two "ID Colleges"—the other being Biola University, where the Mere Creation conference was held—which "offer courses or programs that include an Intelligent Design perspective."[67] (Note: the presence of ID curricula on two campuses fulfills one of the Wedge's "Five-Year Objectives": "Two universities where design theory has become the dominant view.") In a 2000 syllabus, Keas assigned readings in *Of Pandas and People* and required course packets entitled "Science Education Reform" and "Definitions of Positions on Origins." In spring 2002, he required students to read Jonathan Wells's *Icons of Evolution*, as well as Internet articles by Dembski and Behe. He also sometimes makes PowerPoint presentations available to students through his online course schedule. (And of course, sympathetic public school teachers can download these PowerPoint presentations for use in their own classes.) In one, he declares that "Christianity influenced rise of peer review accountability." In a second, he explains Dembski's concept of complex specified information as a competitive scientific theory. In yet another, he cites Jonathan Wells's *Icons of Evolution* among his "Reliable Sources on Human Evolution." If US 312 is a prototype for what other students can expect in college ID classes, they will be thoroughly schooled in the essentials of Wedge doctrine.[68]

Access Research Network: Fighting the "Battle of Gideon." Keas is not working alone in the Wedge's concerted effort to make "educational" materials available on the Internet. The ARN, CRSC's de facto auxiliary website, offers a host of resources, many of which may be downloaded without cost. One of the most heavily promoted has been *Darwinism: Science or Naturalistic Philosophy?*—a videotape of the 1994 Stanford University debate between Phillip Johnson and William Provine of Cornell University. The video is part of a kit that includes a study guide and downloadable PowerPoint "Overhead Masters" for making presentations correlating with the videotape. Also in the kit is a report by Wedge supporter Norris Anderson entitled "Education or Indoctrination: Analysis of Science Textbooks in Alabama," which includes an "insert to be placed in all biology textbooks"—the infamous 1996 Alabama textbook disclaimer.[69] ARN is a treasure trove of ID materials. As a vital arm of the Wedge, it deserves discussion.

ARN evolved from an earlier creationist organization, Students for Origins Research (SOR). One of SOR's founders, Dennis Wagner, is ARN's board chairman.[70] Linked to the CRSC website as one of its "research partners" and headquartered in Colorado Springs, ARN has become a comprehensive clearinghouse of ID resources. Its website offers virtually everything—news releases, publications and multimedia products, a lively discussion forum, and even "Real Science 4 Kids," an elementary science curriculum for both homeschool and public school markets (and which, readers are assured, requires no science background to teach).[71] The site yields a detailed idea of what ARN provides to interested ID supporters, but its "2000/2001 Annual Report" provides more insight into the ambitious goals it shares with the Wedge; it is an important part of the Wedge strategy.

ARN has a small full-time staff whose dedication is exemplified by the fact that Wagner and two office managers gave up their full-time jobs early in 2000 to "devote themselves to the task of growing the organization"; its board of directors and "Friends of ARN" are dominated by CRSC fellows, who also make up most of the editorial board of *Origins and Design*, which is produced through ARN and carried on its website. ARN's mission is "providing accessible information on science, technology and society issues from an intelligent design perspective." Defining ID as "a recent movement of theistic scientists and scholars who have joined together to pursue a credible, and academically rigorous, alternative to the dominate [*sic*] view of naturalism in our culture today," the organization sees itself as engaged in a "cultural battle of scientific worldviews." [72] ARN is eager to promote its message even to those who do not frequent its website; to that end, according to the report, it sent a "sample mailing to approximately 2,000 [public and university libraries] offering them the William Provine vs. Phillip Johnson debate video." One hundred videos, financed by a patron, were sent to public libraries, and ARN expected to send out another 700 in 2001.

Though it does not enjoy the lucrative funding of the CRSC, ARN has a loyal and expanding donor base according to its report. One unnamed foundation gave $10,000, with a commitment of another $6,000 for 2001; roughly a dozen individuals donate more than $1,000 per year. ARN reports a total income for the 2000 fiscal year of $226,648 (a substantial portion of that income was in the form of donated labor and book/video sales), with total expenses of $229,534. Its 1998, 1999, and

2000 IRS 990 forms, available through Guidestar, show some fluctuation in revenues. Possibly in response, ARN announced in its report ambitious plans for 2001 to increase its revenues and broaden its reach:

- Increase of our web traffic from 40,000 . . . to 60,000 visitors/month
- Increase in retail sales from an average of 5 . . . to 8 sales/day
- Increased focus on wholesale distribution of our products
- Increase product and subscriptions sales from $81,979 to $130,000
- Increase individual donations from $29,712 to $100,000
- Increase grant funding from $10,000 to $80,000 . . .
- Recording and release of new video lecture and interview products featuring noted ID academic speakers. . . .
- Increase the frequency and distribution of the *Origins and Design* journal . . .
- Add more resources for high school and undergraduate students.

ARN's budget goal for 2001 was to increase its operating budget from $220,000 to $370,000. The Annual Report appealed to recipients in its cover letter to help ensure its growth by becoming one of the "ARN Gideon 100," drawing from the biblical story of Gideon, who defeated the Midianites with a small army of 300 men divided into units of 100. As an incentive, ARN offered potential donors "an ARN Gideon 100 Certificate" and a videotape of Jonathan Wells's lecture, "The Fall of the Darwinian Empire." Such fundraising efforts are clearly considered vital—ARN is fighting a war: "join us in the battle against the Darwinian Empire. . . . Pick up your trumpet and your torch: the battle intensifies."[73]

Student-Oriented Web Resources: "Future Talent"—IDURC and IDEA. One of the Wedge's goals in Phase II is to nurture "future talent"—to recruit student followers. The Wedge appears to be making good headway toward this goal. There is a link from the ARN website to a student organization, the Intelligent Design Undergraduate Research Center (IDURC), which has also become an arm of the Wedge. Virtually everything on the IDURC site is related to CRSC activities. In "The Coming Revolution: It's Up to Us," IDURC member Jeremy Alder highlights the dependence of ID on youthful supporters: "The intelligent design movement is a movement of the youth. . . . [Its] ultimate triumph or failure will be the direct result of the work done by our generation and the ones following." There is no evidence on the site itself that CRSC established IDURC. However, the organization has been officially taken under the Wedge's wing by being assimilated into ARN.[74] Whether the CRSC established IDURC or not, but especially if it did not, the group's existence and obvious adherence to ID principles indicate that CRSC is gaining influence in the college and high school populations, although such direct influence is probably still small at the moment.

Finally, the Intelligent Design and Evolution Awareness Club (IDEA) was formed in May 1999 by some University of California-San Diego students after they attended a UCSD lecture by Phillip Johnson. Among the group's goals are to "promote, as a scientific theory, the idea that life was designed by an Intelligent Designer," to "educate people about scientific problems with purely natural explanations for the origins and evolution of life," and to "challenge the philosophical assumptions of Darwinism, naturalism, and materialism." With the graduation of its founders, IDEA has matured into the IDEA Center, affiliated with a young-earth Lutheran seminary. Its major goal is high school and college recruitment (more on this in chapter 9).[75] These student movements certainly do not presently represent a tidal wave of awareness among the young about the CRSC and ID, but they do represent a vast potential pool of recruits that the Wedge is cultivating with its usual political energy.

OP-ED FELLOW

Given the number of CRSC members who have published editorials in major newspapers, this project is clearly a team effort. The following op-eds, published in some of the country's leading newspapers, are a small sample of an ever-lengthening list:

1. Michael Behe, "Darwin Under the Microscope," *The New York Times*, October 29, 1996.
2. Phillip E. Johnson, "The Church of Darwin," *Wall Street Journal*, August 16, 1999.
3. Jay Richards, "Darwinism and Design," *The Washington Post*, August 21, 1999.[76]

Despite the shared responsibility for getting the Wedge's viewpoint into the newspapers, CRSC fellow Nancy Pearcey seems to have the primary responsibility for such writing jobs; she produces a steady stream of op-eds for journals and magazines, most prominently the religious *World* magazine. Since 1997, Pearcey has published at least eight *World* articles on ID-related topics, seven in 2000 alone.[77] ID is getting a great deal of publicity from *World* editor Marvin Olasky through Pearcey's frequent op-eds. Olasky is one of the "influential individuals" in the print media who the Wedge Document says the Wedge will "seek to cultivate and convince" in Phase II. He obviously supports the Wedge's work: in the December 4, 1999, issue of *World*, he and Gene Veith, in apparent seriousness, named Behe's *Darwin's Black Box*, Dembski's *The Design Inference*, and Johnson's *Darwin on Trial* among "The Century's Top 100 Books."[78]

World is owned by Bob Jones University and covers current events and politics from a strictly biblical perspective. According to Skip Porteous of the Institute for First Amendment Studies, Olasky is a Christian Reconstructionist who once served as a consultant for Howard Ahmanson's Fieldstead & Company, one of the Wedge's major financial benefactors. (Frederick Clarkson, an authority on the Religious Right, describes Christian Reconstructionism: "Generally, Reconstructionism seeks to replace democracy with a theocratic elite that would govern by imposing their interpretation of 'Biblical law.' . . . [It] would eliminate not only democracy but many of its manifestations, such as labor unions, civil rights laws, and public schools.") Olasky is also an adviser to President George W. Bush, who wrote the foreword to Olasky's book, *Compassionate Conservatism: What It Is, What It Does, and How It Can Transform America*. Through the frequent pro-ID articles of Pearcey and others in *World*, the Wedge has been the beneficiary of publicity in the nation's capital: *every week from January 2000 until the November 2000 presidential election, copies of World were hand-delivered weekly to each member of Congress*.[79]

"Significant Coverage in National Media." One of the Wedge Document's "Five Year Objectives" is "significant coverage in national media," which CRSC divides into four categories:

- Cover story on major news magazine such as *Time* or *Newsweek*
- PBS show such as *Nova* treating design theory fairly
- Regular press coverage on developments in design theory
- Favorable op-ed pieces and columns on the design movement by 3rd party media [Wedge Document]

The CRSC has not yet become the cover story in national news magazines, nor has it been the subject of a PBS documentary (though it *turned down* an invitation to be featured in one, as we discuss later). But it has made significant headway toward its goal of making ID a frequent subject of newspaper articles and op-ed columns written by people other than its own members and supporters.

Phase II of the Wedge Document celebrates DI president Bruce Chapman's "rare knowledge and acquaintance of key op-ed writers, journalists, and political leaders." Chapman's connections have paid off handsomely—the Wedge has received third-party coverage in nationally recognized media outlets, most of it available on the CRSC website.[80] Cultivation of third-party publicity was an early stage of the Wedge's execution of its agenda, exemplified by a "seminar" described in *Origins & Design*: "On April 30, 1996, at the Ethics and Public Policy Center in Washington, D.C., biologists Michael Denton (Department of Biochemistry, University of Otago) and Michael Behe (Biological Sciences, Lehigh University) presented summaries of their new books to an audience of journalists, scientists, educators, and think-tank staffers."[81] The event attracted prominent journalists and other attendees: "Ronald Bailey of New River Media, Tom Bethell of *The American Spectator*, David Brown of the *Washington Post*, historian Gertrude Himmelfarb, Irving Kristol of *The Public Interest*, Anne Morse of Break Point Radio, Joel Nelson of the Johns Hopkins School of Medicine, Paul Nelson of the University of Chicago, Nancy Pearcey of the Wilberforce Forum, Scott Walter of *American Enterprise*, Larry Witham of the *Washington Times*, and science scholar Hubert Yockey."[82]

Some participants were certainly supporters: Gertrude Himmelfarb, Irving Kristol, and Tom Bethell are vocal, but certainly *not* scientifically qualified, critics of evolutionary theory. However, not all attendees were there to show their agreement with the Wedge program. Subsequent to his attendance, Ronald Bailey wrote a hard-hitting article in the July 1997 issue of *Reason*, analyzing neo-conservative support for the ID movement, pointing out quite properly that this neo-conservative enthusiasm for a creationist renascence "comes just as evolutionary thinking itself is shedding considerable light on an array of questions and problems, from brain growth to . . . software design. *Such research is yielding anti-designer results*" (emphasis added).[83] Yet whether the participants attended because of curiosity, on assignment, or to show ideological support, the ability of the Wedge to command such attention is impressive. And the media effort continues, with events such as this: on November 1, 2000, DI held a "CRSC Media Lunch" for twenty-eight people at the Waldorf Astoria in New York City. Discovery Institute billed the event as "an exclusive opportunity for east coast media to meet with key design theorists, William Dembski, David Berlinksi, and Jonathan Wells."[84]

The Wedge considers as favorable *any* coverage recognizing ID as one side in an ongoing evolution-creationism debate; the point is to have the ID movement legitimized in the media as a phenomenon that must be taken into account. Criticism is better than being ignored, as Dembski said in response to devastating criticism from evolutionary biologist Massimo Pigliucci: "A biologist, Pigliucci's sputtering, angry review of *The Design Inference* published in the journal *BioScience* called Mr. Dembski's work 'trivial,' 'nonsensical,' and 'part of a large, well-planned movement whose object . . . is nothing less than the destruction of modern science.' Mr. Dembski loved it. 'If the worst humiliation is not to be taken seriously, at least we're being taken seriously. . . . If we're generating such strong, visceral responses, *we must be doing something right*'" (emphasis added).[85] Judging from the attention paid to ID by the media, the Wedge *is* doing something right. Major, nationally distributed newspapers not only carry their op-eds but report frequently on their activities. Again, we offer only a small sample; the full range of articles is on the CRSC website.

1. Marla Freeman, "Surviving Darwinism," Special to the *San Francisco Chronicle*, December 19, 1997.
2. Laurie Goodstein, "Christians and Scientists: New Light for Creationism," *The New York Times*, December 21, 1997.
3. Mary Beth Marklein, "Evolution's Next Step in Kansas," *USA Today*, July 19, 2000.
4. Larry Witham, "How a Theologian, Two Biologists See Darwin," *The Washington Times*, November 19, 2000.
5. Uwe Siemon-Netto, "It's Perilous to Ponder the Design of the Universe," UPI, December 21, 2000.
6. James Glanz, "Evolutionists Battle New Theory on Creation," *New York Times*, April 8, 2001.
7. Teresa Watanabe, "Enlisting Science to Find the Fingerprints of a Creator," *Los Angeles Times*, March 25, 2002.

In addition to these national newspapers, the Wedge has been the subject of stories in numerous local and regional papers.

The CRSC receives a significant amount of attention from *The Washington Times*. At least part of this attention may spring from its being owned by the Rev. Sun Myung Moon, leader of the Unification Church and spiritual-intellectual father to Jonathan Wells (though the articles on ID by *Times* religion writer Larry Witham are relatively even-handed). Wells is an active Unification Church theologian. As indicated by the UPI article, UPI is another media outlet to which CRSC has access, again probably through Wells's Unification Church membership. Rev. Moon bought UPI in 2000. According to Frederick Clarkson, "This small church has played a pivotal role in the development of the conservative movement . . . [and] the Christian Right—helping, for example, to develop the theme of religious persecution as the response to criticism of their political activities." Through his numerous front organizations, Moon has poured hundreds of millions of dollars into a plan to replace American democracy with a Unification theocracy.[86]

PBS (OR OTHER TV) CO-PRODUCTION

The CRSC has not yet achieved its goal of a PBS production on intelligent design, but it has logged quite a bit of time on PBS and major commercial networks. The Wedge's most high-profile PBS appearance so far occurred when Johnson, Behe, and David Berlinski formed the pro–intelligent design side of a debate on PBS's *Firing Line* in December, 1997.[87] Although it aired two years before the formal beginning of the Wedge Document schedule (1999–2003), this program is the closest the Wedge has come to the objective of "a major public debate between design theorists and Darwinists (by 2003)." More time on public airwaves came when Johnson was featured as a "talking head" throughout PBS's special *Monkey Trial* in February 17, 2002. Johnson took credit for influencing producer Christine Lesiak's "research and thinking from the beginning."[88] He has also appeared on CNN and ABC's *Nightline*. Along with others, he was a guest on CNN's *Talkback Live* following the Kansas Board of Education's deletion of evolution from the state science standards in 1999; his inclusion was clearly a reflection of CNN's recognition of the role Johnson played in assisting pro-creationist forces in Kansas. And the Wedge's continuing role in Kansas maneuverings was recognized again when, along with People for the American Way president Ralph Neas, Johnson was Ted Koppel's guest on *Nightline*, just prior to the August 1, 2000, Kansas Board of Education Republican primary election.[89]

The Wedge came close to being featured on PBS's much-publicized series, *Evolution*, which aired in September 2001. DI was invited to participate but balked at the producers' intention to feature them in the segment on religion, "What about God?"—they wanted to be featured in the science segment as *scientists*. As a result, they were replaced by the young-earth creationists at Answers in Genesis (AIG), a substitution that prompted snide comments: "'We wanted to talk about science, and they wanted us to do Sunday School,' said Mark Edwards, a spokesman for the Discovery Institute. 'The final episode paints a picture that the only critics of Darwinian theory are these guitar-strumming hillbillies in Kentucky who are creationists, and that's just not true. We're glad we're not part of that stereotype.'"[90]

DI subsequently launched a full-court press against the series (as did ARN—*and* the unappreciative "hillbillies" at AIG), creating an extensive website, "Discovery Institute's Critique of PBS Evolution," on which were posted numerous news articles and critical op-eds. They produced a 152-page *Viewer's Guide*, complete with a set of learning objectives to repair PBS's damage to young minds: "Accuracy and objectivity are what we should be able to expect in a television documentary—especially in a science documentary on a publicly funded network. . . . [T]he PBS EVOLUTION series distorts the scientific evidence, omits scientific objections to Darwin's theory, mischaracterizes scientific critics of Darwinism, promotes a biased view of religion, and takes a partisan position in a controversial political debate. . . . It is for these reasons that we have prepared . . . [this viewer's guide]."[91]

"100 Scientists." The Wedge's most publicized move, however, was publication of "A Scientific Dissent from Darwinism," an ad signed by "100 scientists," who publicly affirmed that "we are skeptical of claims for the ability of random mutation and natural selection to account for the complexity of life. Careful examination of the evidence for Darwinian theory should be encouraged." Such a statement could easily be agreed to by scientists who have no doubts about evolution itself, but dispute the exclusiveness of "Darwinism," that is, natural selection, as evolution's sole

mechanism when other mechanisms such as genetic drift and gene flow are being actively debated. To the layman, however, the ad gives the distinct impression that the 100 scientists question evolution itself. Released shortly after the PBS series aired, this craftily worded statement was prefaced with a statement from the Discovery Institute: "The public has been assured, most recently by spokespersons for PBS's *Evolution* series, that 'all known scientific evidence supports [Darwinian] evolution' as does 'virtually every reputable scientist in the world.' The following scientists dispute the first claim and stand as living testimony in contradiction to the second. There *is* scientific dissent to Darwinism. It deserves to be heard." (In response, the National Center for Science Education launched "Project Steve," its "tongue-in-cheek parody of [the] long-standing creationist tradition of amassing lists of 'scientists who doubt evolution.'" In honor of Stephen Jay Gould, NCSE asked scientists named "Steve" to endorse a statement affirming their acceptance of evolution and rejecting the teaching of creationism, including ID, in public schools. As of April 25, 2003, the "Steve-o-meter" registered 320 names, including cosmologist Stephen Hawking and 1979 Nobel laureate Steven Weinberg. Dembski was not amused: "If Project Steve was meant to show that a . . . majority of the scientific community accepts . . . naturalistic . . . evolution, then . . . [NCSE] could have saved its energies. . . . Project Steve is . . . an exercise in irrelevance.")[92]

As it turned out, when the NCSE requested the source of the quote from the "spokespersons," DI could not provide one (and has not to this day, despite assurances by spokesman Mark Edwards that he would try to find it). NCSE found one possible source in an internal PBS memo (discussed later), containing a line almost identical to the quote in the ad, but DI itself had inserted the bracketed reference to Darwinism. On contacting a sample of the scientists who signed the statement, NCSE found no monolithic opposition to evolution but rather "a diverse range of opinions about the role of natural selection in evolution." The ad ran in major publications such as *The New York Review of Books* and *The New Republic*; according to NCSE, the magazines' rate cards suggested that all three ads cost a total of approximately $50,000.[93]

Finally, DI leaked an "internal PBS memo" outlining PBS's production goals, marketing strategy, and intended audience messages. The memo, in DI's words, disclosed "an improper political agenda" behind the series. Bruce Chapman—with stunning irony, in light of what this book shows about the Wedge's agenda—charged that "clearly, one purpose of 'Evolution' is to influence Congress and school boards and to promote political action regarding how evolution is taught in public schools. . . . In fact, 'Evolution's' marketing plan seems to have the trappings of a political campaign."[94] Despite DI's attack on the truthfulness of the PBS series, the Wedge has not managed to break into the public television scene as a star attraction, as members had hoped—and *because* of the attack they are not likely to do so. The campaign to discredit PBS arguably had little real effect except on ID supporters, but it was another index of how far the Wedge is willing to go to become an audible voice in the cultural mainstream.

Although they have not quite become the television presence they would like to be, in addition to the TV time they *have* managed to obtain, Wedge members also receive publicity through frequent radio appearances, including public radio. In April 2000, David K. Dewolf, the Wedge's chief legal strategist, was featured with Dr. Eugenie Scott of the NCSE on NPR's "Justice Talking."[95] Through public radio, however, they reach neither their largest nor their most receptive audience. To reach the conservative Christians the Wedge Document names as the movement's "natural constituency," the Wedge relies on the radio shows of major Religious Right operatives.

Phillip Johnson appeared on James Dobson's *Focus on the Family* radio program as one of Dobson's "favorite authors" in September 1996; Dobson and Johnson talked about Johnson's *Reason in the Balance* and "about naturalism, and how it worms its way into our lives and into the lives of our children. . . . [I]t is very pervasive, and at this time, very influential, and very damaging." *Dobson has a loyal contingent of four million daily listeners*, giving him a bigger audience than Jerry Falwell or Pat Robertson, according to *U.S. News & World Report*. Johnson had been Dobson's guest a couple of years before to promote *Darwin on Trial*. He appeared again in October 1999 in a two-day segment entitled "Teaching Children About Science," which was broadcast again on May 30–31, 2000. Dobson is now selling audiotapes and CDs of these broadcasts.[96]

There have been many more appearances like these. Whether the CRSC arranges these TV and radio appearances itself through its media connections or whether they receive invitations, the air time raises the Wedge's profile. Such appearances demonstrate that the CRSC has come to be viewed by the national media as an available source of interviewees who can appear whenever they are called. The Wedge's plan is that, as their media presence becomes familiar, their message will acquire among American listeners the legitimacy so often associated with regular exposure—independently of its content. Of course, there is nothing new about such "legitimacy."

PUBLICITY MATERIALS/PUBLICATIONS

U.S. Newswire. Discovery Institute cultivates publicity by announcing the ready availability of CRSC fellows and including contact information at the end of all its U.S. Newswire press releases. DI directs these press releases to "National Desk, Science and Education Reporters," as it did in the July 20, 2000, press release "Discovery Institute: Scopes Trial 75 Years

Later: Darwinian Fundamentalists are the Ones Suppressing Dissent." DI also did press releases on September 5 and September 26, 2000, respectively, publicizing Jonathan Wells's "Evaluation of Ten Recent Biology Textbooks" and condemning to the flames the report Lawrence Lerner wrote for the Thomas B. Fordham Foundation, *Good Science, Bad Science: Teaching Evolution in the States*.[97] According to U.S. Newswire's online newsletter, "USN has agreements with thousands of the world's most widely accessed online services and databases, helping your news reach millions of non-media audiences in addition to the targeted reporters and editors which receive your news through USN's comprehensive satellite newswire, fax, e-mail and Web-based distribution network." U.S. Newswire's press releases are permanently archived through Lexis-Nexis, Dialog, Westlaw, and other databases; the Wedge has clearly chosen a very productive source of publicity here.[98]

Internet Advertising. Another source of publicity is the Internet, exemplified by a banner running in June 2000 on the conservative online newspaper WorldNetDaily.com (WND), advertising the videotape *The Triumph of Design and the Demise of Darwin*, featuring Phillip Johnson. According to WND editor and CEO Joseph Farah, *PCDataOnline* weekly statistics showed that WND received 4.235 million page views during the last week of May 2000 alone.[99] And just as CRSC uses the Internet to provide resources for teachers, members use it to influence public opinion about the Wedge's educational agenda where it counts most: with parents. Jonathan Wells has posted to the Internet materials that are not educational resources for teaching ID, but rather are plainly intended to sway public opinion away from evolution and toward ID. One such offering is "An Evaluation of Ten Recent Biology Textbooks: A Report for the Center for the Renewal of Science and Culture."[100] This piece is not so much a report as a *report card*, evaluating some current textbooks according to whether their presentation of evolution is compatible with ID "theory." Textbooks are graded, but none of the books receives an "A." One receives a "B" in one category; another receives a "C" in one category. Otherwise, all ten books receive grades of "D" and "F." Most pages contain nothing more than Wells's explanations of his grading criteria for the evaluational categories. There is no scientific documentation of any kind in the report; no scientific sources are referenced.

Along with his promotion of *Icons of Evolution* at www.iconsofevolution.com, Wells provides other materials, "Tools for Change," designed to undermine the teaching of evolution. He has composed a list of "ten questions to ask your biology teacher about evolution." The questions *themselves* beg a number of questions, incorporating assertions that students are in no position to evaluate, but with which Wells is clearly encouraging them to challenge their teachers: "EVOLUTION A FACT? Why are we told that Darwin's theory of evolution is a scientific fact—

even though many of its claims are based on misrepresentations of the facts?" Wells even provides a page of "print-your-own" warning labels for textbooks: "WARNING: The subject of human origins is very controversial, and most claims rest on little evidence; drawings of 'ancestors' are hypothetical." The warnings are downloadable in pdf format, conveniently laid out for photocopying on self-adhesive labels.[101]

This is not the only publicity material available for parents on the Internet. In a document entitled "Dealing with the Media in the Science Textbook Controversy," the Wedge dispenses advice for parents about how to handle the media. Written by Judith Anderson and John Wiester, whom Nancy Pearcey calls "the heart" of the ID movement, it instructs parents in media manipulation:

> If you are asked whether or not you "believe in evolution", use the opportunity to educate the reporter. Ask him to explain what he means by the term; then teach him the differences between microevolution, macroevolution and Darwinism. . . . When asked about your religious views, explain to the reporter that you have no interest in discussing your religion; you want to discuss the quality of science education. . . .
>
> Be savvy when dealing with the media. . . . Be aware that the reporter needs outrageous sound bites and that your comments may be heavily edited. Be interested in his point of view: does he want the truth, or has he been told to rewrite Scopes? . . . Don't give the media the opportunity to stereotype you as a Bible-thumping fanatic who is trying to suppress academic freedom. If the reporter can see you as the reasonable party, he will be more inclined to understand and communicate your point of view.[102]

The article, remarkable for its blunt teaching of practical subversion and effective concealment of religious motive, comes with a glossary of stipulated definitions—"intelligent design" is carefully differentiated from "young earth creationism (Creation Science, Scientific Creationism)."

General Publications: "Peer-Reviewed" Journals. One "Five Year Objective" of the Wedge Document is to publish "one hundred scientific, academic and technical articles by our fellows." As already shown, there has been no progress toward this goal—Wedge members have so far simply not published anything on ID of scientific or technical significance in professional journals of physics, chemistry, biology, or geology. Indeed, they publish little else on ID in any regular academic outlets.[103] They do, however, issue *Origins and Design*, which, according to the spring 1998 DI *Journal*, is funded by grants from the CRSC. *Origins and Design* was identified by Henry Schaefer, in his foreword to *Mere Creation*, as a "first-rate interdisciplinary journal" that would be part of the effort to advance the agenda set at the Wedge's *Mere Creation* conference in 1996. This publication is a Wedge tool; most of its content is written by Wedge members. All the most recent editors (as of issue 20:1) are Wedge members: Paul A. Nelson, editor; Bruce L. Gordon, managing editor; and

William A. Dembski, Stephen C. Meyer, and Jonathan Wells, associate editors.[104] Most recently, Dembski has launched a new "peer-reviewed" journal, *Progress in Complexity, Information, and Design,* in connection with his new organization, the International Society for Complexity, Information, and Design (ISCID).[105] (We discuss ISCID in the next chapter.)

Despite the Wedge's establishment of these journals, however, they are not its chief publication outlet. Wedge members publish prolifically in outlets aimed at the general public and, of course, in religious journals and magazines. There is also an extensive list of articles and interviews from "Featured Authors" on the ARN website. Of the sixteen (as of this writing) whose articles are on the site, thirteen are CRSC fellows.[106] (The others are supporters.) The "Leadership University" website of Christian Leadership Ministries, where several Wedge members have "virtual offices," also archives writings by Wedge members such as William Dembski, Phillip Johnson, Henry "Fritz" Schaefer, and Robert Koons.[107]

We have now surveyed an important part of the Wedge's effort. There is, unfortunately, much more.

7

Everything *Except* Science II

[The Darwinists] are already fighting with each other. Sometimes what people say is that we are a major defection or two away from total victory. Once we get some undeniably legitimate figures in the scientific and intellectual communities saying, "This is a legitimate issue . . . " that's when the situation will have radically changed. . . . Once [evolutionary scientists] don't dominate the scene, then we have a radically new world.

Phillip Johnson, quoted in *World*, November 22, 1997

Phase II Continued

Courting Public Opinion

The Wedge hopes to have an outsized impact on American public opinion, and they keep a close eye on opinion polls. As part of its campaign for "spiritual & cultural renewal," one of the CRSC's Five Year Objectives is the "positive uptake in public opinion polls on issues such as sexuality, abortion and belief in God." (Presumably, "positive uptake" means an increased alignment between public opinion and the Wedge's own views.) We cannot at this point quantify the Wedge's impact on American public opinion. Certainly, few Americans even know what the "Wedge" is (by that name), although a great many are now familiar with the Discovery Institute and individual Wedge members. But more Americans are aware of the "intelligent design movement" as a result of the Wedge's frequent involvement in public affairs and the consequent news coverage. And this is calculated to have an effect on public opinion.

Creationism is a perennial American problem for two reasons: low scientific literacy (despite the American love of technology) and a high level of religious conservatism, although the mainstream religions made their peace with evolution long ago and have no objections to its being taught in public schools. On the whole, Americans are, by standards prevalent in the West, a very religious people. Asserted belief in God is consistently high, as revealed in public opinion polls over the years.[1] The Wedge watches these polls carefully, especially as they may pertain to the ID agenda, and recent, well-publicized polls are ambiguous enough in certain respects to suggest exploitative opportunities for the Wedge.

Phillip Johnson sees opportunities both in the remarkably large number of Americans who accept the Bible's creation story literally and in those who seem to believe in theistic evolution. (He does not hesitate, for example, to cite in ID's favor the numbers supporting theistic evolution, despite his own religious repudiation of it.) In "The Wedge: Breaking the Modernist Monopoly on Science" (*Touchstone*, July/August 1999), he exults in the resistance of Americans to the message of science:

> Scientists, educators, museum curators, and others have made determined efforts to convince the public, but public opinion polls indicate that the public isn't getting the message. Over 40 percent of Americans seem to be outright creationists, and most of the remainder say they believe in God-guided evolution. Less than 10 percent express agreement with the orthodox scientific doctrine that humans and all other living things evolved by a naturalistic process in which God played no discernible part. These figures, from recent polls, are practically unchanged from previous polls in the early 1980s. The Darwinists . . . have not convinced the masses.[2]

Johnson is apparently referring to the same information as that gathered by George Bishop in his 1998 paper "The Religious Worldview and American Beliefs About Human Origins." This paper is reproduced in part in his summer 1999 *Free Inquiry* article, "What Americans Really Believe (and Why Faith Isn't as Universal as They Think)." Bishop points to a 1978 Gallup poll in which 50% of Americans said they believed in the literal Genesis account of creation, 20% in theistic evolution *with* special intervention in the creation of humans, and 11% in theistic evolution *without* special intervention to create humans. He then cites a 1997 Gallup poll showing that 44% of Americans still believe the literal Genesis account, 39% accept theistic evolution, and only 10% accept the account provided by natural science, "despite the rising percentage of college graduates, a trend which might be expected to have reduced significantly the proportion of adults believing in biblical creationism."[3]

In discussing his *Touchstone* article with an Internet discussion group,

Johnson commented on the consistency of twenty years of poll data showing the resistance of the "citizenry" to recognizing that evolution is a fact. He sees opportunities in such data: "When a high percentage of the public is seriously at odds with elite opinion, one should make a serious effort to understand the grounds of this skepticism sympathetically, and not just dismiss it out of hand as foolish. When educating the citizens on such a subject, one should make an effort to explain the controversy fairly, and not just adopt the propaganda of one side as 'scientific fact.'"[4]

Indeed, poll data are so important to the Wedge that DI commissioned a poll of its own, hoping to counteract any effects on public opinion produced by the PBS *Evolution* series. DI commissioned Zogby International to conduct the poll, which the Wedge then used in its campaign against the series. Wedge members and their supporters have also invoked energetically the results of this poll in their political campaign to influence federal legislation and Ohio science standards (discussed in chapter 8), and they will probably continue to do so. Therefore, before we examine the significance of other relevant polls, we analyze the Wedge's own poll.

THE ZOGBY POLL

Though DI commissioned the poll, a DI press release said that "the survey itself was designed by the Zogby organization." The poll results were made available to the public on September 24, 2001, the same day the television series began. The National Center for Science Education analysis of the survey, like our own, indicates that the results are of doubtful significance. This survey consisted of four questions (we also show the percentage responses):[5]

1. Which of the following two statements comes closest to your own opinion?
   A. Biology teachers should teach only Darwin's theory of evolution and the scientific evidence that supports it. [15%]
   B. Biology teachers should teach Darwin's theory of evolution, but also the scientific evidence against it. [71%] Neither/Not sure [14%]
2. Do you strongly agree, somewhat agree, somewhat disagree, or strongly disagree with the following statement: "When Darwin's theory of evolution is taught in school, students should also be able to learn about scientific evidence that points to an intelligent design of life."

Strongly agree	53%
Somewhat agree	25
Somewhat disagree	5
Strongly disagree	8
Not sure	9

3. Which of the following two statements comes closest to your own opinion?
   A. When Public Broadcasting networks discuss Darwin's theory of evolution, they should present only the scientific evidence that supports it. [10%]
   B. When Public Broadcasting networks discuss Darwin's theory of evolution, they should present the scientific evidence that supports it, but also the scientific evidence against it. [81%] Neither/Not sure [10%]

4. Do you strongly agree, somewhat agree, somewhat disagree, or strongly disagree with the following statement: "The universe and life are the product of purely natural processes that are in no way influenced by God or any intelligent design."

Strongly agree	12%
Somewhat agree	12
Somewhat disagree	13
Strongly disagree	56
Not sure	7

Though at first glance the results clearly favor the Wedge, the National Center for Science Education reached a different conclusion after its own analysis: "The results of the poll may reflect the popularity of the 'fairness approach.' However, the wording of the poll's questions incorrectly assumes that the theory of evolution is exhausted by Darwin's contributions to it and that any scientific evidence against 'Darwin's theory' invalidates the theory of evolution. . . . [T]hese assumptions and implied connections are erroneous."[6] NCSE's analysis is correct: the questions *do* take advantage of Americans' insistence on fairness, and they identify evolution exclusively with Darwin (not acknowledging the massive additions modern science has made to evolutionary thought). Our analysis shows other serious problems in the design of this poll.

First, questions 1B and 3B contain the crucial assumption that there *is* scientific evidence against evolution. Question 2 contains another: that there *is* scientific evidence for intelligent design. Certainly, a sense of fairness would lead most respondents to answer that such evidence—*if* there were any—should be included. There isn't any, but the questions do not allow for such qualification. Those who already believe in divine intervention in the creation of life (83%, according to the 1997 Gallup data cited by Bishop) would certainly want to include any scientific evidence that points to intelligent design—*if* there were such evidence. But again, the questions do not allow for such a qualification. Respondents had to give the Discovery Institute's preferred answer or risk appearing to be deliberately biased against facts!

There is also a grave technical problem with portions of the data the report narrative selects for special emphasis. For example, the narrative highlights the fact that for question 1, of those who selected option B, "78% each of Republicans, residents of the West, and parents of children under 17," as well as "80% of 18–29-year-olds," believe that scientific evidence against Darwin's theory should be taught. It then points out that agreement decreases to 59% in seniors 65 and older (still a majority). The report likewise highlights the fact that for question 2 (which refers to the intelligent design of life, understood by virtually everyone as God), those who selected B "include an average 83% of Republicans, Protestants, parents of children under 17, and people with household incomes of $25,000–$49,999." Similar data are highlighted for questions 3 and 4, except that here emphasis is placed on the fact that those who selected the Wedge's preferred answers overwhelmingly live in small cities and rural areas. Consequently, the report at first glance shows that the bulk of the Wedge's support comes from low to moderate middle-class, religiously affiliated Republicans, both Protestant and Catholic, and young people, with these supporters residing overwhelmingly in small cities and rural areas—in other words, from "America's heartland."

But the report also highlights data from other "subgroups" that, despite "overwhelming support in each subgroup for teaching both points of view," emphasize the sources of *disagreement* with the Wedge's favored responses. For question 1, the report highlights the support for teaching *only* evidence *for* evolution among residents of suburbs in the eastern United States, even though this response represents an average of only 21% of this subgroup. For question 2, the report highlights *disagreement* with teaching evidence *for intelligent design* among very low income people (less than $15,000, implying that the least educated are among those who oppose teaching ID), relatively high income people (with incomes of $75,000 or more), independents, and again, residents of the eastern United States—even though these represent only "an average 18%" of these subgroups. Similar data are highlighted for questions 3 and 4, except that Democrats, Hispanics, and African Americans are included, with an emphasis on residence in large cities as well as suburbs. Such emphasis, viewed uncritically, creates the impression that opposition to ID and support for evolution is concentrated among the inner-city poor, racial and ethnic minorities, and well-off suburban liberals in the eastern United States—definitely *not* America's heartland. These are the only groups singled out in the report as *disagreeing* with the Wedge's preferred positions; yet the percentages of disagreement the report cites for these groups are very low. What's going on? An analysis of the crosstabs provides an answer.

What the report narrative does *not* emphasize, but is evident on inspection of the crosstabs (tables of numerical data that few people are likely to inspect), is that for *every question*, in *every* demographic category, support for the Discovery Institute's preferred responses was high, in virtually every case more than a majority. In only 2% of the figures was there *less* than a majority—definitely not what one would reasonably expect given the diversity—political, religious, educational, and economic—that the poll's subgroups represent. If read uncritically, these results indicate that the Wedge's favored positions were supported by every subgroup in the country—from Nader supporters to Gore supporters to Bush supporters, from people with less than a high school education to those with college degrees, from rich to poor!

Why then would the report narrative *not* emphasize such unexpected and surprising data? Would such data not mean that the Wedge has won from the majority of Americans unqualified support for "teaching the controversy"? The answer is no. With numbers so consistently high and in agreement across all demographic groups, the only possible conclusion is that the poll is grossly flawed methodologically, that the biases in the questions themselves, as posed, nullify any meaning that might be deduced from the poll. The Zogby poll shows only two things for questions 1–3, namely, that Americans favor fairness and want to appear reasonable and that they do (or say they do) have a healthy respect for *the idea* of scientific evidence. Period. With respect to question 4, it emerges that most Americans reject the idea of a purely natural explanation for "the universe and life." But that is already common knowledge.

Although the report narrative emphasizes only data that portray ID as most appealing precisely to the constituency the ID movement seeks specifically to cultivate, for political purposes the numbers themselves actually say more about the inherent desire for fairness in the American people than about their support for intelligent design. The poll tells us nothing useful about that. Sadly, the American respect for fairness is what the Wedge exploits when members argue, as Stephen Meyer recently did, citing the Zogby poll, that "voters overwhelmingly favor this approach. . . . 71% of those polled stated their support for teaching evidence both for and against Darwin's theory of evolution. Only 15% opposed this approach. An even greater majority favored exposing students to 'evidence that points to an intelligent design of life.'"[7] The superficial reasonableness of such statements appeals strongly to fair-minded people because they are not aware that the other side of the "controversy" (ID) has neither scientific nor philosophic substance. Although an "other side" exists, where the science is concerned, it is a blank. Finally, no recent exercise illustrates better than this Zogby poll story the fatuousness of using opinion polls to decide questions of scientific substance.

More Polls—and Their Significance to the Wedge

Even prior to commissioning the Zogby poll, the Wedge was watching polls closely. In a "Special Update" to the American Geological Institute, David Applegate, who attended the Wedge's May 10, 2000, congressional briefing in Washington, D.C. (see chapter 8), demonstrates Johnson's dependence on poll information for Wedge strategy:

> Asked if there was a critical mass yet of ID supporters among scientists at universities, Johnson stated that you do not convince the priesthood but generationally replace them. He argued that demographics are on ID's side—polls show skepticism about Darwinism so the public at large is sympathetic. . . . The people need to be empowered and that is what is happening with the Internet and talk radio, which takes away control from the scientific gatekeepers. Johnson's stated objective was to get thousands of young people in the classroom asking questions of dogmatic professors, and he said that it is already happening.[8]

Johnson is not the only Wedge member citing the poll numbers. William Dembski has done so, putting a favorable spin on the results:

"Gallup polls consistently indicate that only about ten percent of the U.S. population accepts the sort of evolution argued by [Richard] Dawkins, [Michael] Ruse, and [Michael] Shermer, that is, evolution in which the driving force is the Darwinian selection mechanism. The rest of the population is committed to some form of intelligent design. Now it goes without saying that science is not decided in an opinion poll. Nevertheless, the overwhelming rejection of Darwinian evolution in the population at large is worth pondering."[9] The CRSC has also cited polls in public presentations such as their PowerPoint slide show, "Darwin, Science, & Going Beyond the Culture Wars," an "Evening Public Forum for Parents & Concerned Citizens," which says that 79% of Americans favor allowing creationism in public schools and 83% favor teaching evolution. The presentation also specifically points out that the poll did not survey public opinion on ID and its approach to education, possibly implying that there might be room in public opinion for the acceptance of ID.[10]

The figures cited in the CRSC's PowerPoint presentation are from a poll commissioned by People for the American Way (PFAW) that received wide publicity.[11] Released in March 2000, it seems to have provided ammunition for both sides of the evolution/creation controversy. Daniel Yankelovich, chairman of DYG Inc., which conducted the poll, acknowledged that "you can read the poll as half-empty or half-full."[12] In the PFAW poll, 83% of Americans favor teaching evolution in public schools. However, the numbers are ambiguous enough elsewhere in the results to be interpreted by ID proponents as reason for optimism. George Bishop's remark regarding the high percentage of well-educated Americans who still accept the biblical account of creation is pertinent here: it is surprising but true. Of the 79% of respondents favoring the teaching of creationist arguments, 17% favor teaching evolution in science classes and allowing for religious explanations (creationism) in *other* classes; 29% favor discussing creationism in science class, but as a "belief" only; 13% support teaching both evolution and creationism in *science* class as *equally scientific* theories; 16% favor teaching *only* creationism in science classes; and 4% favor teaching both evolution and creationism, without being sure how it should be done. These figures show that exactly *one third* of the American people favor the teaching of creationism in American public schools—or *think* they do. The fact that, of this one third, more than half favor teaching evolution as well is not yet of concern to the Wedge. If ID proponents cannot eliminate evolution, they will settle—*for now*—for recognition of ID as a serious alternative to it.

Other less publicized figures from the PFAW poll should arouse concern among scientists and all scientifically literate persons about the opportunities for the increasingly respectful acceptance of ID by the public, especially if the Wedge can convince enough people that ID is a legitimate, competitive, *scientific* explanation. The PFAW poll points out that, although 45% of Americans have never previously encountered the term "creationism," it is also true that "most Americans are unclear about Evolution's scientific status. . . . [M]ost are not sure that the theory, as they understand it, is fully accurate or proven."[13] Specifically, PFAW points out that many Americans do not understand evolution. Among the 95% of Americans who have heard of the term "evolution," 50% understand correctly one vague, simplistic, or limited definition of it ("human beings have developed from less advanced forms of life over millions of years"). Of the remainder, 34% use a grossly incorrect definition ("human beings have developed from apes over the past millions of years"), and 16% either think it means something else or are simply unsure what it means.[14] This shows that half of all Americans who have heard of evolution lack even an *approximate* notion of what it means. PFAW's finding is consistent with a 1993 International Social Survey finding, related by Bishop, that of twenty-one nations surveyed regarding people's knowledge of the basic fact of evolution, Americans were last—behind Bulgaria and Slovenia.[15]

PFAW found that, beyond false or absent ideas of the principles and meaning of organic evolution, most Americans have specific doubts about its account of the history of the human species. Of the 95% who have heard of evolution, 69% believe either that "you can never know for sure," or that evolution is "mostly not accurate," or that it is "completely not accurate," as opposed to only 29% who believe either that evolution is "a completely accurate account of how humans were created and developed" or that it is "mostly accurate." Forty-nine percent believe that evolution is "far from proven." As expected, among Americans with an educational level of high school or less, 55% believe that evolution is "far from proven." Of college graduates, 42% agree. Even of people with postgraduate education, a full 30%—almost one out of three—opine (on no basis of competent judgment) that evolution is far from being proven.[16]

The most recent Gallup poll, conducted February 19–21, 2001, shows no appreciable change in the 1997 Gallup poll numbers cited by Bishop.[17] Again, Bishop's observation about the surprisingly widespread rejection of the facts of evolutionary science in such a highly industrialized nation as the United States is borne out. Even disregarding the inflated and uninterpretable numbers in the Wedge's Zogby poll, such data as these keep the wind in the Wedge's sails.

Phase III: "Cultural Confrontation and Renewal"

This is the most aggressive element of the Wedge strategy. As the Wedge Document makes plain, through the objectives of this phase, the Wedge hopes to breach the constitutional barrier that has so far prohibited creationism's advancement into the educational mainstream:

- Academic and Scientific Challenge Conferences
- Potential Legal Action for Teacher Training
- Research Fellowship Program: Shift to social sciences and humanities [Wedge Document]

Following this phase of the Wedge conveys a sobering idea of what the Wedge stands to accomplish should its strategy succeed.

Academic and Scientific Challenge Conferences

The Wedge Document's description of Phase III clearly indicates that conferences have been supremely important to the Wedge strategy: they are, as described, the setting in which CRSC fellows are supposed to "challenge" their invited opponents—mainstream scientists and philosophers who (rightly) defend science's methodological reliance on natural explanations for natural phenomena. However, despite the CRSC's assertion in the Wedge Document that this phase will begin when its sponsored research has had time to "mature," the Wedge has obviously not waited for this maturation. (Research that has not been done needs no time to mature.) But conferences have been integral to the Wedge program since its inception. The Wedge movement actually began in 1992 with a conference at Southern Methodist University. CRSC fellows attend many conferences held by others, but the Wedge's own major conferences—six in only eight years, four of them in the Wedge Document's stipulated "significant academic settings"—have been central to the Wedge crusade. Although the conferences may be officially sponsored or cosponsored by non-Wedge entities, their identity as major Wedge events is established by the presence of a core of CRSC members who, given their relationship as a tightly knit group with carefully orchestrated activities, constitute an overwhelming ideological presence at these events. Indeed, the conferences are important to the Wedge's strategizing, providing members, who live across the country from Washington State to Georgia, with valuable opportunities to confer face to face: "[Interviewer:] How do members of the Wedge communicate with each other? [Phillip Johnson:] We communicate primarily by e-mail, and through conferences about once a year."[18]

DARWINISM: SCIENCE OR PHILOSOPHY?—SMU 1992

This conference produced the Wedge's first book of published proceedings. According to Phillip Johnson, "friends who were interested in promoting my ideas arranged an academic conference at Southern Methodist University, to which they invited ten scientists and philosophers, including Behe, to discuss the relationship between evolutionary science and philosophical naturalism."[19] The Wedge's religious support structure is clearly visible in the conference sponsors: Dallas Christian Leadership (the SMU chapter of Christian Leadership Ministries), the Foundation for Thought and Ethics, and the C. S. Lewis Fellowship. The FTE, as already discussed, has for years been closely allied with the Wedge. The C. S. Lewis Fellowship is "dedicated to evangelizing students and professors and equipping Christians to deal with scientific and historical apologetics and current issues." Tom Woodward, its president, in an article on CLM's "Leadership University" website, celebrated one event supposedly produced by this conference: as Woodward tells it, the pro-evolution philosopher of science Michael Ruse experienced a "change of heart" because of his participation and later "shocked his colleagues" at a 1993 American Association for the Advancement of Science conference when he endorsed Phillip Johnson's point "that Darwinian doctrines are ultimately based as much on 'philosophical assumptions' as on scientific evidence."[20] (Needless to say, this is a classic case of creationists using the remarks of competent scientists and philosophers out of context.)

The Wedge's plan to exploit fully the participation of important mainstream scientists and philosophers in its conferences is *not* in doubt. The aim is to use such events and the reputations of people like Ruse to appropriate a legitimacy and a level of public recognition the Wedge cannot achieve on the basis of its own scientific accomplishments. According to Johnson, "[The SMU conference] brought a team of influential Darwinists, headed by Michael Ruse, to the table to discuss this proposition: 'Darwinism and neo-Darwinism as generally held in our society carry with them an a priori commitment to metaphysical naturalism, which is essential to making a convincing case on their behalf.'"[21] With the SMU conference, the Wedge—at least in members' estimation—had scored mightily in its push for academic respectability.

MERE CREATION CONFERENCE—BIOLA UNIVERSITY, 1996

The Mere Creation conference was a crucial one: the crystallization of the Wedge strategy began there. According to CRSC fellow Dr. Ray Bohlin, director of Probe Ministries, the conference at Biola University in 1996 was a "historic conference" that attracted "two hundred participants" and constituted "the backbone of the future direction of the fledgling intelligent design movement."[22] Rich McGee of Christian Leadership Ministries directed the conference: "Our goal was to conduct an academic conference on the origins issue for leading scientists and scholars who reject naturalism as an adequate framework for doing science. We wanted to challenge the paradigm which reigns in the university and culture today—naturalism . . . and explore the possibilities in the concept of intelligent design." He acknowledged his indebtedness to Wedge members: "Particular thanks goes to Bill Dembski and Paul Nelson who served on the Executive Committee with me; we were in almost daily contact for months before the conference. Bill is also the Academic Editor for the conference and is now preparing the papers for publication."[23] The seeds of the Wedge strategy are evident in the "Conference Overview":

1. Preparing a book for publishing with chapters drawn from the conference papers.
2. Planning a major origins conference for 1997 at a large university to engage scientific naturalists.
3. Outlining a research program to encourage the next generation of scholars to work on theories beyond the confines of naturalism.
4. Exploring the need for establishing fellowship programs and encouraging joint research.
5. Providing resources for the new journal *Origins & Design*, as an ongoing forum and a first-rate interdisciplinary journal with the contributions of conference participants.
6. Preparing information usable in campus ministry, such as expanding a World Wide Web origins site (http://www.iclnet.org/origins) and exploring video and other means of communication. [This item is early evidence of the essentially religious goals of the Wedge strategy.]
7. Formulating a brief statement of unity ("mere creation") with a call for a cease-fire on other issues (though not discouraging research in those areas).[24]

NATURALISM, THEISM, AND THE SCIENTIFIC ENTERPRISE—UT-AUSTIN, 1997

In an April 26, 2000, letter to the *Waco Tribune-Herald*, CRSC fellow Robert Koons denies being part of the ID movement: "I am not myself a part of this movement, because my training is in the field of logical metaphysics, not biology or the philosophy of science."[25] Yet he organized an important Wedge conference. The March/April 1997 issue of *The Real Issue* magazine says that the Wedge's NTSE conference was "the brainchild of just two professors, Dr. Robert Koons and Dr. John Cogdell."[26] In this article, Koons says, "The conference was an outgrowth of work that I was doing for the Veritas Forum [a campus Christian organization] at the University of Texas. . . . Although the conference started as a spin-off of the Veritas Forum, it was from the start a very distinct and separate enterprise. It was designed to be an academic conference in a secular setting, and we followed through on that, I think." (The Veritas Forum, a Christian ministry with campus chapters around the country, works closely with the Wedge. See chapter 9.)

Although there was a Veritas Forum event February 17–21, 1997, which featured Koons's fellow Wedge members Walter Bradley and Phillip Johnson, and overlapped with the NTSE conference (February 20–23), what Dr. Koons neglects to say is that the 1997 NTSE conference, for which he wrote a web page containing a call for papers in August 1996, was already on the agenda as part of the Wedge strategy at the November 1996 Mere Creation conference.[27] Item 2 of the Mere Creation Conference Overview is clearly a reference to the NTSE event later hosted by Koons: "Planning a major origins conference for 1997 at a large university to engage scientific naturalists."

The identity of UT-Austin as the location of this "major origins" conference was reported by Scott Swanson, who covered the Mere Creation conference for *Christianity Today* and wrote in the January 6, 1997, issue, "Leaders [of the Mere Creation conference] are planning a spring conference at the University of Texas."[28] The Discovery Institute's summer 1997 *Journal* carried a short piece on the conference: "Design theory was the subject of a major conference at the University of Texas at Austin. 'Naturalism, Theism and the Scientific Enterprise' was sponsored by the [Koons's] Department of Philosophy. Conference papers were presented by Discovery's Center for the Renewal of Science and Culture Fellows John Angus Campbell, William A. Dembski, Stephen C. Meyer, Paul A. Nelson and Jonathan Wells. The conference opened a dialogue between methodological naturalists and intelligent-design theorists and has since been written up in a number of scientific journals."[29] Subsequent to his organization of the NTSE conference for the Wedge, Koons served as a contact person for the next Wedge conference, "The Nature of Nature" conference at Baylor University.[30]

The location of the NTSE conference in Texas highlights the concentration of a hub of Wedge supporters there, especially at UT-Austin, where Koons, Cogdell, and CRSC fellow J. Budziszewski are on the faculty. Cogdell was a signatory to the Ad Hoc Origins Committee letter written in defense of Phillip Johnson (see chapter 1). Also teaching at UT-Austin is Martin Poenie, who participated in the Wedge's Yale ID conference and is now a member of Dembski's new ID organization, the International Society for Complexity, Information, and Design (discussed later). Dembski himself resides in Texas, as does CRSC fellow Bruce Gordon, his partner in the creation of the Michael Polanyi Center (MPC, also discussed later). Other Texas residents who assist the Wedge are Ray Bohlin, who runs Probe Ministries in Richardson, Texas; Jon A. Buell, who runs the Foundation for Thought and Ethics, also in Richardson; and Walter Bradley, who taught at Texas A & M until his retirement. Another Texas connection is ARN "Featured Author" Gordon C. Mills, retired from the UT Medical Branch at Galveston. Mills was one of (at least) six Texas reviewers of *Of Pandas and People*.[31]

THE NATURE OF NATURE CONFERENCE—BAYLOR UNIVERSITY, 2000

The Nature of Nature conference is among the most important of the Wedge's many conferences; it is also the most controversial. According to Phillip Johnson in an interview with Christian Book Distributors, Inc., about his book *The Wedge of Truth*, "The conference so frightened the Baylor faculty that they demanded that the sponsoring Institute [the Michael Polanyi Center] be shut down at once to make sure that nothing of that kind ever happened again!"[32] This conference was extremely important because it marked the Wedge's official entry into the world of mainstream academia following the establishment of the MPC at Baylor University, and the MPC had planned to hold at least two more major conferences: "Plans for the future include . . . [t]wo major conferences for 2001: (1) A one-week conference on the origin of information; (2) An international conference on the significance of Michael Polanyi's thought."[33] The Wedge's involvement in the conference was evident from the Discovery Institute's co-sponsorship, as well as from the presence of fifteen CRSC fellows. Also assisting were *Touchstone* magazine, which had printed a special issue on ID in July/August 1999, and, again, the Foundation for Thought and Ethics.[34]

Phillip Johnson, in the interview mentioned, claimed ownership of the Baylor conference for the Wedge when he bragged that "*we* had a conference at Baylor University in April 2000 to discuss whether the evidence of nature points towards or away from the need for a supernatural creator. It was probably the most distinguished conference in Baylor history, with two Nobel Prize winners and many of the country's most distinguished professors in science, philosophy, and history" (emphasis added).[35] Using the innocuous title "The Nature of Nature: An Interdisciplinary Conference on the Role of Naturalism in Science," the MPC was able to secure commitments from a few world-class scientists and philosophers, among them Nobel laureates Steven Weinberg[36] and Christian de Duve; the chair of Harvard's History of Science department, Everett Mendelsohn; philosopher John Searle; and MIT physicist Alan Guth. The Wedge then lost no time in exploiting the presence of the distinguished attendees. Johnson's CRSC compatriot Nancy Pearcey boasted in *Christianity Today* of the credibility the conference gave to the ID movement: "Baylor University's Michael Polanyi Center, founded by Dembski, held a conference last month on naturalism in science that attracted nationally known scientists such as Alan Guth, John Searle, and Nobel-Prize winner Steven Weinberg. These scientists' willingness even to address such questions, alongside design proponents such as Alvin Plantinga and William Lane Craig, gives enormous credibility to the id [*sic*] movement."[37]

In addition to the Wedge's exploiting the presence of the Nobel laureates to claim credibility among Pearcey's Christian readers, Phillip Johnson exploited it for political purposes at a congressional briefing held by the CRSC in Washington, D.C., on May 10, 2000, less than a month after the conference. (See chapter 8 for information about the briefing.) Explaining William Dembski's very conspicuous absence from the briefing as the result of his need to avoid bad publicity after the conference because of faculty animosity, Johnson pointed out that Dembski had brought together a star-studded group of scientists the likes of which Baylor had probably never seen—a group that included Nobel winner Steven Weinberg (*but please see note* 37). He cited Dembski's absence from the briefing as an example of the prejudice ID proponents face: "You have not seen the snarling lips, the contempt" of the scientific priesthood, Johnson lamented.[38]

Baylor University was the perfect entry point for the Wedge in its effort to counteract what they believe is the naturalism that has become pervasive even in Christian universities, a pervasiveness that Johnson bewails in *The Wedge of Truth* and in the Christian Book Distributors interview about the book:

> We do not wish to honor the true God, and so we turn from the creator to created things, including idols of the mind like the theory of evolution. Of course secular universities are tempted that way, but the sad thing is that similar inclinations are widespread in the Christian academic world. . . .
>
> The most important question is whether God is real or imaginary. . . . The latter is the teaching of evolutionary naturalism, and even many Christian thinkers tacitly assume that position. . . .
>
> Baylor is a Baptist university, by the way, that advertises itself to prospective students as providing a Christian education. On the issue of naturalism the university world is totally closed-minded and fearful. The nominally Christian institutions are particularly fearful because they are understandably worried that they will be accused of betraying their heritage and advertising themselves falsely. But the truth will eventually wear them down.[39]

DESIGN AND ITS CRITICS— CONCORDIA UNIVERSITY-WISCONSIN, 2000

Billed as "a critical appraisal" of intelligent design and held shortly after the Baylor conference, this Concordia University conference generated less controversy. However, it did produce a fairly dramatic confrontation between Michael Behe and Kenneth Miller, author of *Finding Darwin's God*, in which Miller produced a list of scientific publications to answer Behe's charge that there have been none that attempt to explain the mechanism of natural selection at the cellular level. (See chapter 4.) This conference seemed "designed" to quiet critics of the Wedge who have complained of its members' unwillingness to subject themselves to peer scrutiny in the mainstream academic arena: "As a popular movement, what is coming to be known as 'intelligent design' is growing rapidly. Nonetheless, its status as a scientific and intellectual program is increasingly coming under scrutiny, and there are many misgivings, especially in the academy. This conference seeks to articulate the best criticisms of Intelligent Design theory and to allow its proponents to address these concerns."[40] The conference and a call for papers were announced on both the Discovery Institute and Access Research Network websites.[41]

The Design and Its Critics (DAIC) conference was sponsored by the Cranach Institute at Concordia University and by *Touchstone* magazine. The Cranach Institute is "a research and educational arm of Concordia University of Wisconsin devoted to working out the implications of the Lutheran doctrine of vocation and engaging contemporary culture with the truths of the Lutheran confessions." Like the CRSC, Cranach sees its role as enabling Christians to preserve their religious identity in a secular culture in need of Christian renewal: "The Cranach Institute exists to help Christians understand their cultures in a secularist age, so that they can resist the temptations of worldliness, more effectively evangelized [*sic*] the lost, and be equipped for faithful service in the secular arena." When William Dembski was relieved of his duties as director of the Polanyi Center, Concordia's Cranach Institute issued not one but two strongly worded public letters of protest.[42]

The DAIC conference apparently reflects CRSC's close working relationship not only with the Cranach Institute but also with the leadership of the Lutheran Church-Missouri Synod (LCMS), which owns Concordia. In September 2000, Dr. A. L. Barry, then-president (now deceased) of LCMS, published a statement entitled "Make Room for Intelligent Design," in the Fort Wayne *Journal-Gazette* and on the LCMS website. The "Doctrinal Position" of the LCMS states that "we teach that God has created heaven and earth, and that in the manner and in the space of time recorded in the Holy Scriptures, especially Gen. 1 and 2, namely, by His almighty creative word, and in six days."[43] However, Dr. Barry eased the time restrictions a bit to accommodate the old-earth leanings of most of the ID contingent: "What creationists want is simply this: equal time for what they consider a legitimate alternative—Intelligent Design. As much compelling evidence as there is for a young earth and a worldwide hydraulic cataclysm (the Noachic Flood . . .), Intelligent Design, on its own merits, can be argued effectively without a single reference to the Scriptures." Dr. Barry was not arguing that ID be given equal time with young-earth creationism in church-operated Lutheran schools, but that it be given equal time with evolution *in public schools*: "On the blackboards of America's public school science classrooms, the time has come for the words 'evolution,' 'naturalism,' and 'neo-Darwinism' to make room for 'Intelligent Design.'"[44] In gaining such public support from the head of a denomination of almost 2.6 million members, the CRSC apparently met another of the Wedge Document's Five Year Goals: "Major Christian denomination(s) defend(s) traditional doctrine of creation and repudiate(s) Darwinism."[45]

SCIENCE AND EVIDENCE FOR DESIGN IN THE UNIVERSE CONFERENCE—YALE, 2000

Judging from the roster of speakers, the Wedge was not aiming even for the pretense of even-handedness at this conference.[46] The conference brochure advertises the event using the falsehood, documented as such earlier in this book, that "an increasing number of scientists are discovering fresh evidence that the universe and life may, after all, bear the marks of design." There is little indication that this conference was aimed at any audience other than students, sympathetic faculty, and ID supporters. However, the Yale conference gives the Wedge a contact point with a religious ministry in a major university—a constant feature of Wedge operations—and fulfills one of its early goals, which Phillip Johnson stated at the Mere Creation conference in 1996: "We hope to schedule future conferences at major secular universities."[47] Whether "Yale" per se had anything significant to do with this conference (it did not), its name carries great prestige among those who crave academic respectability.

The typical Wedge plan for a book of published presentations was part of the conference announcement from the hosting Rivendell Institute (RI). Co-sponsor of the event along with the Discovery Institute, the Rivendell Institute for Christian Thought and Learning is not a campus academic organization, but a small team of campus ministers established by Campus Crusade for Christ to proselytize in the Yale community: "For the past three years, we have begun to pursue the goals of the Rivendell Institute as we continue to give leadership to the Campus Crusade for Christ ministry at Yale."[48] The Wedge's contact person appears to have been Gregory Ganssle, former director of the Rivendell Institute (succeeded by David Mahan). RI had little visibility at Yale and did not have a website until a week before the conference. Although the conference was publicized extensively on the Discovery Institute and Access Research Network websites, it was not publicized at Yale until the day before the conference, when an ad appeared in the *Yale Daily News*, but a search of its website turned up no campus coverage of the event.

Attendance was heavy at both the daily concurrent sessions, some of which were held in the Becton engineering building, and the evening sessions, held in Yale's law school auditorium. However, according to one observer, the audience appeared to be largely from *outside* Yale, creating an atmosphere "like an old time revival meeting, especially during [Phillip] Johnson's talk or when it seemed like [Jonathan] Wells was going [to] lead them in hymns after he [had told] them about how it was 'their tax money' being spent on bad science." But the observer also reported that "there were lots of nods, and verbal agreement from members of the audience when [Wells] told them to call their congressmen about NSF [National Science Foundation] funding."[49] Wells also distributed bookmarks and instructions for downloading his textbook warning labels. (See "Publicity Materials/Publications" for information about these labels.)

One usefully placed Wedge supporter, Joseph D'Agostino of *Human Events* magazine, explained why the Yale conference contained a lesson that ID's detractors would do well to learn: "[Yale University] demonstrated a healthy open-mindedness when it allowed a conference on Intelligent Design theory . . . to take place on campus. . . . Contrast this tolerance with the situation at Baylor University, an aggressively Baptist school in Texas. On October 19, Baylor dismissed Dr. William Dembski, one of the most prominent scholars in the Intelligent Design movement, as director of the Polanyi Center . . . after the institute sponsored an Intelligent Design conference at Baylor. . . . After its success at Yale, the Intelligent Design movement can no doubt look forward to many more such conferences, with open minds to influence at every one. Baylor could learn a thing or two from Yale."[50] As noted, however, *the conference had little or nothing to do with Yale* and everything to do with cultivating support from religious allies who are more than willing to create an entryway for ID among impressionable students and a few cooperative faculty.[51]

Yet another ID conference, "Research and Progress in Intelligent Design" (RAPID), held on October 25–27, 2002, was advertised by CRSC fellow Guillermo Gonzales in the September/October 2002 American Scientific Affiliation *Newsletter*. The location for this event was Biola University, site of the 1996 Mere Creation conference. The conference title looked promising—as though the latest scientific products of ID research might be unveiled. However, the description accompanying the ad stated that the conference's purpose was to form working relationships among scientists "seeking" to conduct research on "the interface between science and faith, particularly within the context of ID."[52] The conference website contained the schedule and a list of speakers and attendees. Virtually all listed attendees and presenters were either CRSC fellows or familiar ID supporters. A panel discussion on "Building Bridges in the Academy" was scheduled, and, of course, a postconference "strategizing meeting" was scheduled for the last day. The conference opened, however, with Dembski's keynote speech, in which he admitted that ID's scientific research program has not enjoyed much success: "Because of ID's outstanding success at gaining a cultural hearing, the scientific research part of ID is now lagging behind." He outlined steps that would enable ID to become a "disciplined science" (a full decade after its first conference) and urged that the Wedge be seen "as a propaedeutic—as an anticipation of and preparation for a positive, design-theoretic research program that invigorates science and renews culture." After recommending that the ID movement "take a good, hard look in the mirror," Dembski asked his audience, "[W]here do we go from here?"[53] Judging from Dembski's speech, genuine science may have to wait yet longer.

MISCELLANEOUS CONFERENCES

The conferences just discussed are the prominent ones. There have been others—some in which the Wedge has played an exclusive or a dominant role, others in which its members had a less prominent but still important role as invited participants. As with the major conferences, the frequency of the Wedge's participation in even these lesser events is a measure of its zeal in spreading the word about the ID agenda. These conferences have been less publicized, but they have all served (or have been used) as a platform for the Wedge:

- Summer 1995—"Death of Materialism" Conference, Discovery Institute
- August 1996—"God & Culture" Conference, Discovery Institute
- September 1997—Forbes "Telecosm" Conference, Gilder Technology Group
- Oct. 31–Nov. 1, 1997—"Consultation on Intelligent Design," Discovery Institute
- September 1998—"Conference on the Origin of Intelligent Life in the Universe," International School of Plasma Physics, Varenna, Italy (attended by Phillip Johnson)
- November 1998—"Creation Week," Whitworth College
- September 1999—"Science and Evidence for Design in the Universe," Wethersfield Institute
- November 1999—"By Chance or Design? Critiquing Contemporary Evolutionary Theories and Presenting Evidence for Intelligent Design," Atlanta Christian Apologetics Project
- December 1999—"Life After Materialism" Conference, Biola University's Torrey Honors Institute
- June 2000—"God and the Academy" Conference, Christian Leadership Ministries and Ravi Zacharias Ministries International
- November 2000—"Darwinism, Design and Democracy: A Conference on Scientific Evidences and Education," Trinity College of Florida, The C. S. Lewis Society, and The Foundation for Thought and Ethics

"Citizen" Conferences. Conferences held by the Wedge's Kansas supporters also help to educate the public. Phillip Johnson's June 4, 2001, *Weekly Wedge Update* discusses three important components of the Wedge, the academic, the religious, and "the citizen component":

> There is a citizen component, illustrated by events like the upcoming Kansas City conference. . . .

> We also need to build our citizen base, and to educate the very large number of dedicated people in the religious world about how they can more [effectively] challenge the ruling naturalistic definition of knowledge. We won't achieve a breakthrough in science merely by making a scientific case, no matter how good that case is. We also need to build a growing community of educated people, especially students, who know what is at stake and who can't be bluffed by authority figures.[54]

The Kansas City conference to which Johnson referred was the second annual "Darwin, Design, and Democracy" (DDD) conference hosted by the Wedge's Kansas arm, the Intelligent Design Network (IDnet). Organized in the wake of the 1999 decision by the Kansas Board of Education to delete references to evolution and other objectionable items from the state science standards, IDnet has become a well-organized source of problems for not only proscience Kansans, but as of late 2001, for Ohioans as well (we discuss IDnet in chapter 8).

Held since summer 2000, the DDD conferences feature the ID stars, as well as lesser known up-and-comers in the movement. In its announcement of the July 2002 conference, "Darwin, Design and Democracy III: Teaching Origins Science Objectively," the Discovery Institute explained the theme, indicating special attention to high school and college education: "DDD III will give special focus to how origins science can be taught at high school and collegiate levels. DDD III is designed for the public, parents, students, homeschoolers, science teachers, school administrators, and academics."[55] From these conferences, which present full slates of plenary and concurrent sessions on ID, the Wedge recruits followers who will spread the Wedge's political/religious brand of ID activism into other states. The July 2001 conference has borne fruit in Ohio, where IDnet and its Ohio supporters initiated attempts shortly after the conference to influence the writing of the state science standards, hoping to make a place for ID in Ohio public school science classes.

These conferences are designed to provide a "down home" atmosphere for attendees while providing saturation in ID doctrine and networking opportunities with the people who are enlisting their efforts, as exemplified in the 2002 program: "Have Lunch With the Speakers. Each of the speakers will be dining at separate tables in the Rockhurst cafeteria where you can meet them one-on-one and join them for lunch. We will be selling box lunches or you can bring your own brown bag." So attendees were able to make connections over ham sandwiches with Jonathan Wells and Michael Behe et al. and were later instructed on "Speaking to School Boards and Legislatures about Origins Science." Interested teachers could receive instruction on topics such as "Presenting Evolution Objectively: A 'How To' for Tenth Grade Biology" by Wells and "Mouse Traps and Molecules: Teaching Evolutionary Mechanisms and the Challenge of Irreducible Complexity" by Behe.[56]

These symposium sessions prompt earnest questions from attendees, who generally lack the scientific background to judge what they hear, as exemplified in the "Q and A," which followed Roger DeHart's DDD I (July 2000) session. DeHart, a Wedge supporter, was then a controversial high school science teacher who had been teaching ID in his science classes in Burlington, Washington (see chapter 8). He was field-testing the CRSC's PowerPoint presentation, "Science Curriculum Module for Teaching the Cambrian Explosion," (a few critics can be heard in the recordings, but they are a small minority of the attendees):

Questioner: I'd really love to hear this whole thing again, because it was only halfway through that I kind of got where you were going with this. But put this in perspective, now, on the biology science curriculum. We've got a day, of one lecture here. . . . Now put it with the rest of biology. Where does this fit in? Because, you know, if we graduate kids from high school, and they don't understand why the human genome's a special thing. . . . It's predicated on evolution, you know, this idea that "evolution happens here and we've got proof for it"—that's really pretty robust. But right here at this Cambrian explosion, [it] sounds like you've got a pretty good line of evidence here that maybe there was forking in that small area, and maybe there was something more. Is that what you're saying?
DeHart: Well I think your argument right now, though—you're taking cladics, cladistics, from stratigraphics. I mean, you're comparing those two different datings. . . . You're comparing once again gene sequences and saying that they have common ancestry because they have so many genes in common. I think that's what you're referring to there. . . .
Questioner: I don't know what I'm referring to—I'm not a paleontologist.[57]

The sessions show the attendees' lack of understanding about what science is and how it should properly be taught, as evident in audience responses to DeHart's promotion of ID's "top-down" interpretation of the "Cambrian Explosion":

Respondent A: We keep hearing all this "top-down" stuff. Well, sure, you know, uh, chordates are much different than echinoderms. . . . I mean, we're talking about bilateral symmetry vs. spherical—still talking what? Maybe a few million cells at most. I mean—it's not a lot to be placing great distinctions. . . .
Respondent B: In a way, you're saying it could support either theory—creation or evolution, right?
Respondent A: It's certainly a frontier.
Respondent B: It's not like God kept reinventing the wheel. . . .
Respondent A: It's no reason to stop teaching evolution, [but] I think the strength here [in this presentation] is that, "Look guys, we don't have it all figured out. There's still plenty of room for God . . . to work. We can't talk about it in science, because it's limited to naturalism. But you're allowed to believe." . . .
DeHart: I think that's one of the points that this is making. If you're only looking for naturalistic explanations, I think that is specifically at the point of that. But . . . we've decided *that* a priori. . . . That's not what the scientists in Darwin's day and earlier limited themselves to. We've only limited ourselves to that recently because of philosophical assumptions. . . . Because this is how we've defined science. . . .
Respondent C: One more thing. . . . The naturalistic definition of science by

Darwin—again, it's circular reasoning. There has to be the nature for nature to evolve itself. So it's another circular thing that needs to come out of that definition [of science].

Respondent D: Until the designer can work through the laws of nature. . . . *And* he can work through putting new information into the genome. Like a computer programmer, [he can] put new information into the genome.

Another Respondent: Omnipotence! It's cool! . . .

These comments were made in utter seriousness by attendees who were offering DeHart advice about how to improve the CRSC's curriculum module. The commentators believed that they were saying something meaningful.

Not wishing to deny potential ID activists a bit of fun, IDnet arranges entertainment at the end of the day with attractions (which still maintain the anti-evolutionary tenor of the event) such as the video, "I Was a Teenage Darwinist." The 2001 attendees (roughly 400) were regaled with a performance of a song satirizing evolution's "Overwhelming Evidence" (sung to the tune of Marvin Gaye's "Ain't No Mountain High Enough"), written by Jonathan Wells and performed by "The Mutations"—"three fine Christian ladies" from Kansas City "in sequined gowns [who] waltzed onto the stage . . . dancing like the Supremes."[58] But except for such end-of-the-day diversions, the Wedge is not dancing around at these conferences—the ID agenda is what brings in the crowd, and it is serious business.

Academic Lectures at Universities

Academic conferences are clearly a crucial part of the Wedge's thrust into academia, but they are not the only aspect of it. Individual Wedge members maintain a busy schedule of lectures on major university campuses, exemplified by these lectures:

- Under the auspices of the Templeton Foundation and the American Scientific Affiliation, William Lane Craig and Henry F. [Fritz] Schaefer spoke at Montana State University on February 25 and March 10–11, 1999, reported by Pat McLeod:

  - We want to say thanks to the ASA and Templeton for sponsoring this event and hope that we can continue to work together to engage the brightest minds in the academy in critical reflection on the person, values and teaching of Jesus Christ—especially as he relates to the world of science.[59]

- William Dembski and Paul Nelson spoke at MIT on April 7, 1999, and Tufts University on April 8, as recounted by Nelson in the September/October 1999 ASA *Newsletter*:

  - While most of the MIT questioners were skeptically curious, most of the Tufts questioners were skeptically confrontational. The first question, for instance, from someone who identified himself as a high school science teacher, asked why the National Academy of Sciences and the AAAS was opposed to what Bill and I were doing—i.e., along the lines of "Exactly when did you stop beating your wife?" I confess to some impatience with questions of this sort. I responded by asking the questioner if he telephoned the National Academy of Sciences to learn what he ought to think on any open question before he made up his mind.[60]

- Walter Bradley spoke at UCLA on February 1, 2000, at a forum organized by the Campus Crusade for Christ:

    - Approximately 400 people attended the forum, and the event had to be moved to a larger room. . . . Using the concept of design, Bradley illustrated how the laws of nature worked together so perfectly that the creation of the universe most likely involved a being of greater intelligence than science. "It's almost as if the universe knew we were coming," Bradley said.[61]

- William Dembski lectured on "The Scientific Case for Intelligent Design in Nature," on February 11, 2000, at the University of Washington in Seattle, according to an announcement posted by CRSC fellow Jay Richards to Metanews on February 7:

    - Dr. Dembski will be signing copies of his newest book, *Intelligent Design: The Bridge Between Science and Theology*. . . .
    - William Dembski is perhaps the very brightest of a new generation of scholars willing to challenge the most sacred twentieth-century intellectual idol—the notion that all of life can be explained in terms of natural selection and mutations.—Professor Henry F. Schaefer III, University of Georgia[62]

- Paul A. Nelson lectured at the University of Colorado at Boulder on September 20, 2000:

    - Paul Nelson . . . will speak Wednesday evening . . . at the University of Colorado, Boulder. . . . His topic will be "Darwin versus Design: The Case for a New Scientific Toolkit." . . . *The next day Paul will be speaking to a private faculty lunch group at the U.S. Air Force Academy in Colorado Springs*" (emphasis added).[63]

- William Dembski lectured at the University of New Mexico in November 2001 in a session jointly sponsored by the University Honors Program, the Center for Advanced Studies, and the Department of Psychology, as recounted in Phillip Johnson's November 19, 2001, *Weekly Wedge Update*:

    - On Tuesday, November 13th, at the University of New Mexico in Albuquerque about 500 to 600 people attended a remarkable event. William Dembski and Stuart Kauffman had a public encounter in which Kauffman, the preeminent self-organizational theorist of the Santa Fe Institute, publicly admitted that intelligent design was a legitimate intellectual and scientific project. . . . Kauffman was quick to add that he saw the application of design-theoretic methods to biology as unsuccessful. But . . . *Dembski did an admirable job defending ID and questioning the adequacy of self-organizational methods to account for biological complexity* (emphasis added).[64] (Johnson announced in the following week's *Update* that Dembski *himself* had written the self-congratulatory November 19 column.)

- Michael Behe lectured at the University of New Mexico in March 2002, as announced by the Creation Science Fellowship of New Mexico:

  - March 4–6, 2002 Michael Behe . . . will be in the New Mexico area. He starts on March 4th . . . at the Anthropology Lecture Hall . . . UNM. He ends . . . March 6th at Calvary Chapel. . . . His lectures challenge the current thinking in the area of evolution, and all are free.[65] (Roughly three weeks after Behe's visit, the seventy-seven public school science departments in New Mexico received copies of *Darwin's Black Box.*)

The CRSC's lecture circuit shows no sign of declining. Based on Phillip Johnson's early 2001 schedule posted on ARN, he visited Wayne State, the University of Michigan, Loma Linda University, Michigan Technical University, Vanderbilt, William and Mary, and Case Western Reserve University between January 24 and February 28, 2001.[66] Although Johnson has slowed down considerably since suffering a stroke in July 2001 and has even discontinued his *Weekly Wedge Update,* the relative youth of the other Wedge members and the need for fresh recruits ensures that the lecture circuit will remain busy.[67]

Potential Legal Action for Teacher Training

One of the Wedge Document's "Five Year Objectives" is to have "legal reform movements base legislative proposals on design theory." With an eye toward building a legal defense for the incorporation of ID into the public school curriculum, the Wedge has taken steps to lay the legal foundation for a push into public education. Phillip Johnson began giving talks at law schools and legal societies as early as 1994, when he met with the law faculty of Ohio State University, speaking on the legal status of "naturalism and creationism," according to CRSC fellow John Mark Reynolds, who reported the visit (with typical Wedge hyperbole) in *Origins & Design*: "The ACLU and People for the American Way should be warned that, given enough time, Johnson may effect a paradigm shift in the legal profession."[68]

Reynolds's remark is surely optimistic, but it reveals yet another aspect of the Wedge's strategizing. In 1995, Johnson spoke at the Indiana

University School of Law, which publicized his visit: "Has the Promotion of Naturalism Replaced Neutrality in Establishment Clause Jurisprudence? Is the Constitution, as currently interpreted by the Supreme Court, genuinely neutral between scientific naturalism and theism? Should it be? Professor Phillip E. Johnson . . . will address these issues on Thursday, Sept. 21 at Noon in the Moot Court Room. . . . [H]ear this engaging speaker discuss naturalism and its pervasive influence on law and education."[69] It may be difficult to envision Johnson's talk having a major effect on law faculty at Indiana, but it more likely was intended to plant seeds in the minds of students.

In 1998, Johnson apparently gave two lectures at the Annual Christian Legal Society Conference, because tapes of two talks were available for purchase: (1) "Main Session I: What Darwinism Has Done to Law—and What Can Lawyers Do About It?" and (2) "Understanding Darwinism and Scientific Materialism: Discussion of how lawyers can use their talents to expose weaknesses in the naturalistic philosophy that rules our culture. Professor Phillip E. Johnson."[70] These were clearly intended for lawyers.

CRSC's actual legal strategizing, however, appears to have begun in 1998, as announced in a DI *Journal* article, "The Promise of Better Science and a Better Culture":

> The Center for the Renewal of Science and Culture at Discovery Institute is a major factor in the new scientific debate and the examination of its implications for culture and public policy. . . .
>
> One of the most significant encounters may have been a hearing organized by the U.S. Commission on Civil Rights, with Steve Meyer representing Discovery on the issue of academic freedom and the issue of whether scientific arguments against Darwinism may be allowed in the classroom. Arguing on the other side was Eugenie Scott of the National Center for Science Education. The Commissioners seemed especially sympathetic to Professor Meyer's presentation and line of argument. At the hearing and elsewhere Scott has acknowledged the relative difficulty of defending an exclusionary policy when religion is no longer advocated. . . . The Federalist Society thought the hearing was so important that it published the testimony in its Religious Liberty News.[71]

The Civil Rights Commission hearing to which the *Journal* refers was held in Seattle, Washington, on August 21, 1998. Stephen Meyer's opening statement reveals that the legal tack the CRSC plans to use is "viewpoint discrimination." In short, the argument is that the absence of ID from science classrooms in the nation's public schools is a violation of the civil rights of students and teachers:

> During the last 40 years, evidence, much of which was unknown to Darwin, has come to light to support the design hypothesis. . . . All these lines of evidence . . . suggest the prior action of a designing intelligence.
>
> Is any of this evidence discussed in publicly funded science classrooms. Almost never.

> Why does this selective presentation of evidence persist in a nation known for its liberal intellectual traditions. Very simply, the opponents of full disclosure in science education insist, often backed by threat of lawsuit and other forms of social intimidation, that any deviation from a strictly neo-Darwinian presentation of origins constitutes an establishment [of] religion. They insist that the concept of . . . intelligent design . . . is inherently religious. Whereas Darwinism with its denial of intelligent design is a strictly scientific matter. . . .
>
> Biology texts routinely recapitulate Darwinian arguments against intelligent design, yet if these arguments are philosophically neutral and strictly scientific, why are evidential arguments for intelligent design inherently unscientific and religiously charged. The acceptance of this false asymmetry has justified an egregious form of viewpoint discrimination in American public science instruction. . . .
>
> For students and teachers wanting to consider or express a theistic viewpoint on this scientific subject as opposed to advocating religion, and this is a critical legal distinction, the present imbalance in public science instruction represents a clear form of viewpoint discrimination. In many cases, such discrimination has also entailed the abridgment of academic freedom for teachers and professors and the free speech rights of individual students.
>
> I ask the Commissioners to consider such practical measures as they have at their disposal to rectify this situation.[72]

Testifying with Meyer at this hearing was Richard Sybrandy, the Rutherford Institute–referred attorney who represented Roger DeHart, a Burlington, Washington, biology teacher who had been teaching ID for ten years in the Burlington-Edison School District. This testimony was clearly a collective effort by the Wedge: "Meyer had help in drafting the lead of his testimony from Paul Nelson, report and graphics development from Jay Richards, legal briefing from David DeWolf of Gonzaga U. in Spokane, WA, and documentation of non-religiously neutral statements in biology from John Wiester."[73]

Ironically, Meyer's presentation, in which he disputes the claim that ID is "inherently religious," was delivered as part of the U.S. Commission on Civil Rights' "Schools and Religion Project." The transcript is published in *Schools and Religion: Proceedings Before the United States Commission on Civil Rights* (December 1999), in which the Executive Summary explains that "transcripts have been produced as the result of an effort by the U.S. Commission on Civil Rights to study and collect information about religious discrimination in the Nation's public schools" (p. 1). One would think that the Wedge, attempting to preserve its charade of scientific legitimacy, would have avoided this event. Yet, it repeats the mantra of "religious neutrality," arguing that the exclusive teaching of evolutionary theory is not religiously neutral, but rather conducive to atheism in violation of the civil rights of religious students.

The "viewpoint discrimination" argument (highly popular with the academic left as well as the right, and therefore a potential winner) had been used in a 1997 amicus curiae brief to which Phillip Johnson was a signatory: "Brief Amicus Curiae of the Christian Legal Society and the

Union of Orthodox Jewish Congregations of America, in Support of Appellants," in the United States Court of Appeals for the Fifth Circuit (No. 97-30879). Johnson and four other signatories filed the amicus brief supporting the Tangipahoa Parish, Louisiana, School Board in the *Freiler vs. Tangipahoa Parish School Board* evolution disclaimer case.[74] Among the "Other Authorities" cited in the brief's references are Michael Behe's *Darwin's Black Box,* Johnson's *Darwin on Trial,* and two of Johnson's articles. The main point of interest here is one of the angles from which the signatories argue: "In Part I of this brief we show . . . that the Religion Clause mandates not official hostility to religion, but substantive neutrality by government concerning religious matters and nondiscrimination by government among religious beliefs. . . . Recently, the Supreme Court has become more pointed and specific in its teaching about nondiscrimination, requiring that the government treat all viewpoints equally. . . . Thus the *principle of nondiscrimination or equal treatment has become the normative rule of law concerning private speech of religious content or viewpoint*" (pp. 1 and 8; emphasis added). This is an approach Johnson had been cultivating since at least 1995, when he used the U.S. Supreme Court decision, *Lamb's Chapel v. Center Moriches School District* (1993), to support his contention that a professor's arguing in a public university classroom that ID is a "legitimate alternative to naturalistic evolution" is a protected form of academic freedom.[75]

The CRSC's legal catch phrase, "viewpoint discrimination," surfaced again in June 2000—this time used by a sympathetic congressman, Rep. Mark Souder (R-IN), who cohosted a congressional briefing on ID the Wedge held in Washington, D.C., on May 10, 2000. In an address to Congress in the June 14, 2000, *Congressional Record,* Rep. Souder clearly signals agreement with the Wedge's legal strategy, and the direct imprint of the Wedge on his remarks is clearly visible:

> My good friend, the gentleman from Florida (Mr. CANADY), chairman of the House Judiciary Subcommittee on the Constitution has considered holding a congressional hearing on the bias and viewpoint discrimination in science and science education. Ideological bias has no place in science and many of us in Congress do not want the government to be party to it.
>
> Nevertheless, many of us continue to be concerned about the unreasoning viewpoint discrimination in science. . . .
>
> As the Congress, it might be wise for us to question whether the legitimate authority of science over scientific matters is being misused by persons who wish to identify science with a philosophy they prefer. Does the scientific community really welcome new ideas and dissent, or does it merely pay lip service to them while imposing a materialist orthodoxy?
>
> Only a small percentage of Americans think the universe and life can be explained adequately in purely materialistic terms. Even fewer think real debate on the issue ought to be publicly suppressed.
>
> I ask my colleagues to join with me in putting aside unfounded fears to explore the evidence and truthfulness of the theories that are being presented by those on both sides of the issue.

I want to thank Phillip Johnson of the University of California at Berkeley, Robert * * * of Princeton University, and others [who helped in] drafting this response.[76]

The Wedge undoubtedly has plans to implement this legal attack at the nation's highest level. Yet they have also begun preparations for dealing with legal brushfires that might flare up in individual schools. Given the certainty of constitutional challenge should a teacher introduce intelligent design into a public school classroom, CRSC senior fellows David K. DeWolf, a law professor at Gonzaga University, and Stephen Meyer, a philosophy professor at Whitworth College, with Mark E. DeForrest (a Wedge supporter), have written *Intelligent Design in Public School Science Curricula: A Legal Guidebook* (Foundation for Thought and Ethics, 1999). According to Tom Bradford, Jr., of the FTE, this book "will give teachers and school board members who want to add intelligent design to their curriculum the ammunition they need to combat the intimidation of the ACLU."[77] The book has now been supplanted by a lengthy article in the *Utah Law Review* (2000) in which DeWolf, Meyer, and DeForrest argue that the landmark *Edwards v. Aguillard* ruling (1987) that outlawed the teaching of creationism in public schools does not apply to ID:

> Many have assumed that the reasoning in *Edwards* can be extended to cover curricular debates about the admissibility of teaching about design theory. Indeed, many have argued that the theory of intelligent design and creation science are effectively indistinguishable for both scientific and legal purposes. . . . Since the court in *Edwards* ruled that creation science promoted a religious viewpoint, many have concluded that teaching public school students about design theory also illicitly promotes a religious viewpoint in the public schools. . . .
>
> Despite claims to the contrary, design theory and scientific creationism differ in propositional content, method of inquiry, and, thus, in legal status. Recall that in *Edwards v. Aguillard* . . . the Court decided against the legality of scientific creationism because it constituted an advancement of religion. . . .
>
> [T]he Court's decision does not apply to design theory because design theory is not based upon a religious text or doctrine.[78]

DeWolf announced at a public ID event in Washington State in May 2001 (at which copies of the article were made available to the audience) that "we gave it to hundreds of school board members at a recent National School Board Association conference in San Diego."[79]

DeWolf is also prepared to make his legal expertise available to individual teachers who find themselves in constitutional difficulties because of their efforts to broaden their science lessons with ID. On the previously discussed science education website, CRSC assured those who registered for access to their restricted curriculum that "email listserv for registered web curriculum users can include legal advice." With the assurance that "Our Curriculum is Legally Permissible in Public Schools," the CRSC urged, "Don't let legal intimidation squash classroom innovation."[80] In its PowerPoint slide presentation, "In-Service Training Show for Science Teachers: Teach the Controversy," CRSC attempts to forestall teachers' fears of legal action: "One can hardly imagine a credible legal challenge to a teacher who wants to discuss Behe's or Dembski's evidentially-based ideas with students." If such assurance is insufficient to quell their fears, or if a teacher actually encounters legal difficulties, CRSC offers stronger assistance:

> Consult Our Legal Guide on the Internet
>
> - David DeWolf et al., *Intelligent Design in Public School Science Curricula: A Legal Guidebook* (1999)
> - A version of this guidebook accepted in law review journal
> - DeWolf, CRSC Senior Fellow, available for legal consultation
> - DeWolf leads the teacher training program for the CRSC
> - J. D., Yale Law School, law professor at Gonzaga University
> - http://law.gonzaga.edu/people/dewolf/fte2.htm
> - Read it, then email: [David DeWolf][81]

Research Fellowship Program: The Wedge's Shift toward the Social Sciences and Humanities

Science is not the only area that needs an overhaul, according to the Wedge. William Dembski asserts that ID has ethical implications, which means that the CRSC has "designs" on the other side of the curriculum as well:

> Design also implies constraints. An object that is designed functions within certain constraints. Transgress those constraints and the object functions poorly or breaks. Moreover, we can discover those constraints empirically by seeing what does and doesn't work. This simple insight has tremendous implications not just for science but also for ethics. If humans are in fact designed, then we can expect psychosocial constraints to be hardwired into us. Transgress those constraints, and we as well as our society will suffer.[82]

This assertion points to the Wedge's interest in funding research toward one of the Wedge Document's Twenty Year Goals: "to see design theory application in specific fields, including . . . psychology, ethics, politics, theology and philosophy in the humanities." Pursuant to this goal, CRSC has funded research that would extend ID's reach into the social sciences and humanities.

The spring 1998 Discovery Institute *Journal* cites CRSC funding of research for a book by Richard Weikart, a historian at California State University-Stanislaus, "about the impact of Darwinian biology on the development of Nazism," which would cover not only politics but history in the humanities. Weikart's biographical sketch on the CRSC website says that "he is currently completing a manuscript on evolutionary ethics, social Darwinism, eugenics, and scientific racism in Germany, documenting the influence of evolutionary naturalism on the rise of National Socialism." Weikart's book has yet to appear, but he has published a review of several books on Hitler in *Christianity Today's* "Books and Culture," where he dutifully attributes Hitler's evil to his denial of a personal God, his embrace of "Darwinism," and his elevation of evolution "to the status of a moral absolute."[83]

The spring 1998 *Journal* also announced funding for articles by Jeffrey Schloss, a biologist at Westmont College, on "the impact of evolutionary reductionism on ethics." Schloss's CRSC biography indicates that his research "focuses on structures of altruism in nature."[84] Schloss has offered his notions as part of Wedge activities such as the "By Chance or Design?" conference at Georgia Tech in November 1999 (see section on conferences), where he gave a talk entitled "Darwinism and the Enigma of Human Morality: Can Natural Selection Explain Radical Altruism?"[85] Although Schloss is a biologist, his CRSC-related work seems designed to bridge the gap between the sciences and the humanities, specifically between biology and ethics. At Wedge conferences, he has given presentations with titles such as this one, reflecting the CRSC's descriptions of his funded research: "The Problem with Love: Evolution, Design, and the Quandary of Altruism" at the November 2000 Yale conference. Schloss has also addressed gender and sexuality, a concern reflected in the Wedge Document's goal of having books published on "design and its cultural implications (sex, gender issues, medicine, law, and religion)." He presented a talk on gender and sexuality at the December 1999 Life After Materialism conference at Biola University.[86]

The Wedge seems to have all the academic bases covered, with individual fellows whose work collectively covers all aspects of its interdisciplinary agenda. However, for progress toward its Twenty Year Goals and real reach into the spectrum of academic disciplines, something more systematic than the work of individual fellows was needed—something like the Michael Polanyi Center at Baylor University.

THE MICHAEL POLANYI CENTER

The Polanyi Center, established at Baylor in October 1999, was, as Wedge member William Dembski describes it, the "first intelligent design think-tank at a research university."[87] Established with the assistance of Baylor President Robert Sloan on the Baylor campus, without the usual faculty input or consultation, the MPC was the Wedge's first attempt to secure a permanent university base of operations, with CRSC fellows William Dembski as the director and Bruce Gordon as associate director. This undertaking has not, however, been altogether successful. The Polanyi Center as a self-contained research unit has been dissolved. Dembski and Gordon can no longer act autonomously, as they once did; their activities are under direct supervision of Baylor's Institute for Faith and Learning, within which the MPC was technically housed at its inception, and which is subject to faculty scrutiny and judgment.

Controversy surrounded the MPC from the moment Baylor faculty learned of it—from outsiders who discovered the MPC's Baylor website. However, *the MPC's demise as an independent research center was not a result of the objections of Baylor faculty*, although a majority did indeed fear its effect on Baylor's reputation as a research university and implored President Sloan to dissolve it. Sloan had refused to dissolve the MPC—even after the Baylor faculty senate voted 26–2 to request that he do so.[88] Rather, after consistent disregard for expressed faculty concern, Sloan finally appointed an "External Review Committee" of scholars and scientists from *outside* Baylor to study the issue. This committee issued an equable and carefully worded report, recommending absorption of the MPC into the existing Baylor Institute for Faith and Learning and an end to the inappropriate use of Michael Polanyi's name for its activities.[89] Dembski, however, promptly interpreted this report as an unqualified victory for ID. He gloated: "The report marks the triumph of intelligent design as a legitimate form of academic inquiry. . . . My work on intelligent design will continue unabated. Dogmatic opponents of design who demanded the Center be shut down have met their Waterloo." In the face of this strident self-assertion, even before the committee's recommendations could be carried out, faculty anger overflowed, forcing the administration to act. Dembski was relieved of his duties as director of the Polanyi Center in October 2000.[90]

As we write, approximately one year remains on Dembski's contract. He is thus still technically employed by Baylor, benefiting from this affiliation while he continues his work on behalf of ID and the Wedge. He has no administrative authority. Yet Dembski, ignoring the significant qualifications the review committee included in its recommendations, continued to insist, even after being relieved of his duties, that the committee "validated intelligent design as a legitimate form of academic inquiry" (the substance of the committee's report belies this claim).[91] And despite the Wedge's blunting in this episode, both Dembski and Gordon remain at Baylor. Although their relocation with a similar center of their own at another university, one with at least Baylor's status, seems doubtful for the present, an examination of how the Polanyi Center served the Wedge strategy is warranted. This requires a return to the Wedge Document.

That manifesto points unambiguously to what the Wedge intends for the push into academia. The renewed emphasis on fellowships and the "shift to social sciences and humanities" in Phase III surely figured in the plans for the Polanyi Center, which was established to advance ID theory, as stated on its "Purpose of Center" webpage: "Present design-theoretic research holds much promise, but the ultimate significance of design theory remains to be seen. Nonetheless, the MPC sees design-theoretic ideas as a promising resource for understanding the complexity we observe in nature, and is committed to pursuing this avenue of research to see what fruit it will bear." The MPC's plan for "a *research fellowship program* so that the MPC can sponsor a steady stream of top scientists and scholars . . . who will be present and active at Baylor" and one of its "fourfold" purposes—"to support and pursue research in the history and conceptual foundations of the natural and *social sciences*"—are reflected in the Wedge Document's Phase III: "Research Fellowship Program: shift to social sciences and humanities. . . . *With an added emphasis to the social sciences and humanities*, we will begin to address the specific social consequences of materialism and the Darwinist theory that supports it in the sciences" (emphasis added).[92]

The Polanyi Center was exactly the institutionalization of its strategy that the Wedge needed to pursue its "Twenty Year Goals" as specifically outlined in the Wedge Document:

- To see intelligent design theory as the dominant perspective in science.
- To see design theory application in specific fields, including molecular biology, biochemistry, paleontology, physics and cosmology in the natural sciences, psychology, ethics, politics, theology and philosophy in the humanities; to see its influence in the fine arts.
- To see design theory permeate our religious, cultural, moral and political life.

These goals were reflected, with a good deal of elaboration, on the Polanyi Center's "Purpose of Center" webpage:

> The first [purpose of the Michael Polanyi Center] is to promote and pursue research in the historical development and conceptual foundations of the natural and social sciences. . . .
>
> The MPC . . . is dedicated to exploring . . . the conceptual foundations and . . . metaphysical, epistemological and ethical issues raised by . . . sciences like quantum mechanics, relativity, genetics, evolutionary biology, neuroscience, and sociobiology.
>
> The impact of science on the humanities and . . . arts is the second focus of . . . the . . . Center. . . . [H]ow has [science's impact on human life] been manifested in the humanities and . . . arts, and has the . . . effect been positive, negative, or a mixture of both? . . .
>
> [M]odern science has been understood as demonstrating the unreasonability of religious belief and creating a crisis of meaning that has had a widespread impact on literature and literary studies. . . . It has also contributed to . . . widespread acceptance of historicism and cultural relativism in the humanities and social sciences. . . .
>
> Thirdly, the . . . Center has a special interest in . . . interaction between science and religion. . . . [T]here must be a rapprochement between what is true in science and what is true in religion. . . .

> The fourth . . . component . . . is the articulation and application of a specific research tool in the sciences. . . . [Design] may serve to elucidate various phenomena that prove intractable from the . . . neo-Darwinian and self-organizational approaches.[93]

The Polanyi Center's "programmatic" plans and goals mirrored the Wedge's broad agenda as far back as the 1996 Mere Creation conference, the intended results of which were outlined by CRSC fellow Henry F. Schaefer in the foreword of the conference's 1998 publication, *Mere Creation: Science, Faith & Intelligent Design* (see chapter 1). Comparing the Mere Creation conference goals with those of the Polanyi Center amplifies the latter's significance as an important arm of the Wedge:

> [T]he MPC is concerned to see the pursuit of scientific research freed from programmatic external philosophical constraints, particularly those associated with a materialist or naturalist agenda. [MPC "Purpose of Center" web page]
>
> Plans for the future include:
>
> - A technical research journal on complexity, information, and design.
> - Two major conferences for 2001. . . .
> - Bringing in speakers outside the Baylor community to present talks and symposia. [This reflects the Mere Creation foreword's emphasis on "joint research."]
> - A research fellowship program so that the MPC can sponsor a steady stream of top scientists and scholars. . . .
> - Sponsorship of an international professional society focused on complexity, information, and design, and maintaining a full website as well as organizing an annual or biannual meeting. [MPC "Events and Programs" web page][94]

With the establishment of the Polanyi Center, a long-term association with Baylor was expected within which the Wedge would have what it most needs for the successful pursuit of its long-term goals—a physical, unequivocally high quality, research university base of operations, with Baylor providing "general administrative support," as called for in the Wedge Document's list of "activities," as well as the setting from which to engage in "alliance-building" and "recruitment of future scientists and leaders" (Wedge Document). William Dembski's five-year contract, which began in 1999 and remains in effect at Baylor until 2004, is for almost the same stretch of time set by the Wedge for the achievement of its major goals: 1999 to 2003.[95]

Clearly, the Polanyi Center's goals were congruent with the goals of the Wedge. Yet the most revealing connection between the two was this statement on the Polanyi Center's "Purpose" web page: "The successful achievement of these goals, therefore, is a task that the Michael Polanyi Center shares with a network of individual scholars and other established Centers around the world that have similar research projects." The most prominent center with a *really* similar "research" project—indeed, the only one known at this time—is the Discovery Institute's Center for Science and Culture in Seattle.

A "Virtual" Polanyi Center

Deprived at Baylor of the desired physical base for a full ID operation, William Dembski has launched elsewhere a resurrection of the Michael Polanyi Center: the International Society for Complexity, Information, and Design (ISCID). Announced in December 2001, ISCID bears most of the earmarks of what had been planned for the MPC at Baylor—even the background of the ISCID website is the same one used on the now-defunct MPC site. One feature of ISCID is different: it focuses actively on recruiting the young: "[ISCID] offers intensive summer workshops for bright undergraduate and graduate students as well as for exceptional high school juniors and seniors. Students have the opportunity to interact with some of the premier researchers in complex systems for several weeks at a time." The Polanyi Center, in its very early days, planned to offer workshops for students as young as those in grade school.[96] That published goal did not long survive on the MPC website—no doubt it would have appeared inconsistent with the MPC's need to cultivate an image of a university research center that could attract top scholars to its planned conferences and fellowship program.

There are some differences between the MPC and ISCID: one finds at ISCID no references to modern science's creation of a "crisis of meaning," with the latter's consequent engendering of "cultural relativism" and "aleatoric music," as on the MPC site. Rather, ISCID is ostensibly oriented toward natural science, engineering science, and complex systems. In addition, there is no physical establishment. The operation is a "virtual" one, which keeps the overhead low. An April 3, 2002, search at www.register.com revealed that ISCID's administrative and billing contact is Micah Sparacio, a student ID follower affiliated with Access Research Network's Intelligent Design Undergraduate Research Center (see chapter 6). According to information posted on the ARN discussion list, its "suite" at 66 Witherspoon Street in Princeton, New Jersey, is a post office box at Mailboxes, Etc.[97] As of January 8, 2002, a month after its establishment was announced at ARN, ISCID had still not registered its presence with either Mercer County or the state of New Jersey, where any organization handling money must register with the New Jersey State Treasury Department. By March 2003, ISCID had registered with Guidestar.org, and ISCID's website announced that it is a 501(c)(3) nonprofit organization. Two of its fiscal year 2003 goals were to increase its funding base, since 90% of its funding comes from membership fees ($25–$45 a year), and to increase membership from 160 to 400. Donors may contribute up to $250 a month or $5,000 as lifetime benefactors. Guidestar reports that ISCID is "not required to file an annual return with the IRS because its income is less than $25,000." (Yet it offers lucrative monetary prizes and fellowships.)[98]

Despite their differences, however, the similarities between the MPC and ISCID (and, one might add, between ISCID and its earlier progenitor, the CRSC) are strong, especially the methodological one. Like its MPC progenitor, ISCID seeks to carry out its research program "apart from external programmatic constraints like materialism, naturalism, or reductionism" (the website announces the organization's goal of "retraining the scientific imagination to see purpose in nature").[99] So its mission permits, consistent with its ID orientation, an appeal to the supernatural. Through the medium of ISCID's discussion forum, "Brainstorms," Dembski has proposed yet another principle (recall his biblically rooted Law of the Conservation of Information described in chapter 5) to cover any eventualities falling outside the "programmatic constraints" of naturalism:

> One . . . hears . . . that a purely naturalistic origin of life is . . . likely, perhaps not in the observable universe, but when all . . . other universes . . . are factored in. . . . Against such claims, I want to propose . . . the Principle of Inflated Probabilistic Resources [PIPR]: "If a theory requires vastly more stuff than is available in the observable universe to render it plausible . . ., then it is not . . . an insightful scientific explanation and we need to keep looking."[100]

Questioned by a Brainstorms discussant about his response to a critic "who argues that the ID hypothesis itself requires 'vastly more stuff than is available in the observable universe to render it plausible,'" Dembski responded readily:

> ID commits one to . . . generic intelligence, not . . . supernatural intelligence. So it's not . . . clear in what sense ID commits one to "vastly more stuff." . . . [E]ven if the intelligence . . . [is] unembodied or supernatural, . . . [it] would be . . . fundamentally different . . . from physical stuff, so . . . there is no question of quantifying . . . stuff that is supposed to render a vastly improbable event probable. The bottom line may just end up that you've got to choose your metaphysics: a world in which unembodied designer(s) are real or a world in which the amount of physical stuff vastly exceeds anything to which we have empirical access.

It is no secret that the metaphysical alternative Dembski prefers is the real, unembodied designer, making the programmatic guidelines of ISCID clear and convenient: an unembodied intelligence cannot be quantified anyway, thereby eliminating the problem of being *unable* to quantify it while at the same time *using* it as an explanatory principle for highly improbable events. So one is not burdened with the problem of insufficient "stuff" in the universe to render plausible one's *theory* of that universe.

On the basis of such programmatic freedom, ISCID has established a fellowship program, as the MPC had planned to do and as the CRSC does, to pay three stipends of $35,000–$50,000 a year beginning in 2003; it also offers grants for research in complex systems. It already has fifty-eight current fellows, however, eighteen of whom are also CRSC fellows (a number of others like Alvin Plantinga are long-time Wedge allies).[101] (Many ISCID fellows are also among the "100 scientists" mentioned in chapter 6.) Maintaining the MPC's establishment of its identity through the memory of Michael Polanyi, ISCID sponsors the Michael Polanyi Essay Prize, awarding $1,000 for the best undergraduate essay on complexity, information, and design. For graduate students, ISCID awards the $2,000 John von Neumann Prize for essays on the same subject. (The winners of the first Polanyi and von Neumann essay prizes are no surprise: they were awarded, respectively, to John Bracht and James Barham, both of whom are on the ISCID journal editorial board.) A biannual conference was also to be part of the ISCID program; the first, on "The Origin of Information," was scheduled for summer 2003 at Oxford in honor of von Neumann's centennial: "The aim of this conference is not to resolve the origin of information but to lay out a research program within which this problem may be resolved." ISCID's conferences so far, however, have been as virtual as the organization itself, consisting of online affairs such as its October 2002 "e-symposium" and online summer workshops for students.[102]

ISCID has issued its winter 2002 journal, *Processes in Complexity, Information, and Design,* a "quarterly online peer-reviewed" publication. The articles must be approved for publication by at least two of the members of ISCID's editorial advisory board, which consists of ISCID's fellows. The editorial board consists of William Dembski and several of his associates who have previously been involved in ID activities. Two, John Bracht (managing editor) and Micah Sparacio (webmaster), have recently been members of ID student groups, the IDEA Club and IDURC, respectively. James Barham (book review editor) was a presenter at Dembski's 2001 Calvin College conference. Bruce Gordon (associate editor) is Dembski's partner from the Polanyi Center at Baylor, where Gordon is also still employed. The last is Jed Macosko, Dembksi's fellow Wedge member in the CRSC.

Journal articles approved by the referees from the advisory board are published electronically after being in ISCID's "archive" (a moderated discussion group) for at least three months. Five of the eight articles in the first issue are previous works by Wedge members. Behe's, for example, consists of his "Responses to Critics," strung together, which he published on the CRSC site. The six-page article by Paul Nelson and Jonathan Wells (two pages of which are references) is from a poster they presented at an October 2001 conference. One article is by Bracht, one by Barham, and the eighth is by Richard Johns, a contributor to the archive.[103]

ISCID is still in its infancy, to be sure, but so far at least, the *scientific* world, including all the scientific organizations specializing in ISCID's main interests (e.g., "complexity"), has taken no notice of it.

8

Wedging into Power Politics

We are not going through this exercise just for the fun of it. We think some of these ideas are destined to change the intellectual—and in time the political—world.

Bruce Chapman, "Ideas Whose Time Is Coming," Discovery Institute *Journal*, August 1996

It is nearly inevitable that "teach the controversy" will become public policy.

Phillip Johnson, *Weekly Wedge Update*, December 2, 2001

The ID movement has little to do with the advancement of science. Phillip Johnson admitted as much in 1996: "This isn't really, and never has been, a debate about science. . . . It's about religion and philosophy."[1] No area of any science has as yet been affected significantly by ID claims. Nonetheless, ID seeks a secure and prominent place in the cultural mainstream *as good science*. Yet the movement hears and responds to the call of its religious loyalties, and its financial support clearly depends on answering this call. But the cadre marches also to a political tune.[2] And while Phillip Johnson's assurance about the inevitability of the Wedge's triumph in public policy seems too optimistic, the Wedge has for some time been laying the groundwork for major political advances.

Eleven years of scientific nonproductivity, coupled with the Wedge's concentrated public relations campaign, demonstrate that the short-term goal is not to forge a genuine new science but simply to become academically respectable to gain recognition—at least from those in power—in American intellectual life. With such respectability, the Wedge can expect funding from outside benefactors, perhaps even from government—that is, from taxes. William Dembski voiced this anticipation in 1998 in the book *Mere Creation*: "I predict that in the next five years intelligent design will be sufficiently developed to deserve funding from the National Science Foundation."[3] The Wedge pursues its war on "scientific naturalism," however, from *outside* science: acquisition of public funds, according to Dembski, depends on the success of the third prong of a "four-pronged approach to defeating naturalism"—"a cultural movement for systematically rethinking every field of inquiry that has been infected by naturalism, reconceptualizing it in terms of design."[4] Significantly, Dembski does *not* say that public funding success depends on the second prong of his "approach," which is supposed to be "a positive scientific research program, known as intelligent design, for investigating the effects of intelligent causes." The Wedge clearly believes a great deal of *non*scientific legwork will be necessary to secure taxpayer dollars. With such money, a public relations program that really wins elections—or catalyzes cultural revolutions—can proceed apace.

In this chapter, we examine further the explicitly nonscientific accomplishments and political methods of the Wedge. The Wedge's funding sources, its undeniable religiosity, its political techniques and alliances, and its fervent pursuit of academic legitimacy must all be documented. This chapter and the next demonstrate that a thick layer of *politics* surrounds and helps disguise the Wedge's fundamentally religious purpose.

The Politics of the Wedge

In DI's August 1996 *Journal*, DI President Bruce Chapman identifies the CRSC with an explicitly political mission: "The new Center for the Renewal of Science and Culture . . . challenges policy makers—and even our members and sponsors—to stretch their own thinking. . . . It calls upon their imagination to see the world not just as we received it, but as it is becoming and can become. . . . We think some of these ideas are destined to change the intellectual—and in time the political—world."[5]

In the July/August 1999 *Touchstone*, Phillip Johnson said, with considerable presumption, "My sense is that the battle against the Darwinian mechanism has already been won at the intellectual level, although not at the political level."[6] Wedge members may be overconfident about that intellectual battle, but they are clear-headed enough about the political one. The political strategy at the heart of the Wedge has been implemented in earnest, although not without a few setbacks. In an attempt to change the intellectual and political world, the Wedge's efforts have moved to the level at which most political action really occurs—the local level—as exemplified by its activities in Washington State and Kansas. Both states have been breeding grounds for creationist activity. But the Wedge has also cultivated connections in high places—even in Washington, DC.

Washington State

In 1995, William Dembski, addressing fellow theologians, wrote that "design theorists are much more concerned with bringing about an intellectual revolution starting from the top down. Their method is debate and persuasion. They aim to convince the intellectual elite and let the school curricula take care of themselves."[7] But if CRSC members ever did hold this corporate view, their activities show that they promptly abandoned it. Not only do Dembski's statements contradict the Wedge Document itself; they are contradicted by CRSC actions.

In Washington State, the CRSC inserted itself into a controversy surrounding biology teacher Roger DeHart at Burlington-Edison (B-E) High School in Burlington.[8] In the 1987–1988 school year, arriving in Burlington after teaching in a Christian school, DeHart began teaching ID in his B-E classes.[9] For almost a decade, he incorporated material from *Of Pandas and People* into his biology teaching, while omitting the material on human origins from the district's official biology textbook. In 1997, DeHart tested one of his classes on the movie *Inherit the Wind* (a favorite Wedge recommendation for teachers, as we show later); this was part of DeHart's strategy to "teach the controversy." In addition to questions about the movie, he included a subjective question about the issue of teaching both creation and evolution: "Do you think both sides or views should be studied?" One fourteen-year-old girl answered thoughtfully and honestly: "No, I do not. Religion is supposed to be separated from the schools. Evolution, however, should be taught because it is the 'scientific' version of how we came about and has nothing to do with religion. And of course, the beginning of our existence is important. Those who want to form an educated opinion, however, should study many religions, not just the creation story, as well as evolution." The student received credit for the answer, but she also got a comment from DeHart in the margin: "Interesting. Your belief sounds biggoted" [*sic*].[10]

A complaint from another student's parent set the stage for legal action, and in July 1997 the American Civil Liberties Union (ACLU) sent a letter to B-E Superintendent Paul Chaplik. DeHart secured assistance from a local attorney, Richard Sybrandy, recommended by the Rutherford Institute, which took up his cause.[11] According to the *American Bar Association Journal*, DeHart also consulted David K. DeWolf, CRSC fellow and Wedge legal advisor.[12] The CRSC's involvement in this local controversy became known in fall 1998, when Jonathan Wells and DeWolf met with the new B-E school district superintendent, Dr. Rick Jones, and B-E principal Beth VanderVeen. Thus began a close working relationship between DeHart and the Wedge.

In May 1999, VanderVeen, attempting to compromise with DeHart and his attorney, allowed him to use a four-page excerpt from *Pandas,* but only to introduce students to the concept of irreducible complexity. She added that he was not to use the words "God," "creation," "designer," or "intelligent design," and he was required to include for balance a pro-evolution article responding to irreducible complexity: "If he wants to show the ongoing controversy and debate, then he needs to do so in a balanced manner. Both sides are to be discussed openly and fairly." DeHart, in response, stressed his desire simply to provide a balanced view for his students.[13] VanderVeen formalized this agreement with DeHart in a July 1999 memo.[14] And despite the ACLU's objection to this arrangement, the 1999–2000 school year began with a promise from Jones that DeHart's class would be monitored to make sure he complied.

In October 1999, after speaking on the evolution controversy to a colleague's honors English class in which other invited guests had already spoken, DeHart was able to arrange a last-minute visit to the class by Jonathan Wells. Wells's appearance was scheduled hastily for the day after the last speaker, Skagit Valley Community College science professor Val Mullen, defended evolution. Wells also presented a community lecture on October 20, 1999, sponsored by Skagit Parents for Truth in Science Education, a group organized to support DeHart. Wells was clearly there to lend (putative) scientific credibility and to show support for DeHart: when he told the audience that DeHart deserved a round of applause, DeHart received a standing ovation.

In November 1999, Jeremy Means, the parent of a public school student, requested a meeting with DeHart and VanderVeen to voice objections to DeHart's use of ID materials from *Pandas* in his child's class. Accompanied for support by Carl Johnson, like himself a member of the Burlington-Edison Committee for Science Education (a local group not affiliated with the school district), Means met with DeHart and VanderVeen. He secured a commitment from DeHart not to use *Pandas* any more but no corresponding commitment about what he *would* use.[15] However, in January 2000, a public records request from the ACLU to the B-E school district attorney revealed that DeHart had never carried out VanderVeen's instructions to include a pro-evolution article with the *Pandas* material (belying his claim that he wanted "balance" in his classroom): DeHart had not filed the required article with the principal, nor had he kept anything he could present on request by the school district.[16] Meanwhile, DeHart's cultivation by the CRSC continued.

At the November 1999 meeting with Means, DeHart had identified himself as a good friend of William Dembski. In a show of support for DeHart, Dembski gave a talk on ID during a February 2000 book signing for *Intelligent Design: The Bridge Between Science and Theology*. The event was sponsored by DeHart's support group.[17] DeHart, in turn, attended the April 2000 Nature of Nature conference organized by Dembski's (now defunct) Michael Polanyi Center at Baylor University. By July 2000, DeHart was traveling out of state to do the "premier showing" of CRSC's PowerPoint slide show, "Science Curriculum Module for Teaching the Cambrian Explosion," at an ID conference.[18]

Because of the efforts of the Burlington-Edison Committee for Science Education, CRSC failed to push its materials into the public school science curriculum. In May 2000, the district denied DeHart permission to use supplemental materials that included two articles by Jonathan Wells.[19] But CRSC's attempt to cultivate popular support and to discredit the teaching of evolution continued. On October 26, 2000, Wells presented a lecture and book signing in Seattle to promote his new book, *Icons of Evolution*. His remarks to the audience about standard biology textbooks (see chapter 5) were consistent with the Wedge's aims: to plant in the public mind grave doubts about current science and science teaching and thereby to break its hold on the public school curriculum. Wells exhorted his audience: "You're supporting public high schools to use these textbooks. If you doubt this, go in and look. Just go in your local high school and say, 'May I please see your biology textbook?' Bring my book with you. Look in the textbook and . . . I guarantee you that these have probably two-thirds or more of the icons of evolution in that textbook. And there will be not even a hint in there that the icons misrepresent the evidence. I know. I've looked. I've checked for you. But by all means, look for yourself."[20]

On May 16, 2001, DeHart's parent support group sponsored a public meeting entitled "Banned in Burlington," held in the B-E High School cafetorium. Wedge representatives were featured speakers: Bruce Chapman, DI president, who described DI's purpose as challenging the materialistic philosophy underlying Darwinian science; Jonathan Wells, who gave a PowerPoint presentation entitled "Uncensored Facts About Textbook Errors"; Roger DeHart, issuing a series of denials regarding his use of ID materials in his classes; and David DeWolf, DI's legal advisor, who compared DeHart to John Scopes of the 1925 Scopes trial and urged that "the [Burlington-Edison] school board should reject the prospect of a possible lawsuit as the reason not to do the right thing." A notable feature of the printed Banned in Burlington program was the request that DeHart supporters run for two school board positions in the November 2001 election.[21] That is indeed what happened, but in the meantime, events surrounding DeHart took an interesting midsummer turn.

In July 2001, the Burlington school district reassigned DeHart from biology to earth science for the 2001–2002 school year, a move that district superintendent Rick Jones said was for practical rather than political reasons. DI President Bruce Chapman spoke up for DeHart, saying that the reassignment was "not in the best interest of the students," while DeHart charged that it was done for "ideological reasons."[22] In August, DeHart suddenly left the Burlington district to teach in the Marysville district, where controversy over his hiring quickly erupted. Although the August 9 *Skagit Valley Herald* reported that DeHart would teach both life science and earth science, and DeHart himself asserted that he was hired to teach high school physical science and biology, Marysville district officials reassigned him to teach only junior high earth science after they learned of DeHart's involvement in the controversy at Burlington-Edison High School. The school officials claimed not to have known of the controversy surrounding DeHart when they hired him. DeHart, contradicting them, called the reassignment "incredible" and fumed—somewhat indiscreetly—"My 23 years of experience in biology is impeccable, and I'm denied as if I'm black or gay."[23]

Meanwhile, the call for pro-DeHart candidates to run for two Burlington-Edison school board seats was answered by physician Paul Creelman and Jerry Benson, a bed-and-breakfast owner. DeHart made his voice heard in the election even though he was no longer teaching in the B-E district (though he still lived there at the time). When Don Zorn, former B-E assistant district superintendent, spoke in an October 23 letter to the *Skagit Valley Herald* of his concern "that we have . . . candidates who have as their agenda the support of intelligent design instruction in our schools" and urged support for ID opponents Richard Spink and Ken Ricci, DeHart responded publicly with a letter on October 28. Although by this time DeHart had begun his new job in Marysville, his continuing ID loyalties were clear: "Zorn is willing to make bedfellows with anti-theistic philosophy. His stance ensures that only . . . a naturalistic belief system is to be taught . . . without any debate or scientific evidence that would contradict it."[24] Creelman was defeated and Benson elected, though for only one term. (District lines redrawn before the candidate qualifying period were filed with the county only after the qualifying period, resulting in Benson's residence falling outside the district in which he ran and was elected.) DeHart taught both physical science *and* biology, as he was hired to do, at Marysville-Pilchuck High School, but only, as he put it, because of "the Discovery Institute and some legal play."[25]

DeHart's tenure at Marysville was short. He resigned in May 2002, saying that he would be teaching biology at a California Christian school the next year.[26] His tribulations in the B-E school district have been chronicled in a video, *Icons of Evolution*, based on Wells's book. The video "premiered" in May 2002 at DeHart's alma mater, Seattle Pacific University, and, according to John West, CRSC associate director and chair of the SPU Department of Political Science and Geography, will be marketed by Focus on the Family. DeHart, calling himself "a modern-day John Scopes in reverse," is philosophical about his Burlington experiences: "This whole experience has made me a better person."[27]

If the ID creationists hoped to precipitate in Washington State the legal test case they clearly want, they have so far been disappointed. But they succeeded for several years in diverting the energies and consuming the time of school officials and citizens of the B-E school district in trying to pressure the district to "teach the controversy."

Kansas

> The evening began with a standing ovation for Linda Holloway, the chairwoman of the Kansas Board of Education.
>
> Then came the main speaker, Phillip E. Johnson. . . .
>
> Johnson saluted Holloway and the . . . "courageous people" on the . . . board who voted . . . to remove questions from state tests about macro-evolution.[28]

This excerpt from the *Kansas City Star* about the Kansas Board of Education's 1999 decision regarding evolution exemplifies the CRSC's political involvement in arguably the most visible creationism controversy since the Louisiana Balanced Treatment Act culminated in the 1987 Supreme Court decision, *Edwards v. Aguillard*, which outlawed the teaching of creationism in U.S. public schools.[29] Although Johnson's remarks were made *after* the board voted to delete most references to evolution from the state's science standards, CRSC's involvement began before the vote; Johnson's appearance had actually been arranged *before* the vote was scheduled. There was apparently great interest in what Johnson had to say, since a crowd of more than 250 people gathered at the Overland Park Church of Christ to hear him.[30] Moreover, Johnson had already spoken the night before the vote in Topeka, where the Kansas Board of Education meets. His supporters handed out speaker announcement fliers—in the board chamber—on the day set aside for public comment prior to the vote.[31] Johnson's involvement in Kansas was more than mere moral support for the creationist cause. When Linda Holloway ran for re-election to the board in August 2000, a year after the vote, Phillip Johnson donated $500 to her campaign.[32]

Despite such help from Johnson, two of the three creationists who ran for re-election, Holloway and Mary Douglass Brown, were defeated in the August 2000 Republican primary, virtually guaranteeing that pro-evolution moderates would retake the board in the November 7, 2000, general election. A fourth creationist had to resign after establishing permanent residence out of state (while trying nevertheless to retain his Kansas board seat). This seat was filled, as were the other two, by a pro-science moderate. As predicted, Republican pro-evolution moderates, as well as a Democratic pro-science incumbent, were elected to the board of education in the general election. On February 14, 2001, the Kansas board voted 7–3 to approve new science standards reinstating the deleted parts on evolution, earth history, and cosmology.[33]

The electoral defeat notwithstanding, the Wedge's influence in Kansas remains strong. In an analysis for *The Nation* just after the Kansas board's 1999 decision, Larry Witham, who has covered the ID movement often, and Edward Larson, author of the Pulitzer Prize–winning Scopes trial account *Summer for the Gods,* interpreted events in Kansas as fueled by Johnson and the ID movement:

> The first step toward understanding the events in Kansas is to disregard all that we've learned about the Scopes Trial from Inherit the Wind. . . . The Kansas episode reflects the convergence of Johnson's new anti-evolution crusade and old-style biblical creationism. . . .
>
> Johnson's books have sold more than a quarter-million copies, and it is no wonder that his kind of arguments showed up among conservative Christians who voiced their opinions during the science standards hearing in Kansas.
>
> Another "authority" often cited in Kansas was . . . Michael Behe, who enlisted in Johnson's crusade in 1991. . . .
>
> Johnson and Behe . . . propound that intelligent design . . . is apparent in nature. This, they argue, divorced from biblical creationism, should be a fit subject for public-school education. With this argument, they have expanded the tent of people willing to challenge the alleged Darwinist hegemony in the science classroom, and this emboldened the populist uprising in Kansas.[34]

Despite the setback of the election, the Wedge continues its work in Kansas via the Intelligent Design Network (IDnet), a Shawnee Mission, Kansas, creationist organization formed in the wake of the board's decision in 1999. To be sure, the move to change the Kansas science standards was initiated by young-earth creationists led by Tom Willis and his Creation Science Association of Mid-America (CSAMA). Willis entered, however, into an (uneasy) alliance with IDnet and used the CSAMA website to publicize a symposium organized by IDnet in July 2000 (see later discussion). The young-earth creationists remain a presence in Kansas since the defeat of the board of education creationists, but IDnet, the most prominent creationist group, is promoting ID even more vigorously after the election than before.[35] IDnet has become, in a practical sense, the Kansas edge of the Wedge.

THE INTELLIGENT DESIGN NETWORK

As described on its website, IDnet's mission is "to promote evidence-based science education with regard to the origin of the universe and of life and its diversity" and "to enhance public awareness of the evidence of intelligent design in the universe and living systems."[36] Intelligent design is defined in classic CRSC style as "an intellectual movement that includes a scientific research program for investigating intelligent causes and that challenges naturalistic explanations of origins which currently drive science education and research." In a June 8, 2000, news release urging Kansas school boards "to reject National Science Standards proposed by Kansas Citizens for Science," IDnet complains that "the National Standards would limit teaching to only 'natural explanations,' so that a teacher could teach only one side of the controversy about the cause of life and its diversity. The evidence supporting design would be ignored, not because it doesn't exist, but because of an a priori philosophical assumption that natural causes are all there is."[37] And in a letter sent that day to all 304 Kansas school districts, IDnet managing director John Calvert (a corporate, not constitutional, lawyer) hinted at possible legal consequences for schools refusing to permit ID in their curricula: "The scientific method requires that the evidence on both sides of any issue be considered. There is also a legally compelling reason to do so. If your school board censors the evidence of design and permits only a consideration of evidence that life results only from the laws of chemistry and physics without design, then we believe that you will be subverting the neutrality mandated by the First Amendment of the Constitution."

Calvert also issued "Teaching Origins Science in Public Schools: Memorandum and Opinion," assisted by William S. Harris, another IDnet director (and nutritional biochemist who "reviewed and endorsed the scientific and other non-legal matters contained in the opinion"). In this document, addressed to "School Boards, School Administrators, Science Teachers and Others Interested in the Teaching of Origins Science," Calvert assures recipients that "the Design Hypothesis is supported by abundant evidence . . . [which] is easily observed and can be empirically detected using the scientific method and logical analysis" and also that "design detection, the evidence of design and logical inferences of design drawn from that evidence fall within traditional definitions of science." School officials who might reject the inclusion of the "design hypothesis" in public school classes are warned that censorship of "evidence of design in teaching origins science so that only natural explanations may be provided will result in violation of the neutrality required by the Establishment Clause of the United States Constitution." Calvert urges schools to adopt a "no-censorship policy," conveniently provided in an appendix, and refers them to the DeWolf-Meyer-DeForrest article on teaching ID in public schools and the related article "Teaching the Origins Controversy: Science, or Religion, or Speech?" in the *Utah Law Review*. And he informs any school systems that might be favorably persuaded by his legal opinions that "efforts are being made to develop [a] constitutionally neutral curriculum," which "may be available in the near future." He even offers to provide copies as soon as the curriculum is available.[38]

On July 15, 2000, two weeks before the Kansas Board of Education

Republican primary, IDnet held a symposium, "Darwin, Design and Democracy: Teaching the Evidence in Science Education." Featured speakers were CRSC fellows Michael Behe, Walter L. Bradley, David K. DeWolf, and Jonathan Wells, CRSC supporter John Wiester, and Roger DeHart, who presented the CRSC's "Science Curriculum Module for Teaching the Cambrian Explosion," a slide show written by CRSC fellow Michael Keas. The symposium was advertised prominently on the main DI web page and the Access Research Network site in June and July 2000 and, according to the July 16, 2000, *Topeka Capital-Journal*, attracted more than 350 people.[39] Throughout October 2000, just before the general election, IDnet held ID events, including a lecture by Jonathan Wells.[40]

THE WEDGE AND LOCAL POLITICS: PRATT, KANSAS

IDnet's alliance with the CRSC, especially with Wells, took a predictable turn in fall 2000. Realizing after the November 7 election that the board of education would certainly reinstate evolution in the science standards in early 2001, the Pratt, Kansas, local school board voted on November 27 to allow the teaching of ID in the town's high school. Kansas Citizens for Science member Jack Krebs wrote in a November 29, 2000, letter to the *Pratt Tribune*, "This is far more than a local issue. National and state organizations in the 'intelligent design' (ID) movement are looking to Pratt as a possible test case for their goal of challenging the current legal restrictions on teaching creationism in public schools."[41] Those national and state organizations are, of course, DI's CRSC and IDnet, and there is evidence for Krebs's view that they selected Pratt, Kansas, as a testing ground for the political machinery that will include in high school and college science courses "alternative theories" such as ID.

In spring 2000, science teachers in Pratt Unified School District (USD) 382 refused to follow the Pratt board of education's instructions to rewrite the tenth-grade science curriculum to add "balance" to the treatment of evolution. This prompted the board, with the help of two ID sympathizers who are now IDnet managing directors Chris Mammoliti and Ernie Richardson, to rewrite the standards themselves—in the style of Wells's *Icons of Evolution*. Mammoliti and Richardson had participated in the July 2000 IDnet symposium, at which David K. DeWolf explained legal strategies for inserting ID into public schools.[42] In November 2000, the Pratt school board added items to the standards in Benchmark #3, which says that "students will understand major concepts of the theory of biological evolution." The influence of design proponents is apparent in these additions, worded to allow the insertion of design theory into the science curriculum. The "Outcomes" section stipulated that students should understand that "there are different viewpoints of how life originated on earth" and that "there are different scientific perspectives regarding the prevailing textbook evidence used to support the theory of evolution." Also added in the "Process" section were requirements for students to "research and critique the strengths and weaknesses of differing viewpoints identified in the textbook" and "understand and critique the different scientific arguments relating to the evidence presented in the primary textbook." (All of these additions are classic Wedge-speak.) Among the "Suggested Resources" for teachers and students was Elizabeth Pennisi's "Haeckel's Embryos: Fraud Rediscovered," an article that may appear to echo Wells's complaints about the "faked" (Haeckel) embryo drawings (see chapter 5), but in fact is merely a short news report of a recent paper that reminds readers of the century-old argument about Haeckel's text figures and discusses implications for current embryology, in which they play no significant part.[43]

The board approved the changes by a 4–2 vote, meaning that ID would be permitted in the tenth-grade science curriculum at Pratt High School.[44] In a December 6, 2000, letter to the *Pratt Tribune*, Wells congratulated the board: "The Pratt school board deserves to be commended for its courage in resisting organized pressure to indoctrinate students in Darwinian evolution, and for its wisdom in encouraging students to think critically about what they're being told in science classes."[45] As a result of the ID activists' success in shaping the public school curriculum, the citizens of Pratt, Kansas, became as divided over the science standards as its board of education.

The controversy engendered by the ID-influenced segment in the high school science standards produced a larger than usual number of candidates for three USD 382 school board positions in the February 2001 primary. Virtually all candidates made public statements on the science standards. Among the candidates was Mammoliti, as well as other ID supporters. In each of the three districts, however, opponents of the revised standards also ran, among them incumbent and board president Bruce Pinkall, one of the two members who had voted against amending the standards. Pinkall, Theresa Miller, and Michael Westerhaus (a biology instructor at Pratt Community College with a Ph.D. in vertebrate zoology), all of whom opposed the changes in the standards, survived the first primary and won in the April 3 final election. After the election, only two of the pro-ID board members, Willa Beth Mills and Sue Peachey, remained on the board. However, the issue of the science standards remained unresolved.

On February 13, 2001, one day before the Kansas Board of Education restored state science standards, Pinkall had received a letter from the ACLU on behalf of parents who had filed a complaint about the Pratt standards. He said at the time that because of the positions of local board members, he could foresee no change in the Pratt curriculum.[46] So the 2000–2001 school year ended and the 2001–2002 session began with the newly adopted standards in effect—but also with a majority of board members on record opposing them. At the middle of the school year, the standards were being reviewed once again according to Pratt Unified School District Superintendent Kenneth Kennedy: "As of December 2001, the current K–12 science standards are being reviewed by the instructional staff of the district for recommended editing and necessary revision. That process is expected to be completed sometime in January 2002. Any revisions and editing changes will be reviewed by the district's curriculum committee, made up of teachers, administrators, and a board representative. Following review and endorsement by the district steering committee, the board may take appropriate action to approve."[47]

On February 25, 2002, the USD 382 Board of Education did approve revised science standards, applicable to grades nine and ten. Compared to those of the previous year, the new standards are considerably improved, with evolution given prominent and serious attention. Resources listed under Benchmark #1, "Science as Inquiry," include the websites of reputable biologists Scott Gilbert and Kenneth Miller, as well as the high-quality resources of the Access Excellence site.[48] The worrisome item in the Benchmark #3 "Outcomes" section is still there, but the "Descriptions and Examples/Activities" for this item are the Glencoe text, *Biology: The Dynamics of Life* (2000 edition), and the cited websites. Evolution is stressed as "a broad, unifying theoretical framework for biology," and Benchmark #3 says that "evolution provides the context in which to ask research questions and yields valuable insights, especially in agriculture and medicine." Unlike the earlier standards, the February 2002 standards call for instruction in the importance of "presenting research for review by scientific peers."[49]

With the integrity of USD 382 and state standards restored, Christian conservatives again began strategizing to recapture the board seats they lost in November 2000. On April 9, 2001, a fundraising letter went out to supporters of the Free Academic Inquiry and Research Committee (FAIR), a political action committee that began raising money to regain the seats lost to moderates. Organized in March 2000 by the Kansas Republican Assembly as a federal political action committee, FAIR began planning in 2001 to field candidates for the five state board seats open in the 2002 election. FAIR believes it is fighting a battle for the minds of Kansas children. Signed by Linda Holloway, Mary Douglass Brown, and Steve Abrams, the only creationist board member to retain his seat in the 2000 election, the letter invoked the specter of "secular humanistic worldviews," pitted against "the Judeo-Christian values upon which this nation was built." As the most prominent example of the struggle between these two views, it cited "the ongoing controversy surrounding the teaching of monkey-to-man evolution in Kansas schools." Holloway, Brown, and Abrams testified to the generous support they received from FAIR during the election: "F.A.I.R. contributed generously to each of our campaigns, held rallies, staged literature drops, and hosted fundraisers to help elect good people to leadership positions in our educational system." The letter then asked recipients for contributions as large as $5000 to support 2002 candidates, warning that "the struggle over the future of our nation that is playing itself out in our school classrooms is far from over."[50] In the February 14, 2001, *Kansas City Star*, shortly after the state board voted to reinstate evolution in the state science standards, Abrams conceded defeat on the issue, then added, "But it's not going to put it to rest."

The 2002 election showed that the issue may indeed not be at rest. According to the November 6, 2002, *Kansas City Star*, "moderates lost their grip on the Kansas board of education." The defeat of two moderates by conservatives converted the 7-3 pro-science majority to a 5-5 moderate/conservative split. Incumbent John Bacon, who had won in the August primary, said that he wanted to "revisit" the science standards if five members would support him. Even without that support, the next election could reverse the 2001 decision to restore evolution to the standards.

Given Phillip Johnson's financial support of Holloway in the 2000 primary, it is a safe assumption that future creationist candidates will have the Wedge's support. Johnson, in fact, had already begun the post-defeat campaign strategizing in a June 27, 2001, speech at the Southern Baptist Theological Seminary in Louisville, Kentucky:

> Since that time, Johnson has helped strengthen the evolution opposition in Kansas. He said that a coalition of creation science proponents and intelligent design proponents was formed to support any future political movement opposing Darwinian evolution. . . . "When this comes up again—as it will—we will be prepared with a leadership that understands the lessons from the first [battle]," he said. ". . . It doesn't matter . . . if you take a defeat in the immediate sense—provided you end up stronger afterwards. We ended up a lot stronger. . . . We demonstrated it is possible to form a unified movement with the academic[s] . . . who traditionally won't have anything to do with the traditional creationist, and the creationists who traditionally won't have anything to do with the academic[s]."[51]

Taking the Battle to Ohio

The lessons ID proponents learned in Kansas are no doubt being studied for future action there, but the political insights gleaned from the Kansas battle have been exported to Ohio, where in late 2001 they bore fruit as efforts by "Science Excellence for All Ohioans" (SEAO). SEAO reflected the ubiquitous presence in pro-ID efforts of the conservative Christian right. It was a "project of the American Family Association of Ohio," an organization with aims strikingly consistent with those of the Wedge: "The American Family Association of Ohio exists to motivate and equip individuals to restore the moral foundations of American culture. We seek to do this through providing information and education on social and moral issues facing our state, and through encouragement to apply traditional Judeo-Christian principles to all areas of life (a Scriptural world-and life-view)."[52]

Phillip Johnson noted in his January 17, 2002, *Weekly Wedge Update* that "a full scale Kansas-style revolt against the dogmatic teaching of evolutionism has broken out on the Ohio State Board of Education."[53] His announcement does not quite reflect the true scale of the revolt on the full board itself, although there was more than enough cause for concern: two elected board members, Michael Cochran and Deborah Owens-Fink, initiated the Ohio "revolt," and in early 2002 there appeared to be three additional ID supporters—on a board consisting of eleven elected and eight at-large, governor-appointed members. There also appeared to be five ID opponents on the board, with the other eight sitting members giving noncommittal answers to a February 2002 newspaper survey.[54]

However, a more ominous cause for concern was the pivotal position of the state board's standards committee, composed of eight board members, five of whom were the ID supporters. At a January 13, 2002, meeting, these five, led by Michael Cochran, voiced their support for an alternative, ID-friendly draft of the science standards that would refer to evolution as "a theory, or an assumption, but not a fact," with "a clear distinction between the different understandings of evolution as minor genetic variation [microevolution] versus evolution as a single common ancestry [macroevolution]." Cochran having suggested this, Richard Baker, Marlene Jennings, Owens-Fink, and Sue Westendorf voiced their support. Member Martha Wise explicitly voiced her dissent.[55] Since ID supporters outnumbered opponents on the committee, they were in a position to overrule the 41-member writing team that, directed by an advisory committee consisting of scientists and educators, had already written a draft of the new standards incorporating evolution; such a ruling would throw the decision to the full board.[56] As we shall see, ID proponents attempted to hedge their bets on the full board by introducing legislation to ensure that the Ohio legislature must approve all curriculum decisions by the state board before they could be implemented, thus taking control of the science curriculum out of the board's hands altogether.

Supported by SEAO, pro-ID efforts quickly produced a flurry of counter-activity by ID *opponents* as in Kansas, resulting in the formation of Ohio Citizens for Science.[57] As in Kansas, ID supporters in Ohio targeted the science content standards, the review of which began in 2001 with a deadline of December 2002 for acceptance of the final draft. The Academic Content Standards science writing team met in June 2001 with the goal of revising the standards and in November produced a draft, "Ohio's Academic Content Standards for Science: Fall 2001 Review Draft"; this draft thoroughly and properly incorporated evolutionary theory throughout the standards as the backbone of Ohio life sciences instruction. Shortly thereafter, however, SEAO produced its own fall 2001 version of its "Suggested Modifications to Draft Indicators," in which ID figured prominently. In addition to amending the indicators for earth and space sciences and life sciences in grade 10, and for life sciences in grade 12, SEAO's modifications contained a new indicator explicitly calling for the teaching of ID in grade 10: "New Indicator. Grade 10, Life Sciences . . . Know that some scientists support the theory of intelligent design, which postulates that the influence of some form of intelligence is a viable alternative explanation for both the origin and diversity of life. Compare and contrast the evidence that supports the design hypothesis with the evidence that supports the evolutionary hypothesis."[58] This indicator specifically referenced the legal opinion Calvert sent to all Kansas school districts.

HELP FROM THE HEARTLAND: IDNET IN OHIO

The Ohio incident unmistakably reflects IDnet influence, with managing directors Calvert and Jody Sjogren (and William Harris, to some extent) becoming the chief IDnet operatives there. Such influence was acknowledged in Phillip Johnson's January 31, 2002, *Weekly Wedge Update*, most of it written by Calvert: "Luckily the [Ohio] physical chemist who is leading the charge, Bob Lattimer, came to the annual IDnet symposium in Kansas last June. Bob heard Phil, Bill Dembski, Nancy Pearcey, Walter Bradley, Jonathan Wells, Bill Harris, yours truly [Calvert] and a number of other scientists and came away convinced that what he had heard was true. . . . After the IDnet symposium, he submitted a set of proposed Modifications to the first draft of Ohio science standards that also reflect the comments of other ID Netters. He also helped form an organization in Ohio to promote excellence in science education [SEAO]."[59] Sjogren, illustrator of Wells's *Icons of Evolution*, relocated to the Columbus, Ohio, area and became a member of the SEAO Speakers Bureau.[60] On December 14, 2001, SEAO celebrated its formation with a "design seminar" with both Sjogren and Calvert as featured speakers, Sjogren outlining the "evidence" for ID and Calvert sharing his legal insights on the unconstitutionality of "viewpoint discrimination" against ID in public schools.

The primary ID supporters on the state board, Cochran and Owens-Fink, formed an alliance with SEAO member Robert Lattimer, who managed to be appointed to the Ohio science standards writing team. Not surprisingly, none of these individuals has biology credentials. Cochran is an attorney with a master's degree in divinity. Owens-Fink has an undergraduate engineering degree and an M.B.A. and Ph.D. in marketing; she teaches marketing and international business at The University of Akron.[61] According to the January 31, 2002, *Columbus Dispatch*, Lattimer said that he "joined the writing team to make sure intelligent design was considered."[62] He is a chemist in the private sector and board member of the Ohio Eagle Forum (OEF), a branch of the rightist Eagle Forum founded by Phyllis Schlafly. OEF's mission is to "to provide oversight of the state educational system and its various components" and to "influenc[e] educational policy by educating families, educators, legislators, and school board members."[63] His relationship with IDnet secured for him a spot as a presenter on "The Ohio Firestorm of 2002" at the latter's "Darwin, Design and Democracy III" conference in Kansas in July 2002.[64]

Having prepared their suggestions for the board of education, SEAO had Calvert ready to defend them before the curriculum standards committee on January 13, 2002, when he was allowed to speak without rebuttal for thirty minutes, even though scientists in the audience asked to respond. Admitting that no other states presently teach ID, Calvert assured the committee that "we're talking about a groundbreaking paradigm here."[65] He presented standard ID arguments, calling attention to his own (purported) qualifications for involvement in the issue and drawing on his previously written legal opinion. Calvert clearly was attempting (no doubt with the ACLU in mind) to forestall the committee's fears about legal pitfalls surrounding the teaching of ID—and to warn them of legal consequences if they *prohibited* it:

> I was trained to be a geologist and wound up as a lawyer. . . .
>
> What qualifies a lawyer to talk about origins science? Lawyers are qualified because the key issues do not involve issues of fact. They involve issues of logic, issues of evidence and procedure and whether the rules, when applied by the State are consistent with the speech and establishment clauses of the constitution. They are issues that lawyers are particularly qualified to speak to. . . .
>
> The effect of modern origins science is to imbue a belief in naturalism. This has led our government into a practice that has the effect of indoctrinating our children and culture in Naturalism. . . . That is why I am here today. To talk about State sponsored naturalism. . . .
>
> A constitutional issue arises when the State decides to teach origins science. The reason is that Origins science unavoidably takes students into a religious arena. . . .
>
> The question then becomes, when the State decides to enter this religious arena can it choose to use a practice—Methodological Naturalism—to censor the design hypothesis? Can it choose to simply tell teachers to hide the evidence of design. . . .
>
> So if you tell Ohio teachers to go hide the evidence of design . . . are you causing the state to be neutral or are you causing it to imbue Ohioans in a belief in Naturalism—a non-religion?
>
> I think you will be involved in unconstitutional indoctrination.[66]

Calvert's letters and columns in Ohio newspapers were cast in the same mold as his address to the standards committee, and IDnet lent itself generously to the Ohio effort: a page soliciting endorsements from supporters of the ID-friendly changes to the standards was posted on the IDnet website.[67] But more effective help was on the way.

THE WEDGE IN OHIO—FROM BEHIND THE SCENES TO CENTER STAGE

IDnet's high profile in Ohio was just a prelude to the eventual appearance of Wedge members themselves; IDnet and SEAO simply did the advance work, creating the initial platform. Given that Ohio is just another stage in the Wedge's nationwide strategy, and given their constant need for publicity to acquire fresh recruits, the Wedge's public involvement was inevitable; the only question was when it would begin. It happened on March 11, 2002, after the standards committee "determined it wanted to host a special moderated panel discussion among four national expert panelists who will give varied views on science and science education standards."[68]

Since the committee was dominated by ID supporters, it was not surprising that two of the "expert panelists" were Jonathan Wells and Stephen C. Meyer. The other two were evolutionary biologist Kenneth Miller (see chapter 4), and Lawrence Krauss, chair of Case Western Reserve University's Department of Physics, who marveled that such a debate was held: "It's amazing to me that we are even having this debate. We should be working to improve [the] science curriculum and not fighting off some medieval attack on science. . . . [T]hese people are . . . trying to bypass the science community and peer review by going straight to the State Board of Education and legislature." Dr. Russell Durbin of Ohio State University, who observed the event (which attracted roughly 1500 people), recognized very perceptively what this opportunity meant for the Wedge: "First, the very fact that this session occurred is a victory for the ID side. This point was made by Krauss, but I think it may be the single most important impression. The ID side is clearly intent on getting across the idea that there's a real controversy here, and fudges the distinction between social/political and scientific controversy. So to have both sides equally represented on the podium is a real coup for them."[69]

This high-profile appearance was the stage for an unexpected tactical maneuver by the Wedge. Probably because of the opposition of groups such as Ohio Citizens for Science, Meyer proposed a "compromise" to the standards committee: ID proponents would no longer ask for intelligent design to be required, but rather that the board simply *allow* teachers to include it. Meyer later summarized his proposal in the *Cincinnati Enquirer*:

> First, I suggested . . . that Ohio not require students to know the scientific evidence and arguments for the theory of intelligent design. . . .
>
> Instead, I proposed that Ohio teachers teach the scientific controversy about Darwinian evolution. Teachers should teach . . . about the main scientific arguments for and against [it]. . . .
>
> Finally, I argued that the state board should permit, . . . not require, teachers to tell students about the arguments of scientists, like . . . biochemist Michael Behe, who advocate the competing theory of intelligent design.[70]

The implication of this proposal was immediately evident. This compromise, if enacted, would simply push to the local level any decisions about whether to teach ID, a move that would make stopping it much more difficult: it would be harder to put out brushfires all across Ohio, which has 612 local school districts, than to counteract it at the state level. Meyer did not stop there with his proposal, however; he addressed the standards' content by proposing that "Ohio should enact no definition of science that would prevent the discussion of other theories." The intent was clearly to undermine the understanding of science as a method of inquiry requiring natural explanations for natural phenomena—the loophole through which to slip ID. Meyer asserted that this more flexible approach would eliminate the debates over whether ID had passed the test of peer review or whether it was based on religion.[71] In short, it would preserve ID's exemption of itself from the processes of evaluation by which all legitimate science is judged. But there was a point to this supposedly spontaneous suggestion of compromise, which no doubt had been part of the Wedge's advance strategizing: by backing down from the demand that ID be included in the science standards and proposing instead that teachers simply be allowed to "teach the controversy," the new strategy would seem more reasonable to undecided board members who harbored doubts about the initial proposal.

Meyer's proposal *appeared* to make the science standards safer. Accordingly, SEAO made the corresponding modifications to their "Modifications," timing their publication to coincide with the Ohio board's posting of the revised draft of science content standards:

> New Indicator. Grade 10, Life Sciences . . . Discuss how various types of scientific evidence may either support or not support the theory of biological evolution. . . . (NOTE): The consideration of alternative theories, such as intelligent design, is permitted—but not required—under this standard.
>
> Explanation. This indicator is added to reflect the proposal made by Dr. Stephen Meyer at the March 11, 2002, Panel Presentation sponsored by the Ohio Department of Education. . . .
>
> This approach seems reasonable for several reasons:
>
> 1. It calls for coverage of evolution with intellectual honesty (since evidence both supporting and not supporting evolutionary theory is presented).
> 2. It promotes academic freedom for teachers (since they are permitted to discuss various aspects of evolution as well as alternative theories).
> 3. It enhances critical thinking in students (since they are exposed to a variety of viewpoints on the issue).
> 4. It generates student enthusiasm for science (since the controversy is interesting).
> 5. It aligns Ohio with the Santorum language in the federal education law. [See further discussion.]
> 6. It maintains government neutrality on a matter (biological origins) touching on religion.[72]

The preservation of the integrity of the standards, however, was only half the battle facing Ohio evolution proponents. In late January 2002, two bills were introduced by Rep. Linda Reidelbach (R) in the Ohio House of Representatives during the regular session of the 124th Ohio General Assembly. HB 484, introduced on January 24, provided "that before the science curriculum standards that are to be adopted by the State Board of Education prior to December 31, 2002, may be effective, those standards must be approved by a concurrent resolution passed by both houses of the General Assembly." An identical bill, SB 222, sponsored by Sen. Jim Jordan (R), was introduced in the Ohio Senate.[73] These bills, requiring a vote by all 132 members of the General Assembly, were clearly intended to block acceptance of the revised science standards without the ID-friendly changes. HB 481, however, introduced on January 23, was the really important bill:

> BE IT ENACTED BY THE GENERAL ASSEMBLY
> OF THE STATE OF OHIO:
>
> *Section 1*. That section 3313.6013 of the Revised Code be enacted to read as follows:
>
> *Sec. 3313.6013*. It is the intent of the general assembly that to enhance the effectiveness of science education and to promote academic freedom and the neutrality of state government with respect to teachings that touch religious and nonreligious beliefs, it is necessary and desirable that "origins science," which seeks to explain the origins of life and its diversity, be conducted and taught objectively and without religious, naturalistic, or philosophic bias or assumption. To further this intent, the instructional program provided by any school district or educational service center shall do all of the following:
>
> (A) Encourage the presentation of scientific evidence regarding the origins of life and its diversity objectively and without religious, naturalistic, or philosophic bias or assumption;
>
> (B) Require that whenever explanations regarding the origins of life are presented, appropriate explanations and disclosure shall be provided regarding the historical nature of origins science and the use of any material assumption which may have provided a basis for the explanation being presented;
>
> (C) Encourage the development of curriculum that will help students think critically, understand the full range of scientific views that exist regarding the origins of life, and understand why origins science may generate controversy.[74]

HB 481 was carefully crafted mostly to "encourage" rather than "require" and to avoid specific references both to evolution and intelligent design; instead, it uses "origins science," a term apparently intended to convey a tone of scientific seriousness, but which is familiar (and essentially dishonest) creationist code to anyone who follows the evolution/creation controversy. Evolutionary biology is *nothing like* "origins science," the term Calvert used in his legal opinion, "Teaching Origins Science in Public Schools." Other language in the bill reflects the Wedge's influence, especially references to "academic freedom," "naturalism," and the repeated emphasis on avoiding "assumptions." But the bill did *require* disclosure of the "historical nature of origins science" and "any material assumption" used in classroom explanations of "origins," amounting, in effect, to an evolution disclaimer: by stressing the historical nature of "origins science," the bill's champions sought to undermine evolutionary biology by presenting it as a "historical," thus less than rigorously scientific, part of science; by requiring the disclosure of "any material assumption," they were seeking to force teachers to state publicly that science is a methodologically naturalistic enterprise, thereby opening the classroom to challenges to the "evil" of naturalism. The bill's encouragement of a curriculum to help students understand the "full range of scientific views" on origins was the ID loophole.

Fortunately, the bills died in committee. Remarks by Ohio House Speaker Larry Householder (R-Perry County) had indicated the probable death of all three, according to the March 13, 2002, *Plain Dealer*: "The State Board of Education has been given the authority to handle this issue, and unless I see evidence that would suggest otherwise, I think probably we're going to leave the state board to handle it."[75] Yet a significant fact about this legislation is that John Calvert helped write it. Since Calvert directly represents the Wedge, his involvement directly falsifies Phillip Johnson's June 2001 statement: "We definitely aren't looking for some legislation to support our views, or anything like that. . . . I want to be very cautious about anything I say about the public interest, because obviously what our adversaries would like to say is, 'These people want to impose their views through the law.' No. That's what *they* [Wedge adversaries] do. We're against that in principle, and we don't need that." But as we have seen, this was not the first time when Wedge deeds were at odds with Wedge words.[76]

October Showdown. In September 2002, the Ohio Department of Education made available for public inspection a new draft of the science standards. The plan was to allow changes before the October 14 meeting of the standards committee and an October 15 meeting at which the board would vote on an "intent to adopt," signaling approval or disapproval of the draft in advance of the final December 2002 vote. This draft was a major improvement in Ohio's science standards; whereas evolution had been present only as the euphemistic "change through time" in the old ones, it was now the backbone of Ohio biology instruction—*and intelligent design was not included*. The standards reflected the success of pro-evolution advocates, aided by the strong, focused labors of Ohio Citizens for Science.

Taking advantage of the chance to propose changes, Owens-Fink, Cochran, and fellow board member James Turner proposed amending one of the Life Science "Benchmarks" for grades nine and ten to include having students "understand how scientists today continue to investigate and critically analyze aspects of evolutionary theory."[77] To anyone who understands how science is actually conducted, this amendment refers to what scientists do routinely. But to veteran ID watchers, singling out evolution for "critical analysis" was a recognizable Wedge tactic. Moreover, this indicator—legitimate when interpreted scientifically—was nevertheless ambiguous enough to be "spun" as permitting teachers to "teach the controversy." Although the Wedge promotes the latter as good pedagogy, it is a ruse for promoting the "alternative theories" of Michael Behe, Jonathan Wells, et al.

When the standards committee met on October 14, this change was adopted, along with another proposed by the same pro-ID trio. Grade ten indicators for "Scientific Ways of Knowing" had specified that students should "recognize that scientific knowledge is limited to natural explanations for natural phenomena based on evidence from our senses or technological extensions." Consistent with the Wedge's anti-naturalism strategy, Owens-Fink, Cochran, and Turner offered a substitute: "Science is a systematic method of continuing investigation, based on observation, hypothesis testing, measurement, experimentation, and theory building, which leads to more adequate explanations of natural phenomena." This statement is only the *first half* of the Ohio Academy of Science's definition of science; the other half, which was not included, amplifies what is meant by "more adequate explanations of natural phenomena": "explanations that are open to further testing, revision, and falsification, and while not 'believed in' through faith may be accepted or rejected on the basis of evidence."[78] It is not difficult to see why ID proponents eliminated the second half: it specifies criteria that ID *cannot meet*. And though the first half explicitly says that science must include hypothesis testing (and ID has never produced an empirically testable hypothesis), the phrase "more adequate explanations of natural phenomena" gives ID proponents another opportunity for spin, since the Wedge maintains that ID *is* a more adequate explanation of natural phenomena.

On October 15, the board unanimously approved the "intent to adopt," *with* a strong emphasis on evolutionary theory, *without* either intelligent design or SEAO's proposed indicators, and *with* the amendments proposed by the pro-ID board members. This was in reality a significant victory for the pro-evolution side: the strong emphasis on evolution reflected the progress they had made in obtaining board support, despite the Wedge's energetic campaign to muscle its way into the standards. According to Patricia Princehouse of Ohio Citizens for Science, "We won big time here. The creationists have lost. There is more evolution in the standards now than there would have been had they kept their mouths shut." The ID contingent also celebrated: Michael Cochran declared the standards a compromise presenting evolution as a theory, not a fact (the familiar creationist perversion of "theory"). But IDnet managing director Jody Sjogren's response more clearly reflected the Wedge's goals: "All along, we've been pressing for a 'teach the controversy' approach. . . . I think these changes are very good in the sense that the definition of science moves beyond a naturalistic bias." Nonetheless, ID champions Cochran and Owens-Fink steadfastly denied any relationship between their amendments and intelligent design.[79]

However, the amendments, together with the fact that Ohio curriculum *content* is determined by local school districts (the standards are not mandatory), pointed quickly to a disquieting realization, shown in an October 16 *Mansfield News Journal* headline: "Guidelines Open Door for Intelligent Design." Associated Press writer Liz Sidoti indicated that ID might yet find its way into Ohio schools since the standards committee's acceptance of the amended standards "kept the decision on whether to teach alternatives such as 'intelligent design' with individual school districts." Joseph Roman, board member and committee co-chair, confirmed that "what the individual school districts choose to teach beyond this is up to them."[80] The Wedge was denied entrance through the front door, but with the back door opened (if only a crack), the Wedge's PR campaign began instantaneously.

An October 15 "Discovery Institute News" article on DI's website featured Stephen Meyer congratulating the board "for insisting that Ohio students learn about scientific criticisms of evolutionary theory as part of a good science education." Of course, the board had not *insisted* on this, though they made a concession that appeared to permit it. Meyer twisted the amendment wording further by declaring that it "explicitly states that students need to know about scientists who are subjecting evolutionary theory to critical analysis." But understanding *how* scientists critically analyze evolutionary theory is not equivalent to Meyer's preferred interpretation of having students know about *scientists* who do it—an interpretation favoring Behe and Wells as the scientists about whom students should know. (SEAO's postvote statement, "Latest News on the Standards," reflected that interpretation: "The 'scientists' referred to in this benchmark/indicator would include mainstream evolutionists as well as dissenters.") Meyer's rhetoric stretches the truth, and DI's U.S. Newswire release the same day repeated these declarations.[81]

The Intelligent Design Network followed with its October 17 release, containing more overblown acclamations. Congratulating the board for its "handling of a major scientific controversy about teaching biological origins in public schools," IDnet celebrated the vote for "objectivity and academic freedom and against censorship of competing scientific views." (The release did express disappointment that the standards excluded intelligent design.) The *Washington Times*, not surprisingly, related its account in Wedge-speak: "The new language does not limit life sciences to materialism, which some consider a kind of atheism, and says students must learn how scientists 'critically analyze' Darwinism and not just accept it dogmatically." Pamela R. Winnick sustained the pitch in her October 18 *National Review Online* "Guest Comment" (she also made a factual error: the *board* of education voted, not the *department* of education): "In what could turn out to be a stunning victory for opponents of evolution, the Ohio Department of Education voted 17–0 on Tuesday to pass a 'resolution of intent' to adopt science standards that would allow students to 'investigate and critically analyze' Darwin's theory of evolution."[82]

The Wedge had effectively demonstrated how even its splinter of a concession could provide a pile of fuel for its spin machine. And even though the more ambitious parts of its Ohio strategy failed, the fight that had to be waged by evolution supporters consumed untold time and resources of scientists, educators, and other citizens, which could have been more productively spent, just as in Kansas and Washington. And there were signs immediately after the October 15 vote that the Wedge's Ohio strategy had not even then completely played out. SEAO's "Latest News on the Standards," posted immediately after the vote, was already advising its troops on the next phase of the strategy: "[T]hree more steps remain before the standards are adopted. First, a public hearing . . . on Nov. 12 . . . will give citizens another opportunity to make comments . . . on the draft standards. Second, State Supt. [Susan] Zelman will make a presentation on the standards to the Ohio House and Senate Education Committees . . . in late November or early December. Third, a final vote . . . will be taken at the December . . . [board] meeting. The next steps . . . will be the development of (a) a model science curriculum and then (b) state assessments. . . . Citizens need to be actively involved at all stages as the process moves forward." Such anti-climactic strategizing is a clear warning to anti-Wedge activists in other states that the Wedge does not relax its efforts even when pro-evolution victories appear to be won and efforts to stop them begin winding down.

On December 10, 2002, the Ohio Board of Education unanimously accepted the science standards. After several members objected to the critical analysis requirement, the board also, at the last minute, added a disclaimer to life sciences benchmarks and indicators for grade ten: "The intent of this benchmark [and indicator] does not mandate the teaching or testing of intelligent design." The disclaimer gave Ohio teachers and district officials a basis from which to resist pressure to teach ID. But ID supporters immediately began planning to influence the writing of the standards-based curriculum. SEAO called for applicants to the curriculum advisory committee and writing team: "The battle for objective science instruction is far from over. . . . We are now moving into the next phase . . . the development of a science Curriculum Model for the state. This model will be used . . . to implement the new science standards. . . . The deadline for applications is February 21 [2003]." So clearly, ID activity in Ohio was not over at this writing. More ominously, the Wedge's meddling there may be permanent. At IDnet's 2002 Darwin, Design, and Democracy symposium, SEAO's Robert Lattimer announced plans that may forecast the Wedge's effort to profit from Ohio school districts' control over their curricula: "Our group, SEAO, was formed as a temporary organization. We plan to go out of existence at the end of this year. What we're going to do at that time is merge into IDnet of Ohio, which is one of the two subsidiaries now of national IDnet, the other one [being] in New Mexico, which was just announced. And we will become a part of IDnet at that point in time."[83]

Lattimer's reference to New Mexico (NM) is accurate. On July 23, 2002, IDnet announced the establishment of "IDnet New Mexico," saying that "the expansion into New Mexico reflects an interest in New Mexicans in working towards an . . . approach to . . . teaching origins science . . . that goes beyond an 'Evolution Only' curriculum." NM executive director Joe Renick declared, "This affiliation will . . . provide a great foundation for growth of the movement, not only in New Mexico, but in other venues as well." But even before the October showdown in Ohio, the Wedge had advanced into Cobb County, Georgia, where on September 26, 2002, the school board unanimously adopted a policy incorporating the Wedge's buzzwords: allowing teachers to teach "disputed views" on the "origin of the species" for the sake of "a balanced education," "critical thinking," "academic freedom," and "neutrality toward religion." IDnet's John Calvert and William Harris had visited earlier to rally support. Wedge member and Georgia resident Henry F. Schaefer, along with Georgia Scientists for Academic Freedom, had petitioned the Cobb Co. Board of Education to support "careful examination of the evidence for Darwinian theory." After passage of the policy, DI issued a press release in which president Bruce Chapman called the decision "a victory for academic freedom and good science education." And when the Ohio board declared its "intent to adopt," the *Marietta* [*Georgia*] *Daily Journal* saw reflections of the Cobb Co. decision: "In a move that mirrors . . . [the] Cobb County School Board vote . . . the Ohio school board . . . will adopt a science curriculum that leaves it up to school districts whether to teach the concept of 'intelligent design'." (In January 2003, the Cobb Co. board clarified its policy to favor teaching evolution according to state science standards. At this writing, however, the board is being sued by a parent objecting to its August 2002 policy mandating disclaimer stickers in science texts. The disclaimer warns students that the text contains material on evolution and tells them that "evolution is a theory, not a fact.")[84]

Kenneth Miller, who participated in the March 11, 2002, panel discussion before the Ohio standards committee, is not optimistic about Ohioans' success in slowing down the Wedge's momentum: "Ohio is just a skirmish. . . . But it is a rehearsal for what will happen later."[85] The death of House Bills 481 and 484 will not—and the Wedge's defeated attempt to get ID into the Ohio standards *did* not—negate the wider implications and further potential political impact of the Wedge's long-term, national strategy. As we shall now see, the language of the Ohio legislation is directly rooted in the Wedge's political maneuverings in the nation's capital.

Washington, D.C.

"THE BRIEFING"—LAYING THE GROUNDWORK

In May 2000, the Wedge's political strategy took a crucial step toward implementing plans to "cultivate and convince . . . congressional staff . . ." (Phase II, Wedge Document). In a May 8 press release, DI announced that "Discovery Institute will bring top scientists and scholars to Washington D.C. to brief Congressional Representatives and Senators and their staffs on the scientific evidence of intelligent design and its implications for public policy and education, Wednesday, May 10, in the U.S. Capitol Building and Rayburn Office Building." The briefing, entitled "Scientific Evidence of Intelligent Design and Its Implications for Public Policy and Education," was co-hosted by seven members of the U.S. House of Representatives, both Republicans and Democrats, several of whom were on the House Committee on Science and the House Committee on Education and the Work Force: Rep. Roscoe Bartlett (R-MD), Rep. Charles Canady (R-FL), Rep. Sheila Jackson-Lee (D-TX), Rep. Thomas Petri (R-WI), Rep. Joseph Pitts (R-PA), Rep. Mark Souder (R-IN), and Rep. Charles Stenholm (D-TX).[86] Sen. Sam Brownback (R-KS) also attended and warmly introduced several of the speakers to about fifty people who attended (of whom only a small number appeared to be congressional staffers).[87] Yet the CRSC's connections in Washington did not begin with this briefing.

Johnson had already been to Washington on March 24, 2000, to lecture to congressional staffers, apparently laying groundwork for the May briefing. The March 28, 2000, issue of *Baptists & News* recounts this visit, which Johnson made as a speaker for Charles Colson's Capitol Hill Series: "The teaching of creation truths in the nation's classrooms is gaining ground . . . Johnson told some 100 congressional staffers March 24 on Capitol Hill."[88] Johnson distributed to the audience information on polls showing that a majority of Americans favor teaching creationism along with evolution in public schools. But cultivation of pro-ID congressional powerbrokers had begun even several years prior to the Wedge's spring 2000 political events. According to the August 1996 Discovery Institute *Journal,* one of the congressional hosts at the May briefing, Rep. Thomas Petri of Wisconsin, was once a Discovery Institute adjunct fellow while also serving in Congress. Introducing CRSC fellow David DeWolf at the briefing, Petri remarked that he hoped the briefing would lead to a "swelling chorus." As of this writing, Petri sits on DI's "Distinguished Board of Advisors."[89]

This foray into Washington, though small, has paid handsome political dividends: Rep. Mark Souder of Indiana—angered by a letter he received from Baylor scientists defending legitimate science against intelligent design after Souder co-hosted the briefing—subsequently published a defense of ID in the June 14, 2000, *Congressional Record* (H4480); he received assistance in its drafting from Phillip Johnson.[90] David Applegate, director of the American Geological Institute's (AGI) Government Affairs Program, in a "Special Update" sent to AGI members to alert them about the briefing, noted Souder's membership on the House Education and Workforce Committee and Rep. Petri's possible upcoming chairmanship of that committee.[91] In a July 2000 article in *Geotimes*, Applegate warns of what the Wedge's political maneuverings in the nation's capital portend: "The Discovery Institute chose to hold its briefing at the same time that both the House and Senate were actively considering legislation to overhaul federal K–12 education programs. Scientific societies have faced challenge enough to keep provisions in these bills supporting science and math education. If creationists choose to move into this arena and gain support from leading members of Congress, good science will face an even tougher challenge."[92] Applegate could not know then that he would send out another "Special Update" almost exactly a year later.

THE "SANTORUM AMENDMENT"

The May 2000 briefing was clearly the beginning of the Wedge's plan to influence science and science education policy at the national level. The events of June 2001 confirmed this assessment. On June 13, 2001, Pennsylvania Senator Rick Santorum (R) introduced amendment #799 to S.1, the Better Education for Students and Teachers Act. Along with its House companion HR 1, the No Child Left Behind Act of 2001, this piece of legislation was a major revision of the Elementary and Secondary Education Act overhauling federal education programs. Santorum added his amendment to the bill only one day before the Senate was to hold a final vote—after six weeks of debate.[93] Recognized on the floor of the U.S. Senate by Sen. Edward Kennedy, Santorum rose to explain his amendment:

> I rise to talk about my amendment which . . . is a sense of the Senate that deals with the subject of intellectual freedom with respect to the teaching of science in the classroom, in primary and secondary education. It is a sense of the Senate that does not try to dictate curriculum to anybody; quite the contrary, it says there should be freedom to discuss and air good scientific debate within the classroom. In fact, students will do better and will learn more if there is this intellectual freedom to discuss. . . .
>
> It is simply two sentences—frankly, two rather innocuous sentences—that hopefully this Senate will embrace:
>
> "It is the sense of the Senate that—

> "(1) good science education should prepare students to distinguish the data or testable theories of science from philosophical or religious claims that are made in the name of science; and
>
> "(2) where biological evolution is taught, the curriculum should help students to understand why this subject generates so much continuing controversy, and should prepare the students to be informed participants in public discussions regarding the subject." . . .
>
> I will read three points made by one of the advocates of this thought, a man named David DeWolf, as to the advantages of teaching this controversy that exists. He says:
>
> "Several benefits will accrue from a more open discussion of biological origins in the science classroom. First, this approach will do a better job of teaching the issue itself, both because it presents more accurate information about the state of scientific thinking and evidence, and because it presents the subject in a more lively and less dogmatic way. Second, this approach gives students greater appreciation for how science is actually practiced. Science necessarily involves the interpretation of data; yet scientists often disagree about how to interpret their data. By presenting this scientific controversy realistically, students will learn how to evaluate competing interpretations in light of evidence—a skill they will need as citizens, whether they choose careers in science or other fields. Third, this approach will model for students how to address differences of opinion through reasoned discussion within the context of a pluralistic society."
>
> I think there are many benefits to this discussion that we hope to encourage in science classrooms across this country. I frankly don't see any down side to this discussion—that we are standing here as the Senate in favor of intellectual freedom and open and fair discussion of using science—not philosophy and religion within the context . . . of science, but science—as the basis for this determination.[94]

Anyone familiar with Wedge rhetoric will recognize the anti-evolutionary idiom (some would say cant) of the ID movement in the language of the amendment itself, which was added as a "sense of the Senate," and which people unfamiliar with ID regarded as "innocuous." But in quoting DeWolf, Santorum was reading verbatim from the ID legal playbook, "Teaching the Controversy: Darwinism, Design and the Public School Science Curriculum," written by DeWolf, Stephen Meyer, and Mark DeForrest. Sen. Kennedy, presumably unaware at that point of the creationist influence behind the amendment (it is hard to believe that he would simply be indifferent to it), spoke in favor: "[The amendment] talks about using good science to consider the teaching of biological evolution. I think the way the Senator [Santorum] described it, as well as the language itself, is completely consistent with what represents the central values of this body. We want children to be able to speak and examine various scientific theories on the basis of all the information that is available to them so they can talk about different concepts and do it intelligently with the best information that is before them."[95]

Sen. Robert Byrd also supported the amendment, noting his own agreement with the "many scientists who have probed and dissected scientific theory and concluded that some Divine force had to have played a role in the birth of our magnificent universe" and avowing that the point of the amendment was "to support an airing of varying opinions, ideas, concepts, and theories." Sen. Brownback of Kansas, who had supported creationist Linda Holloway in the 2000 board of education race, promoted the amendment as a vindication of the creationist-dominated Kansas board's 1999 decision to defend "scientific fact" rather than "scientific assumption."[96] Amid a series of other votes, the Senate passed the amendment by a 91–8 margin.[97]

That vote prompted a press release from the Discovery Institute indicating the barely concealed delight with which DI viewed the amendment's passage, interpreting it as a green light for the introduction of ID in science classes: "Undoubtedly this will change the face of the debate over the theories of evolution and intelligent design in America. From now on the evidence will be free to speak for itself. It also seems that the Darwinian monopoly on public science education, and perhaps on the biological sciences in general, is ending."[98] Not surprisingly, Phillip Johnson acknowledged his role in crafting the amendment's language, as Larry Witham reported in the June 18 *Washington Times*: "'I offered some language to Senator Santorum, after he had decided to propose a resolution of this sort,' Mr. Johnson said."[99] Such an acknowledgment probably explains another trip Johnson had made to the nation's capital in early June, reported in his June 11, 2001, *Weekly Wedge Update*: "I spent the first part of last week in Washington, D.C., meeting with Senators, Representatives and their staffs." In his June 18 column after the amendment's passage, Johnson announced that a "huge bipartisan majority of the United States Senate has endorsed an intellectual freedom resolution for science education," while also disavowing any "coercive effect" of the amendment: "It does not dictate any curriculum."[100]

Intelligent design supporters, however, quickly saw the amendment's true implications and, in one case, acted upon them. Answers in Genesis, a young-earth organization, headlined a web page "US Senate supports intellectual freedom!" and provided information for readers to "contact your Congressman to express your support of the Senate version of the Education bill that states that evolution is controversial."[101] The conservative religious organization, Family Research Council, announced in its June 22, 2001, *Culture Facts* that "an amendment attached to the education reform bill put the Senate on record as being in favor of opening up science teaching to theories other than solely evolution. While not directly mentioning creationism or intelligent design, Sen. Rick Santorum's . . . 'sense of the Senate' amendment strikes at the widespread practice of ignoring potential flaws in, and alternatives to, the theory of evolution."[102] *World* magazine, which has given the Wedge extensive publicity, announced under the headline "Darwinian Teaching on Trial" that "conservatives are quietly cheering the passage" of the Santorum amendment and highlighted Johnson's role in crafting the language.[103] And, predictably, before the summer ended, New Mexico creationist Paul Gammill invoked the amendment in a letter to the NM board of education requesting the teaching of ID in NM schools:

> As you might know, the United States Senate recently . . . approved Senator Rick Santorum's Amendment No. 799, by an overwhelming vote of 91–8. . . .
>
> Specifically, I request that you develop Life Science Standards for New Mexico that will include a critique of neo-Darwinian evolution, and that will present the theory of Intelligent Design. The supplemental textbook, *Of Pandas and People*, is an excellent Intelligent Design textbook. . . . I have here a copy of the booklet, *Intelligent Design in Public School Science Curricula: A Legal Guidebook*, by David DeWolf, et al. I think every educator ought to have a copy.

Although Gammill stood no chance at that time of succeeding in New Mexico, this incident confirmed that creationists would invoke the Santorum amendment in their pro-ID efforts. (We have indicated in chapter 6 that such efforts may be gathering momentum in New Mexico.) In January 2002, soon after Congress passed the No Child Left Behind Act, Gammill sent another letter to the president of the NM board, as well as to his U.S. and state senators and representatives, other NM education officials, and newspapers, invoking the amendment even more strenuously.[104]

Science and education organizations from all over the country, working for months to have the Santorum amendment deleted from the funding bill, finally succeeded. In early December 2001 the House-Senate conference committee deleted it from the legislation, inserting it instead into the "Joint Explanatory Statement of the Committee of Conference" (also called the "statement of managers"). The wording was slightly revised prior to its relocation:

> The conferees recognize that a quality science education should prepare students to distinguish the data and testable theories of science from religious or philosophical claims that are made in the name of science. Where topics are taught that may generate controversy (such as biological evolution), the curriculum should help students to understand the full range of scientific views that exist, why such topics may generate controversy, and how scientific discoveries can profoundly affect society.[105]

This relocation was a significant move, which means that the Santorum amendment is not part of the legislation and thus is not legally binding, but rather has been relegated to the legislative *history* of the No Child Left Behind Act. However, in a release by the Discovery Institute, Bruce Chapman announced his role in the amendment's survival, stating that he had "helped advise members of Congress in the progress of the Santorum language in the Conference Report of the No Child Left Behind Act of 2001."[106] Chapman erred in locating the amendment in the conference committee report; the joint explanatory statement is separate from the report. But even though the Santorum amendment is now legislative history rather than part of the law itself, according to David Applegate's December 5 Special Update, "its incorporation . . . will be welcomed by evolution opponents."[107]

The survival of the amendment in the joint explanatory statement was a somewhat diluted but still useful victory for the Wedge. Of course, the Wedge would have scored a bigger *political* victory had the amendment remained in the bill as a sense of the Senate, although a sense of the Senate does not, and is not intended to, have the force of law; it is merely an expression of opinion. Moreover, neither house of Congress votes on joint statements.[108] So the amendment's deletion and subsequent insertion into the joint explanatory statement appears to give it even less weight, since neither the conference committee report nor the joint explanatory statement gives any directives regarding the amendment to any state or federal agency. Consequently, from a legal standpoint—since intelligent design falls *outside* anything that could be properly included among the "full range of scientific views"—the "Santorum amendment" does not now, and never did, require teachers to teach anything other than evolution. Prof. Dennis Hirsch of Capital University School of Law in Ohio provided this analysis for Ohio Citizens for Science:

> The Conference Committee did include a watered-down version of the amendment in a separate "explanatory statement" that it issued with respect to the final legislation. To understand the legal significance of this statement, it is important to distinguish between a statute and legislative history. A statute passed by Congress and signed by the President constitutes federal law. Legislative history is merely a record of events leading up to the passage of a law. It is not part of the statute itself, is not voted on by Congress, and is not law, as such. Here, the Santorum amendment was deleted from the statute. It did not become law. At most, the explanatory statement is an expression of the views of a few members of the House and Senate about the law. It forms a part of the legislative history. It does not constitute federal law on the subject.
>
> On occasion, legislative history such as committee reports can be a helpful tool for interpreting the language of a statute. The statement here provides little help in that regard since there is no corresponding statutory language to interpret, Congress having deleted the Santorum Amendment. Moreover, legislative history only serves this interpretative function where it sheds light on the intentions of Congress as a whole. Here, Congress did not support the Santorum Amendment, as evidenced by the fact that it took it out of the final legislation. This suggests that the watered-down version that appeared in the explanatory statement was added at the behest of a special interest group and did not receive the endorsement of Congress as a whole. In such situations, courts give legislative history little weight even as an interpretative tool. They in no way treat it as the considered "federal law" on the subject.[109]

Despite the shaky (or even nonexistent) legal status of the Santorum amendment, the Wedge accomplished what they initially set out to do: they managed to influence the legislative process regarding legislation of signal importance to science education, getting the initial sense of the

Senate passed with an overwhelmingly supportive vote (91–8) before it was removed by the conference committee. (That vote arguably did not at the time reflect genuine understanding or support of the true aims of its ID proponents, though Kennedy later publicly clarified his position, as we indicate later). The thin end of the Wedge has now been inserted at the federal level, giving Wedge supporters and fellow occupants of the "big tent" something to exploit. The scientific and educational organizations that worked to defeat the amendment were forced to recognize the ID movement as the political force it is, and the CRSC can use that initial vote as evidence that ID is more than merely respectable in the nation's capital. (That won't hurt the fundraising.)

Whether or not the conference committee understood the amendment's language as ID code talk, all the original language reflecting the Wedge's aims was left intact in the joint explanatory statement. Although the amendment's opponents view the explanatory statement's version as weaker from a legal standpoint, preservation of the essential part of the wording—and actually the *addition* of a new phrase not in the original—creates additional PR opportunities for ID proponents to spin the outcome to their advantage with boards of education and sympathetic teachers like Roger DeHart: (1) "biological evolution" having been retained in the wording, singled out for special attention as a "controversial" topic (even if only as an example—since it is the *only* specified example), the amendment can be cited to support the Wedge's demand that public schools "teach the controversy" about evolution; (2) consonant with the joint explanatory statement's recognition that "science education should prepare students to distinguish the data and testable theories of science from religious or philosophical claims that are made in the name of science," Johnson can continue to charge that "Darwinism" requires the (anti)religious presuppositions of atheism and philosophical naturalism, while only ID truly explains the data and therefore serves as a testable scientific theory; and (3) consistent with their legal strategy of identifying ID as science in an effort to take advantage of supposed loopholes in Supreme Court rulings, the Wedge can continue to offer ID as one among the "full range of scientific views." All of these opportunities that lie in the wording, as we shall soon see, have already been exploited.

The Santorum amendment may also help the Wedge precipitate the court case they so urgently want. As Mark Edwards of DI said with reference to Ohio, "All we need is one state to stand up and say we are going to permit academic freedom on this issue, a test case."[110] The Wedge has indeed given some indication that, in addition to the Santorum amendment's being invoked before state legislatures and boards of education, it may also be invoked in the courts. Kenneth Miller, who in March 2002 spoke to the Ohio board of education in opposition to Jonathan Wells and Stephen Meyer, disputed the Wedge's interpretation of the amendment's status. He demonstrated to the Ohio board, by conducting an online search of the federal education bill, that the amendment was not there. On April 19, 2002, in a published response to Miller's demonstration, DI claimed that "while the Santorum amendment may not have the 'force of law,' it is a powerful statement of federal education policy, and it provides authoritative guidance on how the statutory provisions of the No Child Left Behind Act (such as state-wide science assessments) are to be carried out. . . . Congress typically provides substantive policy guidance through the language of conference reports. . . . While report language does not have the 'force of law,' it might be said that it has the 'effect of law' because Congress expects it to be obeyed."[111] (As already shown, the amendment is not in the conference report, but in the joint explanatory statement, yet DI consistently fails to distinguish them.)

A week later, Bruce Chapman and David DeWolf issued yet another statement, "Why the Santorum Language Should Guide State Science Education Standards," asserting that "it is especially noteworthy . . . that the Santorum amendment was included as report language for the education act's provision requiring state science assessments." They again invoked the "effect of law" argument and, while admitting that the Santorum amendment "does not legally mandate an even-handed approach to the teaching of evolution" (i.e., the teaching of intelligent design as an alternative theory), still maintain that it is "federal policy."[112] Aside from whatever possible legal nuances DI sees in terms such as the "effect of law," its linkage of the Santorum amendment to the education bill's mandate for state science standards and its insistence on the authoritative status of legislative history with respect to statutory interpretation point in only one direction—the federal courts. And whether there is any legal justification for their position, it is clear that they relish the prospect of testing it.

So the only thing the Wedge must do is find the demographically and ideologically right location, one sufficiently religiously conservative to get ID into the schools, or at least into the science curriculum standards. What the Wedge needs is a compliant board of education with a majority of ID supporters, or some other (possibly legislative?) means of achieving the inclusion of ID in either state or local science standards. And if such inclusion results in a challenge (in which they truly believe the legal arguments would favor their side), the Wedge's lawyers are ready with their strategy. And it is possible that ID proponents will not wait for an anti-ID legal challenge but will themselves file suit against some state board of education if the state refuses to include ID in its science standards or to make any other concessions. SEAO member Robert Lattimer had threatened in June 2002 to do exactly that, as reported in the *Columbus Dispatch*: "'Our job is to convince the state school board that public opinion is strongly on our side.' Failing that, the fight will be taken first to the legislature and then to the courts, Lattimer said."[113]

One of the Wedge's Five Year Goals in the Wedge Document is to have "legal reform movements base legislative proposals on design theory." With the survival of the Santorum amendment in any form, even though less prominently than first envisioned, the Wedge comes close to achieving its first legislative success at the national level. On December 18, the same day that the Senate passed the education funding bill in which the amendment was originally placed, Sen. Santorum began the pro-ID spin with an insertion in the *Congressional Record*:

> Mr. President, I am very gratified that the House and Senate conferees included in the conference report of the elementary and secondary education bill the language of a resolution I introduced during the earlier Senate debate. . . . As a result of our vote today that position is about to become a position of the Congress as a whole. . . .
>
> There is a question here of academic freedom, freedom to learn as well as to teach. The debate over origins is an excellent example. Just as has happened in other subjects in the history of science, a number of scholars are now raising scientific challenges to the usual Darwinian account of the origins of life. Some scholars have proposed such alternative theories as intelligent design. In the Utah law review [*sic*] article the authors state, . . . "The time has come for school boards to resist threats of litigation from those who would censor teachers, who teach the scientific controversy over origins, and to defend their efforts to expand student access to evidence and information about this timely and compelling controversy."
>
> The public supports the position we are taking today. For instance, national opinion surveys show—to use the origins issue again—that Americans overwhelmingly desire to have students learn the scientific arguments against, as well as for, Darwin's theory. A recent Zogby International poll shows the preference on this as 71 percent to 15 percent, with 14 percent undecided. The goal is academic excellence, not dogmatism. It is most timely, and gratifying, that Congress is acknowledging and supporting this objective.[114]

DI quickly issued a December 21 U.S. Newswire press release announcing "Congress Gives Victory to Scientific Critics of Darwinism," treating the Santorum amendment as part of the education bill:

> The new public education bill just adopted by Congress calls for greater openness to the study of controversies in science, including biological evolution, thus giving victory to scientific critics of Darwinism. What began as the "Santorum Amendment," named for its originator, Sen. Rick Santorum (R-PA), says: "Where topics are taught that may generate controversy (such as biological evolution), the curriculum should help students to understand the full range of scientific views that exist."[115]

Shortly afterward, a slightly altered version was issued, which, though citing the amendment's removal to the committee report (which, *again*, is a mistake), still falsely implied that the report, and thus the amendment, were yet somehow part of the education bill itself. The new version also added a pointed statement about the significance of the "new education bill" for teachers, school superintendents, and school boards—precisely the people for whom the Santorum amendment was intended:

> The education bill . . . calls for greater openness to the study of current controversies in science. . . . [T]he "Santorum Amendment" . . . *now resides in report language*. . . .
>
> *The new bill represents a substantial victory for scientific critics of Darwin's theory*. . . .
>
> *There are no penalties in the new bill for failure to observe the Santorum Amendment recommendations.* But as Rep. Tom Petri (R-WI), a lead promoter . . . in the House . . ., noted, "This statement is especially important to make now because H.R.1 . . . requires all students eventually to be tested in science . . . as a condition of aid." . . .
>
> In preparing his amendment, Sen. Santorum followed the criticisms of . . . Phillip Johnson. . . . *The new education bill is likely to give courage to local schoolteachers, superintendents and school boards that have been misled by the ACLU and others to confuse teaching scientific differences with inculcating religion.*[116] (emphasis added)

This last statement might be considered prophetic: in Ohio, less than a month after the DI press release, the Santorum amendment was being exploited precisely as anti-ID critics had predicted and DI hoped. However, given the Wedge's adeptness at long-range strategizing, "calculated" is probably more accurate than "prophetic." It seems likely that the Ohio-Kansas ID coalition was simply waiting to ascertain the final fate of the Santorum amendment in order to put an already constructed plan into action, a plan whose viability was contingent on the amendment's survival.

THE SANTORUM AMENDMENT AS PART OF THE WEDGE'S OHIO STRATEGY

When one surveys the dates on which relevant events unfolded, the conclusion that the Ohio intervention was heavily dependent on the Santorum amendment's fate is almost inescapable; it highlights the Wedge's long-range plans in getting the measure introduced in Congress, thus highlighting further the long-term nature of the Wedge program. The Senate passed the Santorum amendment on June 13, 2001. The Ohio Writing Team Committee, on which Robert Lattimer sits, first met on June 26–27, 2001. Lattimer attended IDnet's Darwin, Design and Democracy II conference on June 29–30, 2001, where news of the amendment's passage was surely discussed and where, as noted earlier, Lattimer became an ID convert. On November 8, 2001, on IDnet's website, John Calvert and William Harris published an essay, "Ending the War Between Science and Religion," providing a telescopic view of arguments Calvert later offered in Ohio. (This paper responded to a February 2000 paper—"The Science and Religion Wars," *Phi Delta Kappan*—by Dr. Mano Singham, a member of the Ohio Advisory Committee for the Development of Science Content Standards.) The first draft of the new science standards, incorporating only evolution, was completed in late November at about the time the writing team met on November 29–December 1, 2001. By December 3, SEAO had registered its website. On December 13, the congressional conference committee for HR 1 published its joint explanatory statement containing the Santorum amendment. On December 18, Congress passed HR 1, and Sen. Santorum entered his celebratory remarks in the *Congressional Record.* On January 8, 2002, President Bush signed HR 1. On January 13, Calvert appeared before the Ohio standards committee, invoking the amendment as federal legislation (discussed shortly).[117]

This chronology shows that the timing of events with respect to the Santorum amendment was, to say the least, advantageous to the Wedge's Ohio strategy. The amendment having been salvaged, supporters invoked it at every turn in their efforts to introduce ID into the Ohio science standards. Like DI, the Ohio and Kansas ID auxiliaries exaggerated the amendment's significance and distorted its status in an effort to derive the political benefit its Wedge progenitors had in mind.

As we have noted, on December 21, 2001, only days after Sen. Santorum's December 18 *Congressional Record* statement, DI issued its press release, implying (wrongly) that the Santorum amendment was still in the education bill. Within days of President Bush's signing into law the education bill itself on January 8, 2002, SEAO posted a web page entitled "Federal Santorum Amendment," echoing DI's misleading slant:

> The Santorum Amendment, a key provision to bolster objective science education, remains in the final version of the federal 'No Child Left Behind' bill (H.B. 1) passed by the U.S. Congress on Dec. 18, 2001. . . . Bruce Chapman, a spokesperson for the Discovery Institute in Seattle, said: "This is a victory for those who believe that academic freedom ought to include the right of students to hear honest accounts of the scientific disputes over Darwinian theory." . . . John Calvert, Director of the Intelligent Design Network, said that the Amendment is "supportive of the legislative mandate that has been suggested in Ohio."[118]

On January 13, Calvert invoked the Santorum amendment in the conclusion of his unchallenged remarks to the Ohio board's standards committee, referring to it improperly as "legislation": "[T]he concept of objective origins science is consistent with recent public school legislation signed by President Bush on Jan 8. You have been furnished with a two page hand out that discusses the legislation. . . . Thus, the Ohio Board is in a unique position. If it follows the Santorum amendment and the Lattimer proposals it can be on the cutting edge of new science standards that will enhance the effectiveness of origins science and insure that it is conducted objectively and consistent with logic, good science, and the law."[119] On January 18, Calvert and Robert Lattimer published an editorial in the *Plain Dealer,* again calling the Santorum amendment part of the education bill.[120] On January 23, Rep. Reidelbach introduced HB 481, which deliberately reflects concepts and phrasing from the Santorum amendment. Reidelbach specifically referred to the Santorum amendment in defending HB 481, saying that her bill mirrored a federal education bill.[121]

On a web page posted on January 25, Ohio Citizens for Science tried to counter the "misinformation campaign waged by creationists concerning the non-existent Santorum 'Amendment'": "Pretending the amendment still exists in the bill signed by the president is nothing more than a sham, an attempt to mislead the public and state legislators into thinking proposed legislation such as Ohio House bill 481 has a precedent of federal law. It does not."[122] ID proponents, either unaware or unmindful of attempts to correct their misrepresentation of the Santorum amendment, continued their campaign, even though SEAO in early February had posted a new version of its "Federal Santorum Amendment" page conceding that "the Santorum Amendment does not appear in the final version of H.B. 1[*sic*]" and, reflecting DI's failure to distinguish between the committee report and the joint explanatory statement, referring to its location in the committee report.[123]

Deborah Owens-Fink, ID's champion on the Ohio board of education, actually referred to the campaign to amend the state's science standards as the *implementation of federal law*, adding her hope that Ohio's legislation would be imitated in other states: "We are very committed as a board to develop the Ohio science standards that are the best in the nation and can serve as a model for other states and, hopefully, the nation as we implement the education law enacted by Congress. This federal bill included the Santorum amendment."[124] And predictably, DI released reinforcing comments from Sen. Santorum's and Rep. Thomas Petri's *Congressional Record* statements:

> *Statement by Senator Santorum Upon the Passage of the Education Act 2001:* Congressional Record: December 18, 2001 (Senate). . . .
>
> Mr. SANTORUM: . . . By passing [the resolution] we were showing our desire that students studying controversial issues in science, such as biological evolution, should be allowed to learn about competing scientific interpretations of evidence. As a result of our vote today that position is about to become the position of the Congress as a whole. . . .
>
> *Speech of Honorable Thomas E. Petri of Wisconsin in the House of Representatives.* Thursday, December 13, 2001. . . .
>
> Additionally, this conference report makes a strong statement that, where Darwinian evolutionary theory or other controversial scientific viewpoints are taught, students should be exposed to multiple viewpoints. Too often, students are taught only one theory where evolution is concerned, *and this language gives support to those at the local and state level* who uphold the value of intellectual freedom in the teaching of science. This statement is especially important to make now because H.R. 1 requires all students eventually to be tested in science on a regular basis as a condition of aid.[125] (emphasis added)

Eventually, the campaign to exploit the Santorum amendment reached a feverish level. On March 15, 2002, Rep. John A. Boehner

(R-OH), chair of the U.S. House Education and Work Force Committee, and Rep. Steve Chabot (R-OH), chair of the U.S. House Constitution Subcommittee, came to the aid of ID proponents in their home state with a joint letter, on congressional stationery, to Jennifer L. Sheets and Cyrus B. Richardson, president and vice president of the Ohio Board of Education. The letter reinforced claims made by ID supporters: "We are writing to comment on recent Ohio School Board hearings regarding the teaching of science in Ohio public schools in light of some recent developments in federal education policy. . . . [T]he Santorum language is now part of the law. The Santorum language clarifies that public school students are entitled to learn that there are differing scientific views on issues such as biological evolution." Attached to the letter were "comments from members of [the] House and Senate [Santorum and Petri] . . . for your information."[126] The attachment was identical to the version of these comments, just quoted, which DI had placed on its website in January. But Boehner and Chabot, eager to point out what the *law* supposedly says, failed to consider what the U.S. Constitution says. Prof. Hirsch, cited earlier, points to Article I, Section 7: "Every bill which shall have passed the house of representatives and the senate, shall, before it become a law, be presented to the president of the United States." Having been deleted from the legislation and therefore never presented to the president, the Santorum amendment cannot be part of the law.[127] Yet the spin continued.

Only the day before the letter was written, Sen. Santorum had published a *Washington Times* column in which he unambiguously invoked Sen. Edward Kennedy's name in support of teaching ID, quoting Kennedy's words when he voted in favor of the Santorum amendment as a sense of the Senate:

> Supporters for a change in teaching standards want the [Ohio] board to include the idea that living things could have been "designed" in some meaningful way. Sen. Ted Kennedy, Massachusetts Democrat, approves of having alternate theories taught in the classroom. He believes children should be "able to speak and examine various scientific theories on the basis of all information that is available to them so they can talk about different concepts and do it intelligently with the best information that is before them."

But Santorum at this point overplayed his hand. Kennedy, perhaps with a more informed view than that with which he had cast his vote in June 2001, responded with a brief letter in the *Washington Times*:

> The March 14 Commentary piece, "Illiberal education in Ohio schools," written by my colleague Sen. Rick Santorum, Pennsylvania Republican, erroneously suggested that I support the teaching of "intelligent design" as an alternative to biological evolution. That simply is not true.
>
> Rather, I believe that public school science classes should focus on teaching students how to understand and critically analyze genuine scientific theories. Unlike biological evolution, "intelligent design" is not a genuine scientific theory

and, therefore, has no place in the curriculum of our nation's public school science classes.[128]

And—of course—there was a response from the Wedge: a press release by William Dembski, remarking on Kennedy's letter with what can only be described as audacity:

> Kennedy is no scientist or philosopher of science, so presumably he has spoken to the experts, who assure him . . . intelligent design is not science. . . . Kennedy himself offers no argument for why intelligent design fails to be a scientific theory. . . .
>
> To dismiss intelligent design . . . is to insulate evolutionary biology from criticism and turn it into a monopoly. . . . Kennedy, who has been so effective at unmasking monopolies in the business world, seems not to realize that they can be just as virulent and oppressive in the scientific world.[129]

At this writing, even without knowing the future uses to which the Wedge may put this "federal law," it is clear that the Santorum amendment was intended at least partly to provide a foundation for legislation below the federal level, as proved to be the case in Ohio. HB 481 was written using verbatim phrasing and terminology. This was not coincidental. On March 5, 2002, Ohio Rep. Linda Reidelbach again stated publicly in her testimony before the Ohio House Education Committee that "HB 481 was modeled on the language of the Santorum Amendment, which is now the guiding language for the implementation of HR 1. . . . House Bill 481 is an effort to align Ohio's standards with the new federal law." In April 2003, Louisiana legislator Ben Nevers filed House Concurrent Resolution 50 at the spring legislative session. HCR 50 encouraged Louisiana public school systems to "refrain from purchasing textbooks that do not provide students with opportunities to learn that there are differing scientific views on certain controversial issues in science." It referred to the No Child Left Behind Act and cited verbatim the language of the Santorum amendment.[130] (Fortunately, it died.)

Debate currently still swirls around the status of the Santorum amendment and probably will continue as long as the Wedge continues to exploit it. And there will almost certainly be other Ohios. With the controversy there reaching its peak in early 2002, Phillip Johnson had already weighed in on the situation in his January 31, 2002, *Weekly Wedge Update*, with a message that, though ostensibly about the Wedge's setback in Kansas, was easily interpreted as support for the Wedge's Ohio brethren—and a promise to those awaiting with confident expectation the Wedge's support in states beyond: "We have raised new forces, and forged an alliance between groups that were formerly suspicious of each other. We will fight again, and eventually the leaders of science will learn that the costs of imposing a pseudoscientific materialism on America are too great for them to bear."[131]

FURTHER D.C. PROSPECTS

The May 2000 briefing and the Wedge's success in getting an ID-friendly amendment inserted initially into a major federal education funding bill and subsequently preserved in the joint explanatory statement are the strongest evidence yet of the Wedge's broad political ambitions. They show that these ambitions are aimed at science, especially biology, at every level of American education. The "briefing" was a splendid opportunity to cultivate congressional contacts: there was a private luncheon as well an evening reception. And the Wedge's success in securing the personal attention of an influential senator indicates how close their influence has now come to the drafting of legislation.[132] Both events increased the chance that the CRSC and its new friends in the nation's capital will soon attempt to make government funding available for ID "research" (or for a more-or-less legitimate scientific research proposal, which, after funding is secured, could be touted as supportive of ID), or at least to ensure hostile congressional scrutiny of all government funding for real evolutionary science.

Jonathan Wells has already promoted such scrutiny—in public. At the October 26, 2000, lecture and book signing in Seattle, he argued that the time has come for precisely such inspections:

> [M]aybe the touchiest subject of all, tens of millions of dollars, ten of billions of dollars every year of your tax money is being used to support research in our universities. Now most of that is good. I am all in favor of supporting honest, useful scientific research. I think our culture relies on it. I am not attacking scientists. . . . I am attacking the use of research money to promote lies and indoctrination. . . . [I]t is time that we begin looking into that. . . . And finally, I think the time is coming . . . when our elected officials have to start looking at how the money is being disbursed, being used. And I fully anticipate that this is going to happen in the next few years.[133]

In *Icons of Evolution,* Wells provides tips for his readers on how to get this political ball rolling, from the national level down to the local:

> One possibility is to call for congressional hearings on the way federal money is distributed by the NIH, the NSF, and NASA. . . . Scientists who deliberately distort the evidence should be disqualified from receiving public funds. . . .
>
> As we have seen, the National Academy of Sciences published booklets that misrepresent the evidence for evolution. . . . Maybe your representatives should look more closely at how your money is being spent. . . .
>
> State legislators might also want to take a look at the Darwinian establishment, to determine whether state taxes are being used for indoctrination rather than education. State and local school boards could be encouraged to take a closer look at the textbooks they buy for public schools. . . . [S]chool boards might want to alert students to their misrepresentations by attaching warning labels. . . .
>
> The next time you get a fundraising letter from your alma mater, you might want to ask where your money will go.[134]

The Wedge's prospects of success in securing either federal funding for ID or increased scrutiny of current research funding remain to be seen. However, Wedge overtures to national politicians involve more than just connections on the Hill; their hopes apparently reach all the way to Pennsylvania Avenue. After George W. Bush made comments in favor of teaching creationism in public schools during his candidacy for president, CRSC fellows began to position themselves to move on multiple opportunities in the event of a Bush victory. In a December 1999 interview with *U.S. News & World Report* after the Kansas Board of Education decision to delete most references to evolution from state science standards, Mr. Bush stated a position consistent with the fondest hopes of the Wedge:

> [Interviewer] I feel like I'd be remiss if I didn't ask a couple of question[s] on some, I guess, controversial subjects of late, one of which is this whole creationism debate in Kansas.
>
> [Bush] I believe in the alignment of authority and responsibility away from the federal government when it comes to issues of governance and schools. Secondly, my own personal opinion is that I believe that it's important for children to understand there's different schools of thought when it comes to the formation of the world. I have no problem explaining that there are different theories about how the world was formed. I mean, after all, religion has been around a lot longer than Darwinism. And I think it's important for people to know what people believe in—but whatever the case, here's what I believe. I believe God did create the world. And I think we're finding out more and more and more as to how it actually happened.[135]

Phillip Johnson lost no time in exploiting this most visible political support for creationism. He gave it special attention in *The Wedge of Truth* (published before the election, in the summer of 2000)—and took notice of Al Gore's by then well-remarked timidity on the subject:

> A Gallup poll conducted in late June 1999 revealed that Americans favor teaching creationism in public schools along with evolution. . . . Given this polling data, it is not surprising that Republican presidential candidates, especially front-runner George W. Bush, endorsed local control of the public schools and said that they had no objection if the local board chose to teach both sides of the controversy. What truly shocked the scientific community, however, was the embarrassing series of waffles by Vice President Al Gore, a liberal who (in the sarcastic words of a *Washington Post* editorial) "has held himself out as an avatar of science education."[136]

Johnson's pre-election positioning was followed on October 21, 2000, by Stephen Meyer in *World* magazine: "George W. Bush's stump speech shows that he recognizes the need to draw sharp differences with Al Gore on the central issues of the campaign. . . . To do that, he must pierce the media's electronic curtain by dramatizing his differences with Mr. Gore in ways the media can't filter or ignore. If he does, he will win the only debate that matters." The tone and theme of Meyer's piece were, perhaps not coincidentally, very similar to that of *World* editor and Bush adviser Marvin Olasky, who on September 23 had written, "To prevent the media from filtering his message, as they did in the weeks after the Republican convention, Mr. Bush must implement a new strategy. He must find ways not only to state, but also to dramatize, his message and his central differences with Al Gore."[137]

Finally, after the election, while the nation awaited the outcome of the legal processes that would determine the presidency, William Dembski and Mark Hartwig on December 7 called on Mr. Gore to concede, comparing Gore's request for a recount in Florida to someone's seeking to win a bet about how many "blue fish" (Gore votes) were in "67 lakes" (Florida counties): "As any statistician knows, Vice President Al Gore's plan to recount votes in Democratic-majority counties has been intrinsically unfair. . . . The only way to get a fair and accurate tally is to count all the fish [votes] in all the lakes [counties]with uniform standards. . . . The only fair thing you can do is forget your scheme, pull out your checkbook and fork over the dough [concede]."[138] Meyer's and Dembski's articles carry their affiliation as Discovery Institute senior fellows. These op-eds were followed by still another, this time by Meyer and DI president Bruce Chapman (who ran the White House Office of Planning and Evaluation and was director of the U.S. Census Bureau during Ronald Reagan's presidency). In "A Plan for Recovery of the Iffy Economy" in the December 28, 2000, *Seattle Times*, Chapman and Meyer offer advice to Mr. Bush to help ensure the success of his proposed economic policies.[139]

Of course, brute politics—eternal, foolish, but pragmatic—is the key to this depressing story. Thus, Vice President Al Gore, Mr. Bush's opponent, a social and political liberal and articulate science enthusiast, was by no means forthright in opposing such crude politicizing of science. In fact, as Johnson observed, Gore "waffled." He adroitly avoided the challenge of direct argument about evolution and creationism in school science teaching. Through spokesman Alejandro Cabrera he opined, "Obviously, that decision [to teach creationism along with the standard science of evolution] should and will be made at the local level, and localities should be free to teach creationism as well." Reminded that the Supreme Court long ago ruled against teaching creationism in public schools, Gore later modified this initial equivocation by specifying that creationism, if taught, should be taught in classes on religion rather than as science.[140] Gore's modification was an improvement, but not the kind that would endear him to the Wedge. Thus, the Wedge made its own political allegiances crystal clear. Yet it is well to remember that these political maneuverings are not an end in themselves; for the Wedge, they reach toward a much more important goal. As for all Wedge activity, that main goal is religious.

9

Religion First—and Last

Christians in the 20th century have been playing defense. . . . They've been fighting a defensive war to defend what they have, to defend as much of it as they can. . . . It never turns the tide. What we're trying to do is something entirely different. We're trying to go into enemy territory, their very center, and blow up the ammunition dump. What is their ammunition dump in this metaphor? It is their version of creation.

Phillip E. Johnson, February 6, 2000, at a meeting of the National Religious Broadcasters in Anaheim, California

In March 1994, Phillip Johnson participated in the conference "Regaining a Christian Voice in the University." In his lecture "The Real Issue: Is God Unconstitutional?" he lamented the prevalence in American universities of "scientific naturalism." That, according to Johnson, is "the established religious philosophy of America."[1] This complaint is the central theme in Johnson's sponsorship of ID; it fuels the mission of getting ID theory into the academic world and into public life as a full competitor of evolution and any other science consistent with evolution. Johnson's mission is urgent: "The bitter debate over whether 'creation' or 'intelligent design' may be considered as a possibility in scientific discourse is no minor matter. Behind it lies one of the most important questions of human existence: Did God create Man, or did Man create God?"[2] Johnson continues to mourn what he sees as the dismissal of God from universities—both religious and secular. In *The Wedge of Truth*, he recounts the tale of Philip Wentworth, who, according to Wentworth's essay "What College Did to My Religion," entered Harvard in 1924 a faithful Presbyterian and emerged a disillusioned convert to "scientific naturalism"—having been taught (in Johnson's words) by infidels.[3] Johnson sees parallels between Wentworth's experience and his own at Harvard thirty years later. They were both victims of an elite who "are particularly skilled at inventing ways to tame God because they desire either to ignore God or to use Him for their own purposes." Wentworth and Johnson were thus defenseless youths in a university that "offered no instructions in how to recognize idolatry."[4] According to Johnson, Wentworth's experiences of "apostasy" at Harvard are "representative of the experience of an entire culture of educated people over more than a century," all because of scientific naturalism.[5]

For the damnable error of naturalism, according to Tim Stafford in *Christianity Today*, Johnson sees nothing short of complete revolution as the remedy: "Phillip Johnson's idea of revolution is not . . . a struggle to control one corner of the ivory tower. He is playing for all the marbles for the governing paradigm of the entire thinking world. He believes evolution's barren rule can be overturned, that it is rip[e] for revolution."[6] But, as we have shown, Johnson and his associates use ostensibly scientific arguments as a façade behind which to mount their revolution. The plan is to establish their religious worldview as the foundation of all cultural life.

Our epigraph states the Wedge's purpose, even though for much of its history Phillip Johnson has tried in public to distance the ID movement from its religious impulse: "ID is an intellectual movement, and the Wedge strategy stops working when we are seen as just another way of packaging the Christian evangelical message."[7] With the political wheels of the Wedge turning fast, however, Johnson is no longer trying so hard to disguise the religious core of the strategy, as demonstrated in *The Wedge of Truth* and reinforced in an interview with Christian Book Distributors:

> [Interviewer]: How would you describe the main purpose of *The Wedge of Truth* in comparison to your other books?
>
> [Johnson]: . . . My prior books argued that the real discoveries of science—as opposed to the materialist philosophy that has been imposed upon science—point straight towards the reality of intelligent causes in biology. . . . [T]here are two definitions of "science" in our culture. One . . . says that scientists follow the evidence regardless of the philosophy; the other says that scientists must follow the (materialist) philosophy regardless of the evidence. The "Wedge of Truth" is driven between those two definitions, and enables people to recognize that "In the beginning was the Word" is as true scientifically as it is in every other respect.[8]

The Religious Essence

Despite Johnson's earlier efforts to present ID as a scientific enterprise, the Discovery Institute, eager to secure funding in the Wedge's early stages, could not conceal from current and potential supporters the religious aims underlying the political ones. Wedge fundamentals had to be established in its formative phase. DI president Bruce Chapman did so in a letter to supporters. When he cited the newly established CRSC's political aims in the August 1996 *Journal*, Chapman acknowledged the integral role of religion in the organization's mission. Invoking the "moral and political legacy" of "the Bible, history and the writing of the American Founders," Chapman declared that "it is fitting" that DI was making a simultaneous announcement of the establishment of the CRSC and the publication of John West's book, *The Politics of Revelation and Reason*, implying that the book would convince skeptical readers of the need to unite religion and politics: "Before you opine about the place of religion in politics, (or why there shouldn't be any), use this scholarly, but very readable, account of religion in early American politics. It will surprise you—and perhaps, it will inspire, too."[9]

An announcement elsewhere in the *Journal* pointed out that "the Center for Renewal of Science and Culture fits well with Discovery's existing programs in high technology and religion."[10] Indeed, on the CRSC's website, a page entitled "Life After Materialism" advertised the organization's religious aim, asserting that "today new developments in biology, physics, and artificial intelligence are raising serious doubts about scientific materialism and re-opening the case for the supernatural."[11]

Anti-Secularism, Religious Engagement—and Christian Dominance

George Gilder, advisor to CRSC along with Phillip Johnson, takes a particular interest in DI's "existing programs in high technology and religion," as Chapman notes in the August 1996 *Journal*: "George Gilder, who heads Discovery's technology program, has a long-standing interest in the interaction of science and culture. . . . [O]ne of his works-in-process deals with the cultural necessity of faith." Gilder's support of CRSC apparently reflects his own religious awakening and rejection of Darwin, chronicled by Laurissa MacFarquhar in *The New Yorker* of May 29, 2000: "He has become much more religious as he has grown older. . . . Gilder is, in fact, sufficiently committed to the idea that divine intelligence permeates all earthly occasions that he has turned away from Darwin and has begun reading writers such as Michael Denton and Michael Behe. . . . To Gilder's mind, most of what secularism gets wrong about the world can be summed up in the phrase 'the Materialist Superstition.'"[12] He shares the CRSC's view that Western culture urgently needs renewal, remarking in a 1994 interview on the necessity of religion for Western civilization: "Religion is primary. Unless a culture is aspiring toward the good, the true, and the beautiful, and . . . really worships God, it readily worships Satan."[13] In fact, Gilder reveals in this interview that he does not believe in the legitimacy of secular society, nor does he have much regard for the public schools into which his CRSC advisees want to insert their ID curriculum: "Secular culture is in general corrupt, and degraded, and depraved. Because I don't believe in secular culture, I think parochial schools are the only real schools."

Religious motivation drives all the CRSC leadership.[14] Indeed, Stephen C. Meyer, the director of the CRSC, professed his attraction to "the origins debate" precisely *because* it is theistic: "I remember being especially fascinated with the origins debate at this conference. It impressed me to see that scientists who had always accepted the standard evolutionary story [Meyer says he was one of them] were now defending a theistic belief, not on the basis that it makes them feel good or provides some form of subjective contentment, but because the scientific evidence suggests an activity of mind that is beyond nature. I was really taken with this."[15]

But that religious passion is not limited to the CRSC advisory level. It colors the thought and argument of the CRSC fellow who most aggressively seeks acceptance of ID in the academic world on grounds of its legitimacy as a scientific research program—William Dembski (whom Gilder, as we noted earlier, has dubbed "God's mathematician").[16]

> A world in which information is not primary is . . . hampered in what it can reveal. . . . [T]he world [modern science] gave us reveals nothing about God except that God is a lawgiver. But if information is the primary stuff, . . . there are no limits . . . on what the world can in principle reveal. In such a world, . . . Jesus can reveal the fullness of God, and bread and wine can reveal the fullness of . . . Jesus' life and death. A world in which information is . . . primary . . . is a sacramental world; a world that mirrors the divine life and grace; a world that is truly our home.[17]

Dembski sees an amalgam of intelligent design "theory," which he insists is the best kind of science, and his fervent evangelicalism.[18] He acknowledges that the reality he asserts can be analyzed scientifically (through the filter of "specified complexity") is "sacramental" and that ID is simply a form of the "Logos theology of John's Gospel":

> Where is this work on design heading? . . . [S]pecified complexity is starting to have an effect on the special sciences. . . .
>
> [D]espite its . . . implications for science, I regard the ultimate significance of this work on design to lie in metaphysics. . . .
>
> The primary challenge, once the broader implications . . . for science have been worked out, is . . . to develop a relational ontology in which the problem of being resolves thus: to be is to be in communion, and to be in communion is to transmit and receive information. Such an ontology will . . .

> safeguard science and leave adequate breathing space for design, but . . . also make sense of the world as sacrament.
>
> The world is a mirror representing the divine life. The mechanical philosophy was ever blind to this fact. Intelligent design . . . readily embraces the sacramental nature of physical reality. *Indeed, intelligent design is just the Logos theology of John's Gospel restated in the idiom of information theory.*[19] (emphasis added)

Statements such as these, made in the July/August 1999 issue of *Touchstone*, betray the disingenuousness of Dembski's November 2000 statement that "design has no prior commitment to supernaturalism."[20]

The Wedge Document enunciates distinctly religious goals, such as setting up apologetics seminars and working to ensure that "seminaries [i.e., seminaries that accommodate evolution and other modern ideas in their theological instruction] increasingly recognize & repudiate naturalistic presuppositions." In *Unapologetic Apologetics: Meeting the Challenges of Theological Studies* (InterVarsity Press, 2001), written primarily for seminary students and apologetics practitioners, Dembski and Jay Wesley Richards present their worldview, which they believe must be the foundation of *all* Christian theological instruction. Richards particularly warns against naturalism: "Through this essay I want to heighten your awareness of an ideology that is frighteningly pervasive, fundamentally anti-Christian and false—namely, naturalism. I hope thereby to 'inoculate' you so that you'll recognize naturalism when you encounter it and won't be led astray by its unwitting advocates. . . . [M]y main purpose is to convince you that naturalism just is not compatible with Christian belief, and any scholarship that claims to be Christian cannot be simultaneously laced with naturalistic presuppositions."[21]

Not only does Dembski and Richards's particular Christian worldview require intelligent design as its first principle (see chapters 13 and 14); it is unabashedly exclusionary. It shows little leniency toward any religion other than Christianity and any *form* of Christianity other than their own.

> As we engage the world with the truth of Christianity, we need to recognize how very high are the stakes. . . . Christianity claim[s] to possess humanity's ultimate truth, but . . . also . . . that this truth is so urgent that a person ignores it at his or her peril. . . . The opportunity to take part in the divine life is regarded by Scripture and church tradition as . . . the best [news] there is. *But Christianity also has a dark side: those who refuse to embrace this truth face separation from that divine life. . . . [T]he Scripture and church tradition are univocal on this point; only this time the picture they present is incredibly bleak. . . .*
>
> People's lives are in the balance. *Not every story will have a happy ending. Everything is not going to turn out all right in the end. Only where God's grace is manifested will things turn out all right. But where God's grace is spurned, things will not turn out all right.* There is a move afoot . . . in theological circles to embrace . . . universalism—that in the end everyone will be saved. This is the teaching neither of Scripture nor of church tradition. . . . Our feel-good pop psycholo-

> gies urge us to think it . . . befitting of God to save everyone. Reality, however, is not . . . determined by what we think fitting. . . . [W]e should be comforted in knowing that the God who decides human destinies is rich in love and mercy. But we must never neglect the holiness and justice of God. . . .
>
> [T]he truth of Christ is . . . glorious and urgent. It follows that Christians have a mandate to declare [it]. . . . *This mandate consists of bringing every aspect of life under the influence of this truth.*[22] (emphasis added)

Dembski and Richards quote approvingly the words of Princeton theologian J. Gresham Machen to bolster their own view that "Christians are called to be salt and light in the world, . . . to stem and overthrow false ideas":

> False ideas are the greatest obstacles to the Gospel. We may preach with all the fervor of a reformer and . . . succeed only in winning a straggler here and there, if we permit the . . . collective thought of the nation or . . . the world to be controlled by ideas which . . . prevent Christianity from being regarded as anything more than a harmless delusion.[23]

While rejecting "quietism" as a response to "false ideas," Dembski and Richards also disavow "imperialism," which is "always a matter of coercion," in favor of polemical "engagement," which makes the believers of false ideas uncomfortable: "The deeper a lie is entrenched, the greater the discomfort when . . . truth finally unmasks it. The Pharisees did not like it when Jesus unmasked their hypocrisy."[24] They assert that false ideas must be exposed and rooted out. But recognition of their own fallibility is no part of their intellectual equipment. So, despite a thoroughly unconvincing disavowal of imperialism, they insist with a kind of gnostic self-assurance that

> [f]alse ideas need to be weeded out. This requires work, patience and diligence. Above all, it requires a willingness to listen and inquire into ideas that oppose the faith . . . [and] to learn from the world. *We must grasp what the world is saying . . . better than it does itself. Only . . . this way will Christ's authority over the life of the mind be reestablished and the doors of faith reopened.*[25] (emphasis added)

Indeed, Dembski and Richards consider themselves martyrs, as shown in this remarkable statement:

> We are to engage the secular world, reproving, rebuking and exhorting it, pointing to the truth of Christianity and producing strong arguments and valid criticisms that show where secularism has missed the mark.
>
> Will we be appreciated? Hardly. The Pharisees of our day . . . reside preeminently in the academic world. The Pharisees killed Jesus and are . . . ready to destroy our Christian witness if we permit it. Nevertheless, this is our calling as Christian apologists, to bear witness to the truth, even to the point of death (be it the death of our bodies or the death of our careers). The church has a name for this—martyrdom. The early church considered martyrdom . . . an honor and privilege, a way of sharing in Christ's sufferings. . . .

> Christian apologetics . . . is a call to martyrdom—perhaps not a martyrdom where we spill our blood (although this too may be required) but . . . where we witness to the truth without being concerned about our careers, political correctness, the current fashion or toeing the party line.[26]

And Dembski has a rather archaic view of *heresy*:

> Within late twentieth-century North American Christianity, *heresy* has become an unpopular word. Can't we all just get along and live together in peace? Unfortunately, no. Peace cannot be purchased at the expense of truth. . . . There is an inviolable core to the Christian faith. Harsh as it sounds, to violate that core is to place ourselves outside the Christian tradition. This is the essence of heresy, and heresy remains a valid category for today. . . .
>
> The Christian apologist is a contender for the faith, not merely a seeker after truth. Seeking after truth certainly seems a less combative and more humble way of cashing out apologetics. Unfortunately, it is also an inadequate way of cashing out apologetics.[27]

So Christian truth and Christian community—as Dembski and Richards apprehend them—must trump tolerance and civic peace. The implications of such a view for a religiously pluralistic, democratic society are chilling.

While Dembski and Richards's statements may reveal accidentally the inflexible religiosity driving the Wedge, other CRSC fellows make an explicit connection between their religiosity and their work for the Wedge. Jonathan Wells's religious life, for example, has dictated his zealous anti-Darwinism. Wells, who is, as we have seen, a devout member of Sun Myung Moon's Unification Church, obtained a Ph.D. in biology—*after* earning a theology degree—*in order* to attack evolution. He explains this on the Unification Church's "True Parents" website: "Father's words, my studies, and my prayers convinced me that *I should devote my life to destroying Darwinism*. . . . When Father chose me . . . to enter a [theology] Ph.D. program in 1978, I welcomed the opportunity *to prepare myself for battle*" (emphasis added).[28] And, in a defense of Phillip Johnson, CRSC fellow Ray Bohlin, executive director of Probe Ministries, testifies to the close fit between the Wedge and Bohlin's own Christian ministry: "What is needed now to widen the crack [made by the Wedge] are larger numbers of theistic scientists, philosophers, and social scientists to fill in the ever widening portions of the wedge exposing the weaknesses of naturalistic assumptions across the spectrum of academic disciplines. Here Johnson's strategy meshes nicely with Probe Ministries."[29]

The religious essence of the Wedge is therefore obvious—Wedge members themselves bring up religion at every turn. Since the Wedge Document itself stipulates that part of Phase II is "to build up a popular base of support among our natural constituency, namely Christians," the Wedge cannot afford to do *too* good a job of concealing its religious goals.

The conclusion that the Wedge is primarily a religious rather than a scientific movement thus becomes unavoidable. And there is abundant evidence of a more pragmatic kind: the identities of CRSC's earliest guardian angels—its major funding sources, Fieldstead and Co., the Stewardship Foundation, and the McLellan Foundation.

CRSC's Most Generous Benefactors

The large contributions of the CRSC's top benefactors indicate that they recognize and support the CRSC's mission. Bruce Chapman celebrates this support: "We are not going through this exercise just for the fun of it. We think some of these ideas are destined to change the intellectual—and in time the political—world. Fieldstead & Company and the Stewardship Foundation agree, or they would not have given us such substantial funding."[30] Larry Witham observes that all three funding sources—Howard Ahmanson's Fieldstead & Co., the Maclellan Foundation, and the Stewardship Foundation—have Christian roots.[31] Indeed, religion is central to the mission of these organizations, and in the case of Ahmanson, to the ominous mission of America's most radical form of Christianity, "Christian Reconstructionism." The generosity of these sources signals their recognition of the Wedge's religious purposes.

The Stewardship Foundation

According to Guidestar (www.guidestar.org), the Stewardship Foundation's total worth in 1997 (the year closest to the CRSC's founding) was $95,770,585. CRSC's connection to the Stewardship Foundation was apparently through C. Davis ("Dave") Weyerhauser, grandson of Frederick Weyerhauser, founder of the Weyerhauser Timber Company. According to the organization's history, "The Stewardship Foundation was created in 1962 by C. Davis Weyerhauser . . . to contribute to the propagation of the Christian Gospel by evangelical and missionary work and to teach the Christian faith as laid down in the Old and New Testaments of the Holy Scriptures." Weyerhauser, who died in 1999, was a "major funder and supporter of the work at ARN," so he clearly was an ID backer. He also served on the "Board of Reference" of the Foundation for Thought and Ethics, which published and markets *Of Pandas and People*.[32] According to the foundation's grant guidelines, "Grants are made to Christ-centered organizations which have as their primary goal to bring people into a relationship with God through faith in Jesus Christ . . . to organizations which address the following Programmatic Themes: Leadership, Poverty, Reconciliation and Justice, Incarnational Christian Witness, and *Cultural Engagement*" (emphasis added).[33]

The Maclellan Foundation

Guidestar lists the 1997 assets of the Maclellan Foundation, Inc., as $42,648,622. Based in Chattanooga, Tennessee, its mission states that it exists to fund religious endeavors: "The purpose of the Maclellan Foundation is to serve strategic international and national organizations committed to furthering the Kingdom of Christ and select local organizations, which foster the spiritual welfare of the community. We will serve by providing financial and leadership resources to extend the Kingdom of God in accordance with the Great Commission." Its "Giving Guidelines" reflect its evangelical mission:

1. Be proactive in funding strategic evangelical, Christian organizations. . . .
2. Strengthening strategic Christian organizations led by Godly leaders acting in a responsible, organizational manner. . . .
3. Encouraging other evangelical donors in their giving to such organizations.[34]

Maclellan's evangelical mission clearly includes opposing evolution; the Maclellan money was given to the Discovery Institute with this understanding: "One Discovery funder, Tom McCallie of the Maclellan Foundation in Chattanooga, Tenn., said the foundation awarded $350,000 to the institute in the hopes that researchers would prove 'that evolution was not the process by which we were created.' He said Darwinism has promoted a materialistic world view that he blames for destroying morals and producing tragedies such as the recent school shootings near San Diego."[35] Other Maclellan grantees are Campus Crusade for Christ (CCC), the Family Research Council, Focus on the Family, InterVarsity Fellowship, Josh McDowell Ministries, Chuck Colson's Prison Fellowship Ministries, and Promise Keepers.[36]

Howard Fieldstead Ahmanson/"Fieldstead & Company"

Phillip Johnson dedicates *Defeating Darwinism by Opening Minds* (1997) to "Roberta and Howard, who understood the 'wedge' because they love the truth." Johnson is recognizing CRSC's most generous benefactor, Howard Fieldstead Ahmanson, and his wife, Roberta Green Ahmanson. Ahmanson is heir to his father's Home Savings of America fortune and serves as president of Fieldstead & Company, through which he distributes, helped by eighteen employees, almost $10 million annually to preferred beneficiaries.[37] He is well known for his support of right-wing organizations such as the Free Congress Foundation and James Dobson's Focus on the Family.[38] According to Business Wire, H. F. Ahmanson and Co., the parent of Home Savings of America, was worth more than $47 *billion* in assets in 1997.[39]

Fieldstead & Company has provided the most critical indirect as well as direct financial support to the Wedge. In March 1994, along with InterVarsity Christian Fellowship, Fieldstead sponsored the conference on "Regaining a Christian Voice in the University," where Phillip Johnson delivered his paper "Is God Unconstitutional?" Fieldstead also committed funding for apologetics seminars (a Wedge Document goal) by Wedge members William Dembski and J. Budziszewski at Calvin College. The first, Dembski's "Design, Self-Organization, and the Integrity of Creation" seminar, was held June 19–July28, 2000.[40] According to Anna Mae Bush, the coordinator of Calvin's "Seminars in Christian Scholarship," this was one of four Fieldstead seminars. Ms. Bush informed Joseph Conn of Americans United for Separation of Church and State that Fieldstead asked for Dembski to lead the first seminar.[41] The Wedge's receipt of $1.5 million dollars from Ahmanson (see chapter 6) arouses the suspicion that Wedge members share at least some of his religious inclinations, which deserve closer scrutiny given the fanatical nature of Reconstructionism, with which Ahmanson has been closely allied for many years.

Characteristic of Ahmanson's interests and activities is his long-time membership (until 1995) on the board of R. J. Rushdoony's Chalcedon Foundation, a well-known, extreme rightist, Christian Reconstructionist organization.[42] Frederick Clarkson reports that Ahmanson and Rushdoony also helped found the right-wing Rutherford Institute (recall from chapter 8 that the institute recommended Roger DeHart's attorney). In his book, *Eternal Hostility: The Struggle Between Theocracy and Democracy*, Clarkson describes the radical form of Christianity with which the Wedge has become entangled by accepting Ahmanson's money:

> Generally, Reconstructionism seeks to replace democracy with a theocracy that would govern by imposing their version of "Biblical Law." As incredible as it seems, . . . democratic institutions such as labor unions, civil rights laws, and public schools would be on the short list for elimination. Women would be generally relegated to hearth and home. Men deemed insufficiently Christian would be denied citizenship, perhaps executed. So severe is this theocracy that capital punishment would extend beyond such crimes as kidnapping, rape, and murder to include, among other things, blasphemy, heresy, adultery and homosexuality. . . .
>
> One conservative Christian scholar [Gary Scott Smith] notes that "The Reconstructionist movement has had substantial influence among . . . fundamentalists and evangelical Christians." . . .
>
> Reconstructionist theologian David Chilton succinctly describes this view: "The Christian goal for the world is the universal development of Biblical theocratic republics, in which every area of life is redeemed and placed under the Lordship of Jesus Christ and the rule of God's Law." . . .
>
> Rushdoony and fellow Chalcedon director and funder Howard Ahmanson were among the seven founding directors of the Rutherford Institute. . . .
>
> Harvey Cox, a professor of Divinity at Harvard, acknowledged the centrality of Reconstructionism and dominion theology to the Christian Right in an im-

portant article in the November 1995 issue of the *Atlantic Monthly*: "The thought of Rushdoony's disciples gaining governmental power," Cox concluded, "qualifies as the real nightmare scenario presented by the religious right."[43]

> A key, if not exclusively Reconstructionist, doctrine uniting many evangelicals, is the "dominion mandate," also called the "cultural mandate." This concept derives from the Book of Genesis and God's direction to "subdue" the earth and exercise "dominion" over it. . . . [T]he commitment to dominion serves as a unifying principle while people debate the particulars. . . .
>
> As recently as a few years ago, most evangelicals viewed Reconstructionists as a band of theological misfits without a following. But Reconstructionism has come of age, along with the Christian Right political movement it engendered. Neither evangelicalism nor American politics will ever be the same. . . .
>
> Among those Reconstructionists who have achieved significant power and influence [is] . . . philanthropist Howard Ahmanson, (who has contributed over $700,000 to Chalcedon).[44]

Whether or not Ahmanson shares the most extreme views of Christian Reconstructionism, given the nature of the organizations to which he has devoted his time and resources, there is no need to speculate about the future he wants his money to buy: "My purpose is total integration of Biblical law into our lives."[45] Clearly, he considers the CRSC's agenda fully compatible with his own, otherwise, as Bruce Chapman has said, "Fieldstead . . . would not have given us such substantial funding."[46]

Other Religious Associations

William Dembski, the Wedge's leading intellectual, appears to belong to no secular professional organizations. Although he seeks mainstream academic recognition and approval, he apparently shuns mainstream secular academic and professional societies. A mathematician and philosopher by training, he does not list membership in important philosophical and mathematical organizations such as the American Philosophical Association, the American Mathematical Society, or the Mathematical Association of America. Instead, he has listed at different times during his professional life the following "Professional Associations":

Access Research Network
American Scientific Affiliation
Discovery Institute—research fellow
Evangelical Philosophical Society
Interdisciplinary Biblical Research Institute
Pascal Centre
Trinity Institute
Origins and Design–Associate Editor
Princeton Theological Review–editorial board
Wilberforce Forum
Foundation for Thought and Ethics
Torrey Honors Program, Biola University–advisory board

International Society for Complexity, Information, and Design (ISCID)[47]

All of these organizations have religion as integral, and usually foundational, structural components. This applies even to ISCID, the very description of which incorporates the supernatural: "a cross-disciplinary professional society that investigates complex systems *apart from external programmatic restraints like materialism, naturalism, or reductionism*" (emphasis added) (see chapter 7).

CHRISTIAN LEADERSHIP MINISTRIES

From its inception, the Wedge has worked closely with religious organizations. One of the most prominent is Christian Leadership Ministries (CLM). CLM's Southern Methodist University chapter, Dallas Christian Leadership, co-sponsored the 1992 Wedge conference, Darwinism: Science or Philosophy? CLM also sponsored the important 1996 Mere Creation conference at Biola University, devoting a special edition of *The Real Issue* to "A Special Report" on it. Rich McGee, CLM's director of research and publications, directed the Mere Creation conference in 1996 and the Wedge's Consultation on Intelligent Design in 1997.[48]

CLM is the faculty ministry of the Campus Crusade for Christ (CCC), which has in recent years mounted an energetic proselytizing effort on American campuses. According to the CLM website, the aim is to "disciple and mentor Christian professors" using "specialized strategies, resources, and training" because "our portion of the harvest field is the universities of the world." To enable professors to "continually saturate the campus with the gospel, and fully integrate their faith into their academic discipline and culture," CLM's goal is "500 Faculty Affiliates" (faculty who promote CLM aims), with "at least three Faculty Affiliates at each of the 100 largest universities in America and at least one Faculty Affiliate at each of an additional 200 universities and colleges . . . by the year 2000." The program's purpose is to reach students: "In a recent poll of all the major Christian student organizations, it was discovered that altogether only 30 percent of the entire student population is being exposed to the gospel of Jesus Christ even once in an academic year. . . . A Christian professor sharing a brief Christian testimony the first day of class may be the ONLY Christian witness that many students will hear while they are in college."[49] Clarkson notes that CCC founder Bill Bright has himself embraced radical religious views. For a 1988 "Washington for Jesus" rally in the nation's capital, which Clarkson calls "one of the key events in the politicization of Pentecostalism," Bright helped draft a declaration (which was dropped after being leaked to the press) claiming that "'unbridled sexuality, humanism, and Satanism are taught [in the schools] at public expense'" and calling for "'laws, statutes, and ordinances that are in harmony with God's word.'"[50]

VERITAS FORUM

In addition to CLM, the Wedge is actively assisted by the Veritas Forum (VF), a campus ministry founded at Harvard in 1992 which, according to its website, is "emerging in universities around the world." VF apparently works closely with Christian Leadership Ministries. An example of the association between VF and the Wedge is the February 1997 five-day event at the University of Texas-Austin, "The Search for Truth," organized by Robert Koons. There, Walter Bradley delivered an address entitled "Scientific Evidence for an Intelligent Designer," Phillip Johnson explained "Why Darwinism Is Doomed," and Wedge supporter Alvin Plantinga presented "An Evolutionary Argument Against Naturalism." [51]

One of the chapters apparently most active in assisting the Wedge is at the University of California-Santa Barbara (UCSB). The UCSB VF has launched "Veritas Forum TV," on which Wedge members dominate the programming schedule; the video *Icons of Evolution*, based on Wells's book, was added to the lineup in March 2002. Links take viewers to Access Research Network, where videotapes of CRSC fellows' lectures can be purchased. The University of Florida VF chapter has also hosted both J. Budziszewski and Michael Behe, as well as Reasons to Believe creationist Hugh Ross.[52] Veritas Forum, in fact, advertised its treatment of "origins" issues and hosting of Wedge figures as selling points in a July 7, 2000, fundraising letter to potential supporters, citing topics of its previous events: "Examples of Questions Raised at Veritas Forum Events: What are our origins? . . . Don't faith and science conflict? What about evolution, chaos, biogenetics?" Among the thirty hosted speakers named in the letter, (at least) ten were creationists. Of those ten, seven were CRSC fellows, two were Wedge allies, and the tenth was Hugh Ross.[53]

INTERVARSITY CHRISTIAN FELLOWSHIP

Another contributor to the Wedge agenda is InterVarsity Christian Fellowship (IVCF), which sponsors many of Phillip Johnson's appearances. IVCF, along with Fieldstead & Company, sponsored the 1994 conference "Regaining a Christian Voice in the University." Along with CCC and other campus religious organizations, it sponsored a May 1999 talk on ID by Johnson at the University of California-Davis. InterVarsity Press (IVP) has published almost all of Johnson's books. In fact, it maintains a web page devoted completely to *The Wedge of Truth*, where customers can learn more about Johnson and read book excerpts. There is a similar page for Dembski and Richards's book, *Unapologetic Apologetics*. IVP has published a total of eleven books—more than any other single publisher—written by the Wedge and its allies (see chapter 6).[54]

TRINITY COLLEGE

Trinity College of Florida, also with close Wedge ties, sponsored the November 2000 conference, "Darwinism, Design and Democracy: A Conference on Scientific Evidences and Education," at which CRSC fellows Paul Chien and Scott Minnich were featured speakers. Its Center for University Ministries is run by Tom Woodward, as is the affiliated C. S. Lewis Society. CRSC fellows Robert Kaita and Charles Thaxton are on the center's board of advisors, as is Jon Buell (Foundation for Thought and Ethics). Through its *Princeton Chronicles* videos, Trinity promotes Kaita and Michael Behe. In early 1999, Woodward hosted the Chinese scientist whom the CRSC had cultivated because of his connection with the Chengjiang fossils.[55]

Alliances with Theocratic Extremists

A more overtly political aspect of the Wedge's religious alliances is its close ties with powerful Religious Right figures such as James Dobson, D. James Kennedy, and Beverly LaHaye. In October 2001, William Dembski and Mark Hartwig (who also works for Dobson's Focus on the Family) were guests on Dobson's radio program in a segment entitled "God's Fingerprints on the Universe."[56]

JAMES DOBSON

Dobson is an ID supporter who also featured Phillip Johnson in September 1996. His invitation to Dembski was a direct response to the PBS series, *Evolution*, which ran the week before Dembski's visit. Dobson opened his show by expressing concern over the series: "A week ago . . . we expressed great concern and some irritation here on Focus on the Family over a PBS television series that was airing at that time. . . . It was little more than evolutionary propaganda because no new evidence for Darwinian naturalism was presented. It was just a rehash of the old argument, but done in a way to manipulate the minds of kids. . . . This series will be seen in public schools all across the country and probably throughout the [W]estern world. . . . We felt it needed to be answered, and we needed to provide some information and some understanding so that parents of kids in public schools can prepare them for what's coming."[57]

Dobson held a question-and-answer segment featuring Dembski and Hartwig in his "welcome center," attended by about 175 people, to educate listeners on the problems with evolution and their solution—

intelligent design: "Wouldn't it be interesting to hear a discussion of evolution from a more objective point of view? . . . We have brought Dr. Hartwig . . . and Dr. William Dembski, who is associate research professor in the conceptual foundations of science at Baylor University. He has Ph.D.'s in both philosophy and mathematics. He's a very bright man. We've become friends, and I invited him to come here . . . to talk about the evidence, talk about our understanding of scripture, and to talk about this entire subject from a scientific point of view." The rest of the interview consisted of Dembski's (and, to a lesser extent, Hartwig's) fulminations on the failure of evolutionary science and the promise of ID as its replacement.

Americans United for Separation of Church and State devoted much of its May 1998 issue of *Church and State* to Dobson. According to this organization, his radio show is broadcast on over 4,000 stations worldwide, 1500 of them in the United States and Canada. Gil Alexander-Moegerle, a former staffer who helped found Focus on the Family but has broken with Dobson, asserts that Dobson is a political operative who conceals his undertakings behind the façade of what he promotes as a family ministry: "Alexander-Moegerle notes that a good chunk of Dobson's daily radio broadcast is political in nature, as is the monthly *Focus on the Family* magazine."[58] Promoting the Religious Right propaganda that church-state separation is a myth, Dobson sees America as gravely threatened by "Secular Humanism" and has fashioned his ministry to meet this threat. Alexander-Moegerle argues that Dobson wants America to be a fundamentalist theocracy: "I believe it is accurate to say that Jim wants theocracy. . . . Jim really believes . . . America is on it last legs . . . gasping its last breaths. Jim is desperate and fearful. And he really believes that only a theocracy can save us."[59]

D. JAMES KENNEDY

D. James Kennedy, leader of Coral Ridge Ministries, is another Religious Right operative who champions ID. Phillip Johnson spoke at his 1999 "Reclaiming America for Christ" conference, at which Americans United reports that "several speakers discussed strategies for injecting fundamentalism into the schools, and one outlined an ambitious agenda for using public school students to spark a revival in America."[60] Kennedy featured Dembski on his "Truths that Transform" radio broadcast on February 25, 2002, at which listeners could "learn how this scientist has had to fight for a voice against those who would stifle science."[61] He had earlier featured Michael Behe and Johnson: "Both of these gentlemen . . . have been very helpful in developing what is called 'intelligent design theory,' or the intelligent design movement. Our guest today [Dembski] has done more than his part in dispelling the myth of Darwinist philosophy. He has experienced the ire of what he calls 'Darwin dogmatists.'"

Americans United reports that Kennedy's TV ministry, the "Coral Ridge Hour," is carried by more than 500 stations. "Truths that Transform" is also heard on more than 500 radio stations. Calling church-state separation "diabolical," a "false doctrine," and "a lie," Kennedy established the Center for Christian Statesmanship on Capitol Hill, which conducts Bible studies and other evangelical activities for members of Congress and their staffs. He is an ardent foe of evolution, condemning it in his TV sermons. In his 1994 book *Character and Destiny: A Nation in Search of Its Soul,* he writes, "Every new advance and every step taken by science confirm not evolution but the Genesis account of creation. Yet evolution still continues to be taught as fact. . . . Thus, the honorable place that had been given to human beings by God is surreptitiously aborted, and they are dragged down into the slime" (p. 178).[62]

BEVERLY LAHAYE

Beverly LaHaye, founder with her husband Tim LaHaye of Concerned Women for America (CWA), has supported the Wedge through her magazine, *Family Voice*. Tim LaHaye is co-author with Jerry Jenkins of the popular *Left Behind* novels, based on LaHaye's vision of the coming of the Antichrist as described in the New Testament's "The Revelation of St. John the Divine." He is described by Americans United as "a fundamentalist extremist who hates church-state separation, seeks a government-enforced 'Christian nation' and who has a long track record of attacking other religions and promoting bizarre conspiracy theories." LaHaye believes public schools should be "turned into centers for fundamentalist indoctrination with daily prayer, promotion of the Ten Commandments and creationism firmly ensconced." Like Jonathan Wells, both LaHayes have had connections to the Rev. Sun Myung Moon:

> LaHaye was . . . damaged [in the 1980s] by revelations that he had accepted money from the Rev. Sun Myung Moon. . . . Bo Hi Pak, a longtime Moon operative, gave ACTV [LaHaye's American Coalition for Traditional Values] $10,000, and LaHaye subsequently agreed to serve on the board of directors of Moon's own Religious Right group, Christian Voice. LaHaye also joined the board of another Moon front, the Council for Religious Freedom (CRF), which was formed primarily as a vehicle to protest Moon's 1984 imprisonment after he was convicted of filing false tax returns and obstructing justice.
>
> The LaHaye-Moon tie was laid bare after a fawning letter surfaced that LaHaye had penned to Pak thanking him for the $10,000. LaHaye . . . tried to distance himself from Moon, but the damage was done. . . . Despite the flap, LaHaye never did sever all ties to Moon. Beverly LaHaye spoke at a Moon event in Washington, D.C., as recently as 1996.[63]

The Wedge is featured in the September/October 2001 *Family Voice*. Dembski is highlighted as a "leader in the ID movement" whose "credentials are impressive." Also featured is IDnet's John Calvert, who, citing

Romans 1:18–32 and speaking for God, informs readers that "God is using [the ID debate] to bring people to know Him better." *Family Voice* indicates how readers can help the ID movement:

> *Pray* design theorists will find open doors to the science community to publicize their research.
>
> *Praise* God for the wisdom He has given researchers to discover His hand in our world.
>
> *Act* Talk to your children about what they learn in science class; supplement their education with materials on design theory.[64]

These alliances mean that the Wedge is associated with some of the most extreme factions of the Religious Right network. Because of their radical views, Frederick Clarkson devotes a separate section in *Eternal Hostility* to just these three and their respective organizations, which he places "in the forefront of the Christian Right's drive for political power": "[T]hey are increasingly ideological, animated by what is generally referred to as a 'Biblical Worldview.' Although it is usually only vaguely defined even by those who use the term, it is often a code word for theocracy."[65]

Intelligent Design: Mere Creationism

William Dembski angrily denies the charge that ID is "stealth creationism": "Ask any leader in the intelligent design movement whether intelligent design is stealth creationism, and they'll deny it."[66] But such remarks, made in the moderate, more mainstream atmosphere of the Metaviews discussion list, are contradicted in Dembski's writings such as *Intelligent Design: The Bridge Between Science and Theology* that he aims at his religious audience. Here, in chapter 8, "The Act of Creation," he dare not hide his creationism, which ties him to his supporters:

> [T]he Christian tradition plainly asserts that God is the ultimate reality and that nature itself is a divine creative act. . . .
>
> The aim of this chapter then is to present a general account of creation that is faithful to the Christian tradition, that resolutely rejects naturalism. . . .
>
> Now within Christian theology there is one and only one way to make sense of transcendent design, and that is as a divine act of creation. I want therefore next to focus on divine creation, specifically on the creation of the world. My aim is to use divine creation as a lens for understanding intelligent agency generally. God's act of creating the world is the prototype for all intelligent agency. . . . Indeed all intelligent agency takes its cue from the creation of the world. . . . God's act of creating the world is thus the prime instance of intelligent agency.[67]

Likewise, Phillip Johnson, in an interview with Ted Koppel on ABC's *Nightline* on July 27, 2000, made it crystal clear that ID *is* creationism:

Ted Koppel: Mr. Johnson, . . . Mr. [Ralph] Neas [of People for the American Way] was making the point that . . . the theory you are citing, far from being more modern than the Darwinian theory, actually pre-dates it by about 50 years. . . . I'm asking you whether this modern theory that you were citing to me is in fact 50 years older than the Darwinian theory?

Phillip Johnson: The idea of divine creation, which is what I'm talking about, [is] that there is an intelligence behind life.[68]

Dembski's denial that ID is creationism rests upon his identifying "creationism" exclusively with "young earth creationism," as in this November 2000 post to Metaviews:

> By creationism one typically understands what is also called "young earth creationism." . . . Given this account of creationism, am I a creationist? No. I do not regard Genesis as a scientific text. I have no vested theological interest in the age of the earth or the universe. . . . Nature, as far as I'm concerned, has an integrity that enables it to be understood without recourse to revelatory texts. That said, I believe that nature points beyond itself to a transcendent reality, and that that reality is simultaneously reflected in a different idiom by the Scriptures of the Old and New Testaments."[69]

He avows, in other words, that he is not a creationist but merely a believer in the intelligent design of the world, by which is meant the act of creation by a "transcendent" intelligent agent. For strategic reasons, most ID creationists resist being pinned down on their views regarding the age of the earth, as Robert Pennock notes:

> Intelligent-design theorists have learned a few lessons from the failures of their predecessors and have devised a more sophisticated strategy to compete head on with evolution. One of the main things they have learned is what not to say. A major element of their strategy is to advance a form of creationism that not only omits any explicit mention of Genesis but is also usually vague, if not mute, about any of the specific claims about the nature of Creation, the separate ancestry of humans and apes, the explanation of earth's geology by a catastrophic global flood, or the age of the earth—items that readily identified young-earth creationism as a thinly disguised biblical literalism.[70]

CRSC fellow Ray Bohlin provided a wonderful example of Wedge double-talk to avoid specifying the age of the earth when he answered an e-mail that was sent to Probe Ministries: "I just read over your article on the 'Age of the Earth' [written with former Probe research associate Rich Milne] to get Probe's stand on the issue. Apparently, the official stand is no stand. I was wondering after I read this statement of yours, 'Biblically, we find the young earth approach . . . to make the most sense. However, we find the evidence from science for a great age for the universe and the earth to be nearly overwhelming.'" Bohlin answered in "Wedgespeak," saying that the affirmation that a young earth "makes the most sense Biblically" does not mean that he and Milne accept a young earth as "the 'clear' written revelation of Genesis 1." He asserts that he would accept a young earth regardless of the scientific evidence if Genesis were clear on this point. Bohlin next declared it significant that "most young earth geologists and physicists" acknowledge that radioactive dating *does* show Earth's layers graduating from older to younger as one surveys from bottom to top, and that while young-earth scientists do not accept the dates yielded by radioactive dating, "the sequence seems real." Bohlin concluded with studied ambiguity: "Therefore the dating methods are not totally without merit. This is more than just suggestive."[71]

Although most ID proponents are of the old-earth kind, there is a significant contingent of young-earth creationists in both the CRSC and the ID movement in general, despite Dembski's public attempts to distance ID from them. Paul Nelson and John Mark Reynolds are well known young-earth creationists. And there are others: Percival Davis (coauthor of *Pandas*), Nancy Pearcey, Ray Bohlin (his evasiveness notwithstanding), Charles Thaxton—all are listed on Henry Morris's "A Young-Earth Creationist Bibliography."[72] Siegfried Scherer, a CRSC fellow in Germany, is also one. He believes that since "the structure of the intermediate sedimentary layers clearly indicates their formation due to a catastrophe" (the Biblical flood), coal "from the Carboniferous and Permian periods" was formed within 6–10,000 years "entirely from floating forests," which "had to live on the water surface next to each other prior to the Flood."[73] Clearly, Johnson's "big tent" is cozier than most people realize—and ID is much more conventional in its creationist views than most people realize.

A historical look at the ID movement demonstrates its traditional creationist ancestry, and a comparison of ID with young-earth "scientific creationism" reveals close kinship in tactical approach, rabid anti-evolutionism, and the actual content of the respective belief systems.

Continuities and Commonalities

ID's historical continuities and current congruences with creationism make it clear that "intelligent design" is simply a new emphasis on the main element of the old creationism—a new title for an old argument.

HISTORICAL CONTINUITIES

The Wedge's eagerness to disconnect ID, at least formally, from creationism and religion has usually been quite pronounced. Responsible for this, at least in part, is their entry into the national spotlight since the Kansas episode and publication of their legal strategy in the *Utah Law Review*, where, to advance their case judicially, they must for tactical reasons argue that ID is *not* creationism: "Recall that in *Edwards v. Aguillard* . . . the [Supreme] Court decided against the legality of scientific creationism because it constituted an advancement of religion. . . . [T]he Court's decision does not apply to design theory because design theory is not based upon a religious text or doctrine."[74] The reason for such a disconnection was not lost on D. James Kennedy during his February 2002 interview of Dembski:

> Kennedy: Now, I understand that the Darwinian advocates use the term [intelligent design] differently than those that are actually associated with the movement. How is it used by Darwinians? What's the difference?
> Dembski: Well, the phrase "intelligent design," you mean. I think the Darwinians will want to identify intelligent design with creationism. And I think there's a crucial distinction to keep in mind. Intelligent design is certainly friendly to the Christian doctrine of creation, but they're not identical. . . . [W]e're just saying there's an intelligence there, we can know it, and we do science with it.
> Kennedy: Well, it seems to me that there are several advantages to that. . . . First of all, it keeps them [Darwinian advocates] from throwing intelligent design out, say, out of the courts as being merely religious, which they have frequently done, in Arkansas and the Supreme Court, saying creation is religious and evolution is scientific.[75]

Notwithstanding Dembski's disavowals, we learn much by looking at earlier days, both before the formal beginning of the movement and shortly after, when the Wedge was less in the public eye and consequently less cautious about disclosing its creationism.

NEW WORDS FOR (VERY) OLD IDEAS

Walter Bradley, a retired mechanical engineering professor, is not one of the higher-profile members of the Wedge. Yet he co-authored, with Charles Thaxton and Roger Olsen, a seminal creationist book, *The Mystery of Life's Origin* (1984). A 1984 paper by Bradley, also co-authored with Olsen, "The Trustworthiness of Scripture in Areas Relating to Natural Science," is available on Christian Leadership Ministries' "Origins" website. The paper is a reprint from an anthology entitled *Hermeneutics, Inerrancy, and the Bible* (Zondervan, 1984). In this paper, which is religious apologetics, Bradley and Olsen distinguish two types of creationists: "mature" creationists, who "see God working only through fiat miracle," and "progressive" creationists, who see "God working . . . through a combination of miracle plus [natural] process."[76] They identify themselves as progressive creationists, but it is clear that everywhere, except concerning the age of the earth, the Bible is the ruling framework for their arguments:

> In this paper we . . . focus on the interpretation of the Hebrew words "yom" and "bara/asah" . . . used in . . . Genesis to describe the time frame and mechanism of creation. . . .
>
> The goal is to first define the latitude of permissible interpretation of the biblical account of origins. Then God's revelation of his world as perceived through . . . science will be used to identify the best . . . interpretation of origins *within the prescribed boundary. This methodology allows the authoritative position of Scripture to be maintained* while taking advantage of insights from sci-

entific studies to supplement our understanding where . . . Genesis 1 . . . is ambiguous. (emphasis added)

They never use the term "intelligent design," yet elements of what is now called intelligent design are found throughout the article, and elements of what Bradley and Olsen call "progressive creationism" are omnipresent in Dembski's work (and certainly in the work of other ID creationists), even though Dembski *disavows* progressive creationism in a 1996 paper, "What Every Theologian Should Know About Creation, Evolution and Design," on the same website: "Nor can it be said that design theory endorses progressive creation."[77] Moreover, the 1984 Bradley paper rejects theistic evolution, which Dembski, too, rejects (e.g., in the cited paper).

As described by Robert Pennock, progressive creationism accepts much of what science shows about the universe's development, including the idea that most of this development proceeded according to natural laws. With respect to other things, however, especially life on earth, progressive creationists believe God intervened directly at crucial points, all of this creation taking place over millions of years. Progressive creationism shares much with other forms of old-earth creationism, and it is sometimes attacked by young-earth creationists for departing from the young-earth view.[78] Bradley's "progressive creationism" is essentially old-earth creationism that requires the special creation of humans and the historicity of Adam and Eve. Dembski, likewise, in "What Every Theologian Should Know," accepts an old earth but asserts "that human beings were specially created."

Bradley and Olsen also reject macroevolution: "[P]rogressive creationism suggests that God created the major types of animal and plant life at various times in geological history in a miraculous way and then worked through process . . . to develop the . . . variety of plant and animal life we see today. To accept the compelling evidence for geological age should not be equated with accepting the general theory of evolution (macroevolution). . . . Some . . . mechanism more efficient than chance must be invoked." Again, Dembski echoes the same views twelve years later: "No, I don't believe in fully naturalistic evolution controlled solely by purposeless material processes [thus he accepts miraculous intervention], and Yes, I do believe that organisms have undergone some change in the course of natural history (though I believe that this change has occurred within strict limits)," meaning that he rejects macroevolution.

Last, just as progressive creationism is a combination of major miracles and relatively minor natural processes, so Dembski identifies this as his preferred view:

Finally, my own metaphysical preference is to view creation as an interrelated set of entities, each endowed by God with certain inherent capacities to interact with other entities. In some cases these inherent capacities can be described by natural laws. Nevertheless, no logical necessity attaches to these laws, nor for that matter to the inherent capacities. On this view God freely bestows capacities and can freely rescind them, not least the capacity to exist. *In performing a miracle, God overrules the inherent capacities of an entity, endowing the entity with new capacities.*[79] (emphasis added)

In short, for Dembski, as for Bradley and Olsen, God works through "miracle plus process."

However, not only are the traditional elements of Dembski's "new" creationism evident in the Bradley article, but two other elements commonly thought unique to ID are here as well: Dembski's argument that "complex biological information" cannot increase is prefigured here, and Behe's special emphasis on "molecular machines" is also prefigured in Bradley and Olsen's "metabolic motors." Here is complex biological information as it shows up in the Bradley/Olsen article:

Having noted the lack of clear support for macroevolution in the fossil record, we turn our attention . . . to the proposed mechanism for evolution: . . . mutation/natural selection. Mutations are the result of replicating mistakes along the DNA chain. Such alterations change the information coding implicit in the base sequence along the chain. . . .

The important question, however, is . . . whether mutations can give rise to . . . increasing information content along the DNA chain resulting in increasing complexity of the organism. . . .

Not only is the model of random mutations along the DNA chain untenable statistically as a means of increasing the information content . . . , it is without experimental verification.[80]

In other words, natural processes cannot increase the information content of DNA, as Dembski claims to have discovered, eighteen years later, in *No Free Lunch: Why Specified Complexity Cannot Be Purchased Without Intelligence*: "Algorithms and natural laws are in principle incapable of explaining the origin of CSI [complex specified information]. To be sure, algorithms and natural laws can explain the flow of CSI. Indeed, algorithms and natural laws are ideally suited for transmitting already existing CSI. . . . [W]hat they cannot do is explain its origin."[81]

Next, Behe's irreducible complexity as it applies to "molecular machines" is found in progressive creationism as well, in the form of "metabolic motors"—there is even an unsuccessful literature survey. Bradley and Olsen discuss the prebiotic conditions necessary for the beginning of the metabolic processes, or pathways, essential to life:

Increasing chemical complexity . . . represents the single biggest challenge to explain by natural processes. . . .

In the formation of the basic macromolecules essential for the biochemical function of living systems (e.g., DNA, protein, etc.), two types of "work" must be done. First the chemical energy of . . . macromolecules must be increased over that of the simple building blocks from which they are constructed. Second, the . . . blocks must be arranged in a . . . specific sequence to achieve proper function in the system . . . the same way letters must be arranged in . . . specific sequences . . . to give a useful function, i.e., communicating information. We will call the first type of work chemical work and the second type of work coding work.

One of the authors . . . recently surveyed the literature on attempts to synthesize DNA and protein in the laboratory. . . . The conclusions from this survey were . . . that no successful experimental results have . . . been achieved, and . . . that the coding work is far more difficult to do than the chemical work. . . .

A "metabolic motor" is involved in the initial conversion of solar energy into chemical energy and a second "metabolic motor" is required to allow the combustion of these chemicals . . . [so] that . . . energy released may be coupled to the . . . chemical and coding work required by the organism. . . . [T]he "metabolic motors" in various plants and animals . . . have . . . common components such as DNA . . . and proteins . . . that channel . . . energy through the system and regulate the rate at which it is released. Apart from such a . . . motor, coupling of energy flow through the system to do the necessary ordering work would not be possible.

This is . . . why prebiotic . . . synthesis experiments fail. "Raw" energy does not seem . . . capable of doing the . . . work necessary to construct the complex macromolecules of living systems. . . . [T]he uniform failure of . . . thousands of . . . attempts to synthesize protein or DNA under prebiotic conditions is a monument to the difficulty in achieving . . . high . . . information content, or order from the undirected flow of energy through a system. . . .

[Michael Polanyi] has correctly noted that the laws of chemistry and physics as presently understood cannot explain the existence of machines or living systems. Though . . . living systems may be described in terms of . . . chemistry and physics, it is in . . . fixing of the boundary conditions (e.g., making the first metabolic motor) that the necessary function . . . becomes possible. Man designing and building an . . . engine constitutes . . . a fixing of . . . boundary conditions for the chemical energy in gasoline to be converted into useful work. In a similar way the metabolic motor common to all living systems allows maintenance of the . . . systems in . . . highly ordered conditions. . . .

[O]ne *cannot* dismiss the possibility of ordering simple chemicals into complex . . . organisms by an appeal to the Second Law of Thermodynamics since the earth is an open system. However, an open system without some mechanism to couple the energy flow through the system . . . is equally unacceptable as a model for the origin of life. Either there is some . . . undiscovered energy coupling mechanism or self-ordering mechanism, or . . . God accomplished this part of his creation in a supernatural way.

In other words, just as there must be a machine, a motor, to do the work of converting gasoline into usable energy for a car, there must be a "metabolic motor"—Behe's "molecular machine"—to do the energy conversion work in a cell, the essential condition for life.

As for how this process takes place, there are only two alternatives: either it occurs naturalistically or by divine intervention. The correct alternative here is not in doubt for Bradley and Olsen, just as it is not in doubt for Behe when he discusses metabolic pathways:

> To build a house you need energy. Sometimes the energy is just in the muscles of the workers, but sometimes it is in the gasoline that powers bulldozers or electricity that turns drills. The cell needs energy to make AMP [adenosine monophosphate]. . . .
>
> To make AMP from the ingredients that the cell uses we also need very high-tech equipment: the enzymes that catalyze the reactions of the pathway. . . . The point is that even if adenine or AMP can be made by simple pathways, those pathways are no more precursors to the biological route of synthesis than shoes are precursors to rocket ships. . . .
>
> AMP is required for life on earth: it is used to make DNA and RNA. . . . There may be some way to construct a living system that does not require AMP, but if there is, no one has a clue how to do so. The problem for Darwinian [naturalistic] evolution is this: if only the end product of a complicated biosynthetic pathway is used in the cell, how did the pathway evolve in steps? . . . On their face, metabolic pathways where intermediates are not useful present severe challenges to a Darwinian scheme of evolution.[82]

And Behe's characterization of cellular structures as machines and motors is now a cliché: "What has biochemistry found that must be explained? Machines—literally, machines made of molecules. . . . The flagellum is an outboard motor. . . . It consists of a rotary propeller, motor, and stationary framework. . . . Darwin's theory is completely barren when it comes to explaining the origin of the flagellum or any other complex biochemical system."[83] So what is the alternative to the barrenness of Darwin's theory? As we know the alternative chosen by Bradley and Olsen, so we know it for Behe, as well.

Just as Dembski's and Behe's ID concepts of information and molecular machines are prefigured in Bradley and Olsen's 1984 progressive creationist concepts of information and metabolic motors, so *all* are prefigured in Henry Morris's 1974 work, where we find Morris speaking of the inability of mutation and natural selection either to generate information or to provide the "marvelous motor" needed to produce the chemicals necessary for cell replication. Morris concedes that there *are* systems in which order appears to increase, a fact that seems to violate the Second Law of Thermodynamics. He adds that, for such systems to "supersede the Second Law locally and temporarily," two criteria must always be met:

> (a) *There must be a program to direct the growth.*
>
> A growth process which proceeds by random accumulations will not lead to an ordered structure. . . . Some . . . pattern, blueprint, or code must be there to begin with, or no ordered growth can take place. In the . . . organism, this is the intricately complex genetic program, structured as an information system into the DNA molecule for the particular organism. . . .
>
> (b) *There must be a power converter to energize the growth.*
>
> The available environmental energy is of no avail unless it can be converted into the . . . forms needed to organize and bond the components into the complex and ordered structure of the completed system. . . .
>
> The evolutionary process, if it exists, is . . . the greatest growth process of

all. If a directing code and specific conversion mechanism are essential for all lesser growth processes, then surely an infinitely more complex code and more specific energy converter are required for [evolution]. . . .

But . . . no such code and mechanism have ever been identified. Where in . . . the universe does one find a plan which sets forth how to organize random particles into particular people? And where does one see a marvelous motor which converts the . . . solar . . . energy bathing the earth into the work of building chemical elements into replicating cellular systems . . . ? . . .

[M]utation and natural selection are . . . inadequate for such a gigantic task. Mutation is not a code, but a random phenomenon. Neither can it assimilate energy into a more . . . organized form. . . . Natural selection is not a code which directs the production of anything new; it . . . merely . . . sieves out unfit variants and defective mutants. It certainly is not an energy conversion device.[84]

The absence of response until very recently from serious scientists to such arguments is not evidence of their having ignored them. On the contrary, the arguments have been examined and found unworthy of the time and effort needed to reply in kind and at length. ID does, however, exist quite obviously in a "big tent," not only with progressive creationism but with young-earth creationism as well. The Wedge itself is a microcosm of that tent, populated, as we have seen, by a significant number of young-earth creationists. And none of these commentators on science has learned anything new about the science upon which they comment.

We may look at an early essay by Phillip Johnson, "Evolution as Dogma: The Establishment of Naturalism," which predates Johnson's first book, *Darwin on Trial* (1991), by a year. Here we find all the same cavils and complaints about evolution and naturalism as in his later work but never the term *intelligent design*. The terms *artificer* and *designer* are used, but that is not a distinguishing feature. "Scientific creationists" have always invoked "design." "Evolution as Dogma" is a defense of creationism, and no distinction is made between creationism and intelligent design theory per se, since the latter term is not yet a part of Johnson's terminology. Rather, the distinction is between creationists who are biblical literalists and those who are not. But according to Johnson, a creationist is "any person who believes that God creates"; and he specifies that, with respect to biblical fundamentalists, "I do not myself think that such advocacy groups should be given a platform in the classroom." However, his urging in this 1990 article is for creationist solidarity. The following, a defense of the creationists who have so far lost in the courts and an indictment of the "Darwinism" that has defeated them, is an example. It is indistinguishable from Johnson's later speechmaking on "intelligent design."

Victory in the creation-evolution dispute . . . belongs to the party with the cultural authority to establish the ground rules that govern the discourse. If creation is admitted as a serious possibility, Darwinism cannot win, and if it is excluded *a priori* Darwinism cannot lose. The point is illustrated by the logic

> which the Natural [*sic*] Academy of Sciences employed to persuade the Supreme Court that "creation-scientists" should not be given an opportunity to present their case against the theory of evolution in science classes. Creation-Science is not science, said the Academy, because "it fails to display the most basic characteristics of science: reliance upon natural explanations. . . ."
>
> Creationists are disqualified from making a positive case, because science is based by definition upon naturalism. The rules of science also disqualify any purely negative argumentation designed to dilute the persuasiveness of the theory of evolution. Creationism is thus ruled out of court and out of the classroom—before any consideration of evidence. . . .
>
> With creationist explanations disqualified at the outset, it follows that the evidence will always support the naturalistic alternative.[85]

In *Darwin on Trial* (1991), we see the same defensiveness. Though Johnson has "very little" to say about "creation-scientists" in the book,[86] he uses "intelligent design" only four times and never defines ID, though he takes pains to define both "creationism" and "naturalism." In "What Is Darwinism?" in 1992, we see much the same stance as in the 1990 article and *Darwin on Trial*, except that intelligent design is a more prominent element of the discussion. But the key terms Johnson wants specifically to define are "creationism, evolution, science, religion, and truth," *not* "intelligent design theory," and the solidarity with creationism remains: "In a broader sense, . . . a creationist is simply a person who believes in the existence of a *creator*, who brought about the existence of the world and its living inhabitants in furtherance of a *purpose*. Whether the process of creation took a single week or billions of years is relatively unimportant from a philosophical or theological standpoint. Creation by gradual processes over geological ages may create problems for Biblical interpretation, but it creates none for the basic principle of theistic religion."[87]

A survey of the historical continuities between the current version of ID and its earliest manifestations shows that, despite its adherents' wordy protestations, the differences are of terminology only, not substance. Finally, the historical link to ID as creationism comes most directly from the Wedge's own Jonathan Wells. In 1996, when he was already one of the first CRSC fellows, he published an article in *The World and I* (a Moon magazine); this article was also reprinted on an earlier CRSC web page. Here he distinguished between creationists and evolutionists but notably did not list a separate category for ID:

> Not only are there many distinct issues, but there are also many parties to the debate. For example, there are young-earth creationists . . . ; there are other creationists who accept the geological time scale but reject the notion that all living things are descended by natural mechanisms from a common ancestor; there are theistic evolutionists . . . ; there are evolutionists who see no need for God's involvement but question one or more aspects of Darwin's theory; and there are other evolutionists who substantially agree with Darwin's theory and its atheistic implications. Even these five categories (which . . . are somewhat artificial) do not do justice to the complexity of the controversies, but they will help to illustrate it.

Wells allowed no separate category for intelligent design because it was not then important to him to do so. The CRSC was newly instituted; the important task at that point was to establish its creationist identity with its most important constituency. But there is more:

> *The most vocal advocates of design in the creation-evolution controversies, however, are creationists* rather than theistic evolutionists. The creationists' arguments sometimes resemble Paley's; that is, design in living organisms is used to prove the existence of God. Evolutionists object that such arguments are unscientific because they invoke the supernatural. . . .
>
> *Creationists such as law professor Phillip Johnson* object that using biological evidence to argue against design or a designer is no more scientific than arguing for design or a designer, and charge evolutionists with misusing science as a platform for their atheistic beliefs.[88] (emphasis added)

We could not have said it better.

CURRENT COMMONALITIES

An analysis of American creationism of all varieties reveals a number of shared characteristics: (1) belief in the creation of the universe by a supernatural designer and (usually) the designer's *continuing intervention* in the creation; (2) implacable anti-evolutionism, stemming from opposition to the scientific consensus on the evolution of the universe and life, such opposition being based on theological, moral, ideological, and political, but never scientific grounds; (3) criticism of all or most methodologies underpinning current scientific evidence for the evolution of life, without presenting for peer review any competing theory of origins; and (4) the most fundamental aspect of creationism: the explicit or implicit grounding of anti-evolutionism in religious scripture.[89] And, of course, there is the indefatigable *political* effort to influence and ultimately to rewrite school science curricula.

ID meets all these criteria, but rarely is the opposition of ID to evolutionary theory and evidence so clearly acknowledged as in Phillip Johnson's statements in a 2001 interview with the *Australian Presbyterian*: "Again, I think it's fair to say that the questions [*sic*] of whether Darwinism leads to immoral consequences is a side issue too. Please, don't get me wrong here. It *has* led to many immoral consequences. . . . But the real question is whether it's true or not; not whether Darwinism has had undesirable results. If it's true, it's still true, even if it has had undesirable effects. *But the important thing is that evolution is not true*" (emphasis added). Even more recently, in a 2003 interview with Hank Hanegraaff, Johnson asserted flatly that "evolution is basically a hoax."[90]

When we look at specific characteristics of "scientific creationism" and ID, we see that the major planks in the ID platform are continuous with those in the older, more traditional creationism. Robert Pennock has commented on these with respect to Dembski's work:

> [A]ll the key elements of Dembski's argument . . . were previously articulated by Norman Geisler, who was one of the expert witnesses for the creationist side in the famous "balanced treatment of creation-science and evolution-science" case in Arkansas in the early 1980s. Today the ambiguous term they use is "designer" instead of "creator," but otherwise the game plan of the "design theorists" is little different than that of the "creation scientists." . . .
>
> One might occasionally get the impression that Dembski does not reject evolution, as when he says . . . somewhat vaguely, that he accepts "that organisms have undergone some change in the course of natural history." However, that impression is dispelled when he immediately qualifies this and asserts that "this change has occurred within strict limits." Dembski's position is indistinguishable from that of Henry and John Morris and other creationists, who put the point in exactly the same terms.[91]

The first commonality between "creationism" and the current ID is in the definitions of creationism offered by "scientific creationism" founding father Henry Morris and by Phillip Johnson; they are indistinguishable in substance. Morris defines "the creation model" as "(1) supernaturalistic; (2) externally directed; (3) purposive, and (4) completed."[92] Johnson's most detailed definition calls creationism a belief that "[1] a supernatural Creator not only [2] initiated this process [of the evolution of life forms from simple to complex states] but [3] in some meaningful sense *controls* it [4] in furtherance of a purpose." Lest this latter give the mistaken impression that it *includes* theistic evolution, Johnson adds that "'evolution' (in the contemporary scientific usage) excludes not just creation-science but creationism in the broad sense."[93] Like Morris's definition, Johnson's includes supernaturalism, external direction (control), and purpose. Since Johnson rejects common ancestry and "macroevolution," one could say that his definition implies that the creation of life is "completed" as well.

We also find commonalities of content. In one statement, Dembski touches all of the major areas of agreement: "The following problems have proven utterly intractable not only for the mutation-selection mechanism but also for any other undirected natural process proposed to date: the origin of life, the origin of the genetic code, the origin of multicellular life, the origin of sexuality, the absence of transitional forms in the fossil record, the biological big bang that occurred in the Cambrian era, the development of complex organ systems and the development of irreducibly complex molecular machines." To anyone reasonably familiar with the modern *science* of any of Dembski's "natural processes," his statement is nonsense. He is saying that neither natural selection *nor any other natural* process can account for any of these things. Creationist polemicists Duane Gish and Henry Morris agreed, in books from 1978 and 1974, respectively:

Origin of Life

[M]ost evolutionists insist that creation be refused consideration as a possible explanation for origins. Creation has not been witnessed by human observers, it cannot be tested experimentally. . . . The general theory of evolution also fails to meet . . . these criteria, however. It is obvious, for example, that no one observed the origin of the universe, the origin of life.

Gish, *Evolution: The Fossils Say No!* (p. 13)

There is no slightest *scientific* evidence that life can come from non-life. The creation model emphasizes the unique origin of life, at the creative word of a *living* Creator. The scientific law of cause and effect requires the First Cause of life to be living!

Morris, *Scientific Creationism* (p. 51)

Genetic Code

[T]he concept of special creation does not exclude the origin of varieties and species from an original created kind. It is believed that each kind was created with sufficient genetic potential, or gene pool, to give rise to all of the varieties within that kind that have existed in the past and those that are yet in existence today.

Gish, *Evolution: The Fossils Say No!* (p. 40)

[E]ven if such a [protein] molecule could ever be formed by chance, it could never reproduce itself. The fact that the DNA molecule is necessary for reproduction and that it can only operate in the presence of proteins which it had previously specified and organized seems to be an impenetrable barrier to this vital phase of evolution. . . .

[H]owever, this is no problem to the creationist. The creation model predicts that life can come only from life.

Morris, *Scientific Creationism* (p. 49)

Absence of Transitional Forms

While transitions at the subspecies level are observable and some at the species level may be inferred, the absence of transitional forms between higher categories (the created kinds of the creation model) is regular and systematic.

Gish, *Evolution: The Fossils Say No!* (p. 150)

All orders and families (as well as kingdoms, phyla and classes) appear suddenly in the fossil record, with no indication of transitional forms from earlier types.

Morris, *Scientific Creationism* (p. 79)

The Cambrian Era

What do we find in rocks older than the Cambrian? *Not a single, indisputable, metazoan fossil has ever been found in Precambrian rocks!* Certainly it can be said without fear of contradiction that the evolutionary ancestors of the Cambrian fauna, if they ever existed, have never been found.

Gish, *Evolution: The Fossils Say No!* (pp. 62–63)

There is obviously a tremendous gap between one-celled microorganisms and the high complexity and variety of the many invertebrate phyla of the Cambrian. If the former evolved into the latter, it seems impossible that no transitional forms between any of them would ever be preserved or found. A much more likely explanation for these gaps is that they represent permanent gaps between created kinds.

Morris, *Scientific Creationism* (p. 81)

Irreducible Complexity

> To believe that the incredibly complex functions necessary in the bombardier beetle came about as a result of genetic accidents, is, at best, pure fabrication.
>
> Gish, *The Amazing Story of Creation from Science and the Bible* (p. 101)

> The creationist maintains that the degree of complexity and order which science has discovered in the universe could never be generated by chance or accident.
>
> Morris, *Scientific Creationism* (p. 59)

Then, there are the commonalities in political tactics, one example of which will suffice. Both Morris and the Wedge's *Utah Law Review* legal strategists argue that teaching only evolution violates the civil rights and academic freedom of teachers and students.

> [T]here are serious objections and harmful aspects to the present practice of teaching evolution exclusively as the only acceptable explanation of origins. . . . It is discriminatory and unfair to those children and parents who, for whatever reason, believe in creation. . . . It is contrary to the principles of civil rights. . . . It is inimical to the principle of academic freedom for those teachers who desire to teach creationism but are inhibited from doing so by fear of academic reprisals.
>
> Morris, *Scientific Creationism* (pp. 14–15)

> While public schools are not public *fora* per se, they are publicly funded places where ideas are exchanged. . . . Thus, if public schools or other governmental agencies bar teachers from teaching about design theory but allow teachers to teach neo-Darwinism, they will undermine free speech and foster viewpoint discrimination. At the very least, the government has no affirmative duty to censor teachers who attempt to present alternative viewpoints on scientific issues. Instead, the Constitution prohibits such censorship or the regulation of speech "in ways that favor some viewpoints or ideas at the expense of others." David K. DeWolf, Stephen C. Meyer, and Mark Edward DeForrest, "Teaching the Origins Controversy: Science, Or Religion, Or Speech?" *Utah Law Review* 39, no. 1 (2000): 106.

Morris and the Wedge have even come up with the same marketing approaches for their textbooks:

> [W]e acknowledge frankly that *Scientific Creationism* is a book designed to emphasize the creation concept of origins. However, it is scientific and objective in its treatment. . . . It is not designed as a "neutral" textbook on origins, but solely as a supplementary textbook or reference handbook for teachers, supplementing the regular textbooks which emphasize evolution, thus enabling any course to be offered with a good balance between the two models.
>
> Morris, for *Scientific Creationism (Public School Edition)*, 1974 (p. v)

> *Of Pandas and People* is not intended to be a balanced treatment by itself. We have given a favorable case for intelligent design and raised reasonable doubt about natural descent. But used together with your other text, it should help you to balance the overall curriculum. (Introduction, p. ix)
>
> In the spirit of good, honest science, *Pandas* makes no bones about being a text with a point of view. . . . By using this text in conjunction with your standard basal text, you will help your students learn to grapple with multiple

competing hypotheses and to maintain an open but critical posture toward scientific knowledge. (A Note to Teachers, p. 154)

Percival Davis and Dean H. Kenyon, for *Of Pandas and People*, 1993

Finally, Duane Gish, Morris, and Johnson *all* make exactly the same tactical disavowals of the role of religion in their thought on the history of life:

> I've never claimed to have won a debate [with evolutionists], but the evolutionists themselves said that the creationists have won almost every debate. I say, if they have the scientific evidence, and we're just a bunch of religious fanatics, it's a strange result. I never mention the Bible. I never mention the book of Genesis. I talk about thermodynamics. I talk about probability related to the origin of life. I talk about the fossil record. I may present the metamorphosis of the monarch butterfly and challenge them to explain how a caterpillar could change step by step by step by a bunch of genetic mistakes into a chrysalis where now he's a mass of jelly, and then change that mass of jelly into a butterfly. Nobody's been able to explain it yet. That's the kind of evidence I talk about.
>
> Gish, interview in *Creation Matters*[94]

> The purpose of *Scientific Creationism* (Public School Edition) is to treat all of the more pertinent aspects of the subject of origins and to do this solely on a scientific basis, with no references to the Bible or to religious doctrine.
>
> Morris, *Scientific Creationism* (p. iv)

> Religious fanatics versus open minded inquiry. That is a totally false picture of what is going on and we have to escape from it. And that is why I say we have to discuss the science. We have to discuss what science does and does not show without getting the Bible mixed up in it. Because, you see, the Bible just distracts everybody. They say, "Oh, we know why you are asking these kind of questions. You don't want to believe the scientific evidence because you are prejudiced in favor of the Bible." So we want to say no, let's just put that issue entirely aside and let's ask what does the scientific evidence really show?
>
> Johnson, April 2000 speech[95]

One marvels that the Wedge would attempt to disclaim ID's character as creationism in light of what even a brief survey of these commonalities reveals. Yet its members (and followers) regularly do, and Dembski is particularly adept at maintaining to the doubtful that ID is distinct from creationism, that ID is disinterested science while creationism is not. Attempting to distance ID from the usual brand of Christian fundamentalist creationism, he has tried to steer between the Scylla of special creation and the Charybdis of theistic evolution, yet insists at every opportunity on the *logical* compatibility of *both* with ID, as he did when introducing the concept of ID in *Cosmic Pursuit* in 1998, in "The Intelligent Design Movement": "What has emerged is a new program for scientific research known as Intelligent Design. . . . Intelligent Design is theologically minimalist. It detects design without speculating about the nature of the intelligence. . . . Logically speaking, Intelligent Design is compatible with everything from the starkest creationism (i.e., God intervening at every point to create new species) to the most subtle and far-ranging evolution (i.e., God seamlessly melding all organisms together in a great tree of life)."[96] Yet while ID is not the typical fundamentalist fare of the past, when Dembski explained ID in 1996 in the *Princeton Theological Review*—directing his explanation at "mainstream theologians" who "perceive design theorists as theological greenhorns"—not only did he disavow young-earth creationism ("First off, design theory is not young earth creationism") but he also discounted the compatibility of ID with theistic evolution:

> [I]t's hard to imagine how design theorists could be identified as . . . fundamentalists. . . . [N]othing in design theory . . . requires a narrow hermeneutic for interpreting scripture. Indeed, design theory makes neither an explicit nor an implicit appeal to scripture. Nonetheless, design theorists are frequently accused of being, if not fundamentalists, then crypto-fundamentalists. What lies behind this tendency to lump them with fundamentalism as opposed to placing them squarely within the mainstream of American evangelicalism? The answer . . . is quite simple: *Design theorists are no friends of theistic evolution*. As far as design theorists are concerned, theistic evolution is American evangelicalism's ill-conceived accommodation to Darwinism. What theistic evolution does is take the Darwinian picture of the biological world and baptize it, identifying this picture with the way God created life. When boiled down to its scientific content, theistic evolution is no different from atheistic evolution. . . .
>
> As far as design theorists are concerned, theistic evolution is an oxymoron, something like "purposeful purposelessness."[97]

Such flat contradictions are routine in Dembski's oeuvre. Still, despite his contradictory assertions (1) in 1996 that design theorists are "no friends" of theistic evolution, and (2) in 1998 in *Cosmic Pursuit* that ID is compatible with the idea that God guided evolution (i.e., with theistic evolution), it is easy to discover where his feet are actually planted. As always, Dembski locates in the supernaturalist camp when speaking to his creationist compatriots, even while proclaiming his alignment with the time-tested conventions of science. As for his November 2000 assertion on Metaviews that ID need not invoke scripture, Dembski had once again given the lie to his own statements, this time in Strasbourg, France, in 1998, at a Millstatt Forum presentation ("Christian Thought for Science, Art & Society"). There he joined other American creationists (Hank "Bible Answer Man" Hanegraaff, Russell Humphreys and John Baumgardner of the Institute for Creation Research, and Phillip Johnson). Dembski invoked not only scripture on that occasion but specifically a transcendent creator in the same breath with which he proclaimed the scientific nature of ID:

> Theism and naturalism provide radically different perspectives on the act of creation. Within theism any act of creation is also a divine act. Within naturalism any act of creation emerges from a purely natural substrate—the very minds that create are . . . the result of a long evolutionary process that itself was not

> created. The aim of this talk . . . is to present a[n] . . . account of creation that is faithful to the Christian tradition, . . . resolutely rejects naturalism, and . . . engages contemporary . . . science and philosophy. . . .
>
> [T]he claim that transcendent design pervades the universe is no longer a strictly philosophical or theological claim. It is also a fully scientific claim. There exists a reliable criterion for detecting design—the complexity-specification criterion . . . [which] detects design strictly from observational features of the world. . . . *[T]he complexity-specification criterion demonstrates transcendent design*. . . .
>
> This empirically-based criterion reliably discriminates intelligent agency from natural causes. . . . [W]hen applied to cosmology and biology, it demonstrates not only the incompleteness of natural causes, but also the presence of transcendent design. . . .
>
> God's act of creating the world is the prototype for all intelligent agency. . . . [A]ll intelligent agency takes its cue from the creation of the world. . . . God's act of creating the world makes possible all of God's subsequent interactions with the world. . . . *God's act of creating the world is thus the prime instance of intelligent agency.* (emphasis added)

Dembski also made an implicit appeal to John 1:1—"In the beginning was the Word, and the Word was with God, and the Word was God"—the same scriptural passage Phillip Johnson had invoked as justification for ID as early as 1996:

> Let us . . . turn to the creation of the world . . . in Scripture. . . . God speaks and things happen. . . .
>
> God, in speaking the divine *Logos*, not only creates the world but renders it intelligible. . . .
>
> On the view that creation proceeds through a divine spoken word, not only does naturalistic epistemology have to go by the board, but so does naturalistic ontology.

Dembski next made it clear that the "information" that he insists points *scientifically* to an intelligent designer—using his "specified complexity" criterion—is one and the same as the divine Logos:

> Information . . . is just another name for *logos*. . . .
>
> [T]he information that God speaks to create the world, . . . that continually proceeds from God in sustaining the world and acting in it, and . . . that passes between God's creatures . . . is the bridge that connects transcendence and immanence. All of this information is mediated through the divine *Logos*, who is before all things and by whom all things consist (Colossians 1:17). The crucial breakthrough of . . . intelligent design . . . has been to show that this great theological truth— that God acts in the world by dispersing information—also has scientific content. All information, whether divinely inputted [*sic*] or transmitted between creatures, is in principle capable of being detected via the complexity-specification criterion. . . .
>
> The fine-tuning of the universe and irreducibly complex biochemical systems are instances of specified complexity, and signal information inputted into the universe by God at its creation.
>
> Predictive prophecies in Scripture are instances of specified complexity, and

[they] signal information inputted by God as part of his sovereign activity within creation.[98]

The Wedge's Creationist Supporters. We have dealt in chapter 5 with such extraordinary claims of Dembski's in information theory and other branches of serious science; they are, to say the least, not supported by *any* informed opinion in *any* of the relevant fields. Therefore, not only is the Wedge strategy founded on and fueled by religious zeal; it is also the currently successful (evolved!) variant of old-fashioned American creationism. Accordingly, other creationist and religious individuals and organizations monitoring CRSC's activities and publications have come to recognize it as such, and to applaud:

Hal Ostrander (Associate Dean and Associate Professor of Christian Theology, Boyce College, Southern Baptist Theological Seminary). Ostrander, an ardent Wedge supporter, certainly does not hesitate to label ID as creationism. In a February/March 1998 article in the Southern Baptist Convention's *SBC Life,* he shows intimate, firsthand familiarity with the Wedge strategy, the documentary form of which did not surface publicly until February 1999. Speaking with what sounds like the knowledge of an insider, Ostrander refers to "behind the scenes" preparations for a "potential anti-Darwinian *coup d'etat*" and to the Wedge's strategy, citing its phases as they must have been described in the earlier version, "slated for 1996–2001." Ostrander names Wedge members and describes them as "Creationists, one and all."

Creation Research Society (CRS). In its January/February 2000 *Creation Matters*, the Creation Research Society announced the Wedge's Design and Its Critics conference, including contact information, in its "Creation Calendar," along with other creationist events such as "*Creation vs. Evolution* by Dr. Kent [Dr. Dino] Hovind," and "*Workshop: Winning Debates Against Evolutionists* with Dr. Duane Gish." In its July/August 1997 issue, creationist David Buckna compiled a chronicle of events, articles, and books entitled "Wider Coverage in Mainstream Press for Anti-Darwinians and Proponents of Creation/Intelligent Design: A Progress Report," suggesting that "God may be using the various creationist groups as well as the secular anti-Darwinists to begin moving the origins debate into the scientific, educational, and cultural mainstream." Of the forty items Buckna listed, twenty-two were by or about Wedge members. The January/February 1997 *Creation Matters* contains "Mere Creation: A Report on the Origins Conference held at Biola University, Nov. 14–17, 1996," by Todd Wood, who attended and reported, "The promotional material also encouraged us to put aside our differences and find a common creed that all creationists can hold to." The "CRS Books" site sells a number of Wedge books.[99]

Creation Science Fellowship. The lead article in the March 2001 *Origins Insights*, the newsletter of the young-earth Creation Science Fellowship in Pittsburgh, Pennsylvania, announced its March meeting topic: "The Wedge of Truth: The Case Against Naturalism," describing the ID movement as "the most visible arm of modern creationism in America." Comparing Phillip Johnson to Robin Hood, that is, a person whose efforts are necessary even though his methods make one uncomfortable, CSF Vice Chairman Bob Harsh urged members to embrace Johnson—"a hero"—and his message, despite doctrinal differences:

> Let's get a glimpse of what Phillip Johnson is up to. . . .
>
> In American culture today the real issue is who shall have the right to determine policy. . . . If you can successfully define knowledge in such a way that your convictions are knowledge and those of others are not, you get to determine policy, to direct human life. . . .
>
> I . . . have felt that modern creationism has made outstanding progress. *Then . . . I was reminded that I am still on the losing side!* . . .
>
> A hero with the promise of victory, even in the far distant future, is refreshing to me. I didn't hear about the "Intelligent Design Movement" until a couple of years ago. . . .
>
> [F]or the sake of argument, those in the ID movement want to set aside (temporarily) questions about, say, Genesis and the age of the Earth. It is not that such questions are not important; it is just that they are being saved for another place and time. . . . [Marsh here quotes the Wedge Document extensively.] By the way, I discovered this document, interestingly enough, as a link at an anti-creationism web site! . . .
>
> *Read the goals of the Wedge and think about which of the goals you can support and which ones you are opposed to.* . . . [Marsh cites the Wedge's "Governing Goals," "Five Year Goals," and "Twenty Year Goals."]
>
> I like them all, don't you?!! . . .
>
> My hope is that you will . . . consider giving your support to the important work of the Wedge Project. . . .
>
> *I think our success is all but inevitable.*[100] (emphasis in original)

Institute for Creation Research (ICR). Despite doctrinal and "scientific" differences with intelligent design, young-earth creationists recognize enough commonality in content to ally themselves with the movement. In a 1998 *Impact* article, Henry Morris recognizes differences between his young-earth creationism and intelligent design. (The main disagreement is the age of the earth.) Another difference, supposedly (Morris cites Eugenie Scott here), is ID proponents' effort to distance themselves from the Bible. But Morris notes that "this approach is not new . . . each [ICR] debate is framed [for tactical purposes] to deal *only* with the 'details of Genesis,'" and that "this [ID] creationism is really not very new, except perhaps for the terminology."[101] Despite this tactical similarity, Morris does complain about the ID movement's disavowal of biblical literalism and its unwillingness to identify the creator as the God of the Bible in its "intellectual arguments." Yet there is an obvious working alliance, however uneasy, between the young-earthers and ID. The following year, 1999, Ashby Camp, in the Creation Research Society's *Creation Matters*, acknowledged the ID movement as "an ally," because "the cultural elites at present fear the ID movement more than they fear biblical creationism." Although he asserts that "the method of attack has nothing to do with Scripture (at a formal level)," indicating that he recognizes the connection to the Bible at what one might call the material level, he clearly sees ID as helpful to young-earth creationism's case: "If the science establishment can be forced to acknowledge the scientific case for intelligent design, theism will become part of the 'post-Christian' cultural air. In that philosophical environment, a new set of options will open for people, one of which will be biblical creation. . . . If ID is successful in changing the culture, the presumption against the supernatural will be eliminated."[102] And of course "changing the culture" is precisely what the Wedge seeks to do.

By November 2001, we find at least one article in ICR's *Impact* that is openly and unreservedly supportive of ID. Gregory Brewer writes that "intellectual honesty will soon force many scientists to abandon Darwin's theory of the evolution of species in exchange for intelligent design or outright Biblical creation" and that "the death of Darwinism will be a hard pill to swallow because it requires replacement by intelligent design, a paradigm outside the box of naturalism that many scientists embrace."[103]

Most recently, in May 2002, Henry Morris's son and successor, John D. Morris, acknowledged the common social agenda of ICR and the ID movement; indeed, he is realistic about the greater chances of ID's success, even though he still complains that ID is not religious enough (at least overtly, though we have amply disconfirmed that complaint):

> Evolution's dam is starting to crack. Soon its stranglehold on education may loosen. . . .
>
> Biblical creation will not replace evolution in public school[s]. . . . The courts have decisively ruled that "religion" doesn't belong in public schools. . . .
>
> We [ICR] do provide information and counsel . . . as we did in Kansas and . . . other states . . . [and] have seen a tremendous ground swell of opposition to naturalistic evolution and a desire to see better education with less "religion" result from our efforts.
>
> Creationist organizations haven't done it alone. . . . A . . . major player is the "Intelligent Design" movement. . . . [T]his is a better fit for . . . public schools, since it is scrupulously non-religious and pro good science teaching.
>
> As a side note, ICR is not against the Intelligent Design movement. We are not part of it, for we are . . . openly Christian. [As this study has shown, so is the Wedge, in front of the right audience.] We support the . . . ID group, but feel it doesn't go far enough.[104]

Morris then applauds the Santorum Amendment and its use in Ohio, speaking hopefully of its ultimate effects: "In reality, the deck is stacked against creationists in Ohio, and they may not succeed, but the dam has begun to break. Things will never be the same."

"Truth and Science"—and Carl Baugh. The Wedge has a new source of support, although its organizational identity is unclear. The website "www.truthandscience.com," the only identifiable entity, is built completely around ID and William Dembski's work. It is support that, from a public relations standpoint, the Wedge should certainly have avoided, but that Dembski recognized with a public appearance at an April 2002 "Truth and Science Seminar on Intelligent Design" in Belton, Texas. He shared an opening night platform with none other than Carl Baugh, whom Australian anticreationist John Stear, on "The Carl Baugh Page," calls "a charlatan second only to 'Dr.' Kent Hovind" (a flamboyant Florida "creation science evangelist").[105] Two other of the five speakers were Baugh's employees at his Creation Evidence Museum, so Baugh's organization was the predominant presence. According to information on the tickets, part of the proceedings benefited the museum.

Citing the New Testament books of Romans and John, Dembski began his talk at the seminar as though delivering a sermon. He explained a "driving force behind what I've been doing all these years," recounting an experience at the age of twenty, when he offered his Christian witness to two students whose faith was "washed away" at the mainstream seminary they attended. Their loss of faith was the result of the seminary's teaching them that the Bible was not divinely revealed and was at odds with evolution. After warning of the dangers of such concessions to naturalism, Dembski invoked the Christian alternative: a vision of creation by a "Creator God who by wisdom creates the world." He progressed to a defense of specified complexity's superiority to natural selection as an explanation of natural complexity and diversity. He also touted ID's attractiveness to people with a "spiritual longing," including those at *UFO* magazine, who had done a story on "space abductions" and who "like intelligent design." He predicted that ID would ultimately "win out" in the culture, and he criticized "state-regulated schools" and "mainstream academies" that "educate us out of . . . seeing the design that God—the wisdom of God—manifest[s] in the world." He concluded by predicting that this would change. Baugh followed him, sanctioning ID with his unqualified support: "Intelligent design is an honest, bona fide, unbiased, academic research program into life origins." After discussing the anthropic principle, Baugh concluded: "If we really want answers for the data available, intelligent design embraces the potential for eternity."[106]

Baugh is notorious even among other young-earth creationists because of his less-than-sterling academic "credentials" and his blatant misrepresentation of fossil "evidence," which he claims shows a human footprint contemporary with dinosaur tracks in the Paluxy riverbed in Glen Rose, Texas, where he built his museum. Baugh's website also contains ID material: "Design in Creation—Part Two" cites *Darwin's Black Box* and *Of Pandas and People*. He has a page listing talks in his "Director's

Lecture Series"; his topics mirror exactly chapters in Jonathan Wells's *Icons of Evolution*.

Given the Wedge's emphasis on educational pedigrees, not to mention its disavowal of creationism, Dembski's public appearance with someone like Baugh is incredible. However, the significance of the event is that Baugh and the sponsors of the seminar and website—whoever they are—see ID as an ally in their own creationist program.[107]

Some discussion is in order concerning the relationship between the Wedge and one of its chief rivals, Answers in Genesis (AIG, the "guitar-strumming hillbillies," as DI spokesman Mark Edwards called them). There is some strain between AIG and the Wedge, yet in our view, the tension is relatively superficial and should not be overestimated. AIG-style creationism has no chance of inclusion in public school classes, given its openly Biblical affirmations. Thus, the only chance of creationism's gaining any degree of acceptance in public schools is under the guise of intelligent design. AIG's statements about ID reflect that awareness.

AIG's Australian CEO, Carl Wieland, has written a position paper on ID, "AIG's Views on the Intelligent Design Movement."[108] (Ken Ham is the CEO of the American branch.) Wieland explains—in a surprisingly evenhanded way—the differences and commonalities between AIG's young-earth view and ID's predominantly (but not exclusively) old-earth view. Wieland notes tactical differences, such as ID's aversion to making public statements about the Genesis account of creation, even though "some who are prominent in the [intelligent design movement] appear . . . sympathetic to the Bible's account." He points to some substantive differences, charging that ID's refusal to endorse the literal Genesis account leaves it without a coherent "story of the past." However, if AIG were truly alarmed about ID's façade of secularity, it would campaign openly and actively against it, as it does against evolution. Most of Wieland's criticisms are tactical, and he stops well short of burning any bridges between AIG and the Wedge. And never does Wieland deny that intelligent design is creationism. Very few of Wieland's criticisms are new, and he opens and closes with conciliatory remarks: "Many in the creation movement, including AiG [*sic*] and me . . . , have friendly relations with, and personally like, some of the people . . . in the IDM [intelligent design movement]. . . . Where we can be natural allies, . . . we want to be. . . . [W]e neither count ourselves a part of this movement nor campaign against it." And Wieland's analysis of ID contains an important acknowledgment: "We have always felt that teachers should . . . be able to critically examine arguments for and against evolution, and we don't think that the constitutional arguments in the USA supposedly preventing that have ever been strong. . . . What reasonable person could logically defend the notion of shielding any scientific theory or idea from all critical analysis?" The Wedge is waging its "teach the controversy" campaign on precisely this platform.

Although AIG has not hesitated to criticize the Wedge, it posts pro-ID material and sells Behe's *Darwin's Black Box* on the AIG website. AIG defended Roger DeHart's teaching ID in his Burlington, Washington, science class. And it has kept a close and supportive watch on the Wedge's efforts in Ohio. Royal Truman, who reviews books for AIG, has written favorably of Dembski's work. He writes that *Intelligent Design: The Bridge Between Science and Theology* "can be used profitably by those wishing to refine how apologetic arguments can be more subtly presented." He sees *The Design Inference* as "an important piece in [ID's] logical arsenal" and asserts that "the attempts to free academia and research funds from the stranglehold of methodological materialism can only help the creationist movement." He pays special homage to the overt religiosity of Phillip Johnson's *The Wedge of Truth*: "This book does not hide the author's conviction that the Bible is God's inspired message." Finally, an unsigned review of Johnson's *Reason in the Balance* reveals why AIG is unlikely to make a definitive break with ID: "Johnson's important book will help force the issues out on to the table, ultimately assisting the vital, ever-expanding task of biblical creation ministries such as Answers in Genesis."[109]

Many websites of creationist organizations include items by and about the Wedge. Despite Dembski's disavowals of young-earth creationism, the Wedge promotes an account of life on Earth that also advances the aims of all fundamentalist groups. The CRSC is not itself a specifically fundamentalist organization (it is more accurately identified with evangelicalism); but its inclusion on fundamentalist websites is important: if the CRSC were not advancing the creationist agenda in a way that also advances the aims of such organizations, their work would not be included. Johnson's political "big tent" strategy of unifying Christians around ID, despite doctrinal differences, therefore seems to be working. He gloats, "We're unifying the divided people and dividing the unified people."[110]

According to Johnson, the scientific "creation myth" must be replaced by the *true* account of human existence:

> The proper metaphysical basis for science is not naturalism or materialism but the fact that the creator of the cosmos not only created an intelligible universe but also created the powers of reasoning which enable us to conduct scientific investigations. . . .
>
> [T]he materialist story thrives only as long as it does not confront the biblical story directly. In a direct conflict, where the public perceives the issues clearly, the biblical story will eventually prevail. . . .
>
> What we need is for God himself to speak, to give us a secure foundation on which we can build. . . . So it is of the greatest importance that we ask the question: "Has God *done* something to give us a start in the right direction, or has he left us alone and on our own?"
>
> When we have reached that point . . . , we will inevitably encounter . . . Jesus Christ, . . . who has been declared the incarnate Word of God, and

> through whom all things came into existence. This time *he* will be asking the question . . . in the Gospels: "Who do men say that I am?" . . .
>
> When the naturalistic understanding of reality finally crashes and burns . . . the great question Jesus posed will come again to the forefront of consciousness. Who should we say that he is? Is he the one who was to come, or should we look for another? . . .
>
> We are not talking about some mere revision of a particular scientific theory. *We are talking about a fatal flaw in our culture's creation myth, and therefore in the standard of reasoning that culture has applied to all questions of importance.* . . .
>
> The basic story of the Incarnation—that God has taken human form . . . is more equivalent to the scientific truth that apples fall down rather than up.[111] (emphasis in original)

There is no doubt that Johnson wants readers of *The Wedge of Truth* to hear the message that *true* science, built on supernatural creation, is marching gloriously toward (or back to?) Jesus—and simultaneously into the heart of American academic and cultural life.

Wedging into Academia

Phillip Johnson has little respect for his fellow academics in American universities. Says he, "If you want to be a real, closed-minded dogmatist, be a top-level university professor. . . . That's where you can get away with it. . . . I speak as one who knows!" He has little respect even for his fellow Christian academics: "Too many of our Christian academics at Christian colleges have been educated in the secular academies. They've learned theistic evolution. They really don't *understand* the issues. They don't know what's wrong with the scientific evidence. They just follow blindly along where the secular world is going. And so they're *seriously* mis-educating our students." Johnson says that Biola University (where several Wedge members are faculty), with its Torrey Honors Institute program, does a great job of educating students, "but by and large, the Christian intelligentsia has let us down."[112]

Johnson's dissatisfaction with not only secular but Christian universities explains why a major presence in American higher education is one of the Wedge's most ambitious *religious* goals; posing as an academic movement serves the Wedge's fundamental religious purpose and illuminates original hopes for the Michael Polanyi Center. Nancy Pearcey is optimistic: "The design movement shows promise of winning a place at the table in secular academia, while uniting Christians concerned about the role science plays in the current culture wars."[113] Wedge strategists do not expect to establish a numerically large presence—indeed, as noted in the Wedge Document, they do not believe it is essential: "A lesson we have learned from the history of science is that it is unnecessary to outnumber the opposing establishment. Scientific revolutions are usually staged by an initially small and relatively young group of scientists who are not blinded by the prevailing prejudices and who are able to do creative work at the pressure points, that is, on those critical issues upon which whole systems of thought hinge."[114] What is most important is establishing an aggressive, high-profile presence *inside* the academic establishment and the cultural mainstream. Tragically, this is slowly but surely taking place.

Nothing is more important to the Wedge than the academic respectability that comes from earned degrees and real teaching positions at respected (or at least respectable) universities. One of the Wedge Document's goals was to have ten CRSC fellows teaching at major universities by 2003. Restricting "major universities" to nationally recognized public and private universities, we find that by 2000 the Wedge had more than realized that goal in the persons of fourteen CRSC fellows (although two have since retired, and one taught only part-time for a short period):

Michael Behe—Lehigh University
Walter Bradley—Texas A & M (retired, now at Baylor)
J. Budziszewski—University of Texas-Austin
John Angus Campbell—University of Memphis
Robert Koons—University of Texas-Austin
Paul Chien—University of San Francisco
David K. DeWolf—Gonzaga University
Bruce Gordon—Baylor University
Phillip E. Johnson—University of California-Berkeley (retired)
Robert Kaita—Princeton University
Dean H. Kenyon—San Francisco State University (retired)
Scott Minnich—University of Idaho
Henry F. Schaefer—University of Georgia
Richard Weikart—California State University-Stanislaus[115]

In the PowerPoint show, "Darwin, Science, and Going Beyond the Culture Wars," (March 2000), CRSC boasts that it has more than thirty fellows at major universities, so they are interpreting "major" *very* broadly and not limiting the list to holders of proper tenure-track positions or even teaching positions.[116] This allowed them to add the following fellows, putting the total (as of 2000) at exactly thirty.

Michael Keas—Oklahoma Baptist University
Stephen C. Meyer—Whitworth College
Jeffrey P. Schloss—Westmont College
Francis J. Beckwith—Trinity International University
Jack Collins—Covenant Theological Seminary
Robin Collins—Messiah College
William Lane Craig—Talbot School of Theology, Biola University
J. P. Moreland—Talbot School of Theology, Biola University

Joseph Poulshock—Tokyo Christian University
Pattle Pak-Toe Pun—Wheaton College
John Mark Reynolds—Biola University
Siegfried Scherer—Technical University of Munich
Charles Thaxton—The Charles University, Prague, Czech Republic (visiting assistant professor)
Guillermo Gonzales—Univ. of Washington (now at Iowa State University)
William Dembski—Baylor University
Douglas Axe—Medical Research Council, Cambridge University

Even with the two retirements, the Wedge is close to its boast of thirty fellows at "major" universities. However, when one adds known ID supporters and auxiliaries, the Wedge has more than achieved its goal. It would not be farfetched to say that there are supporters at universities in every state. And Mary Beth Marklein's comment in *USA Today* creates exactly the impression that the CRSC *wants* the public to have: "From the intelligent-design movement, advanced by scholars at respected universities, is emerging what could become a battle in science research."[117]

Of course, the presence of CRSC fellows on the faculties and in university research laboratories does not *necessarily* mean they carry their religion and their distorted notions of science into the classroom or their scholarly work; but some, and perhaps many, especially in religion-based schools, unquestionably do so. One of the Wedge's Five Year Goals is "two universities where design theory has become the dominant view." The advance toward that goal apparently consists so far only of Biola University's Torrey Honors Institute, where CRSC fellow John Mark Reynolds is director, and fellow Michael Keas's Oklahoma Baptist University, where "the general science, science, education, and chemistry programs . . . take a strong Intelligent Design advocacy position."[118] But with fellows and supporters in mainstream universities, CRSC creationists are in a position to try to achieve their desires:

1. *The Wedge wants unquestioned academic legitimacy.*

Phillip Johnson realizes that academic legitimacy is the Wedge's first hurdle:

> The conference [Southern Methodist University in March 1992] brought together as speakers some key Wedge figures, particularly Michael Behe, Stephen Meyer, William Dembski, and myself. It also brought a team of influential Darwinists, headed by Michael Ruse, to the table to discuss this proposition: "Darwinism and neo-Darwinism as generally held in our society carry with them an *a priori* commitment to metaphysical naturalism, which is essential to making a case on their behalf." . . . [T]he amazing thing was that a respectable academic gathering was convened to discuss so inherently subversive a proposition.[119]

Wedge members and supporters are doing their part to help achieve this legitimacy in the academic mainstream. According to abstracts in the *Journal of the Mississippi Academy of Sciences*, Robert Waltzer (Ph.D., Ohio State), an associate professor of biology at Belhaven College in Mississippi who presented a paper at Dembski's 2001 Calvin conference, presented pro-ID papers in the history and philosophy of science segment at the 2000 and 2001 meetings of the Mississippi Academy. In February 2002, an apparently *anti*-ID paper was presented, but Johnson would see even this as a victory. Opposition is not a deterrent to the Wedge, but an energizer.[120]

In March 2002, Wedge members William Lane Craig and Robert C. Koons, who, unlike philosopher Dembski, belong to the American Philosophical Association (APA), held a meeting of the Philosophy of Religion Group in conjunction with the APA (Pacific Division) meeting in Seattle. Chaired by Craig, the session was entitled "Detecting Design in Nature," with Koons, CRSC fellow Robin Collins, and Wedge supporter Del Ratzsch presenting papers. Many organizations hold meetings at the APA conventions. Although it exercises no supervisory responsibility, the APA provides space as a courtesy if the session topics might be of interest to philosophers. The groups' names usually indicate clearly their organizational identities, such as the Society for Skeptical Studies, which met the day after the Philosophy of Religion Group. Although its oddly generic name gives no indication of its identity or affiliation, the Philosophy of Religion Group is apparently part of the Evangelical Philosophical Society, which solicited papers for the group's meeting at the APA's March 2003 Pacific Division and April 2003 Central Division conventions: "This is a Call for Papers for the 2003 APA Central Division and Pacific Division meetings of the Philosophy of Religion Group sponsored by the Evangelical Philosophical Society." William Lane Craig was listed as the person to whom abstracts should be sent.[121]

These Philosophy of Religion Group presentations can be seen as an attempt by the Wedge to expropriate for ID some of the American Philosophical Association's prestige; the same might be said of Waltzer's Mississippi Academy of Sciences presentations. They are perhaps small beginnings, but to be offered at a meeting of a state academy of sciences and in conjunction with a national philosophical society meeting represents, for creationism, an advance that only a few years ago would have been unthinkable.

In an interview with *Communiqué: A Quarterly Journal*, Johnson acknowledges the difficulty of acquiring legitimacy in the wider academic world and his resolute commitment to achieving it:

> CJ: That seems to be behind the idea of driving 'the wedge' into the scientific community—that you'd just encourage them [students and faculty] to get behind guys like Behe and join that momentum.
>
> Phil: Yes, the idea is that you get a few people out promoting a new way of think-

ing and new ideas, it's very shocking, and they take a lot of abuse. . . . You have to have people that talk a lot about the issue and get it up front and take the punishment and take all the abuse, and then you get people used to talking about it. It becomes an issue they are used to hearing about, and you get a few more people and a few more, and then eventually you've legitimated it as a regular part of the academic discussion.[122] And that's my goal: to legitimate the argument over evolution . . . as a mainstream scientific and academic issue. . . . We're bound to win. . . . We just have to normalize it, and that takes patience and persistence, and that's what we are applying.[123]

2. *The Wedge wants to influence high school and college students, most of whom are ignorant of genuine science, and to recruit them to the Wedge movement.*

In *Touchstone* (July/August 1999), Johnson updates the Wedge's progress. His remarks indicate that universities are fertile recruiting ground: "[M]any . . . college students are reading our literature, and are responding very favorably. . . . The most talented of these will be the Wedge members of the future."[124] Colleges and universities, even high schools, are the logical source of the "future talent" that, according to the Wedge Document, the CRSC seeks "to cultivate and convince."

Books like Wells's *Icons of Evolution* are designed specifically to reach American students. In a recent radio interview with Hank Hanegraaff, Johnson agreed with Hanegraaff that since *Icons'* publication, students can now challenge their biology professors:

We now have more and more students that are going into [biology] classes who know the score. . . . [Evolutionary biology professors] are getting increasingly put on the spot, and they hate it. We're getting a lot of panicky reactions, and that's why you'll see the entire scientific establishment go off into an absolute panic when, for example, the Kansas State Board of Education last year made a really minor change, downgrading evolution a bit in terms of the required state standards. They were frightened that once any revolt against this officially imposed dogma gets started, the whole thing will unravel.[125]

CRSC fellow Mark Hartwig is doing his part to ensure that science teachers get put on the spot, as he announced during his and Dembski's October 2001 interview with James Dobson:

Hartwig: We've been training our kid all along. I mean, this year my daughter confronted her eighth-grade teacher on evolution, on a couple of facts. She did it very politely—

Dobson (interrupting): So you're homeschooling now! [Laughter.]

Hartwig: Uh, well, no! [Laughter.] We have, but we do school her at home [on challenging her teachers]. We teach her what to watch for. We'll just say, "Honey, you just keep doin' that in college."

Dobson: And so you're preparing her to cope with whatever is thrown at her.

Hartwig: Exactly. And she'll go in with the knowledge because of my experience and my wife's experience with a private Christian college. Don't go in with your guard down, you know. Question things.[126]

Dembski recently indicated hopes for ID recruits from high schools and colleges: "My commitment is to see intelligent design flourish as a scientific research program. . . . To do that, I need a new generation of scholars willing to consider this, because the older generation is largely hidebound. So I would like to see textbooks, certainly at the college level and even at the high school level, which reframe introductory biology within a design paradigm."[127]

The recruits may not be long in coming. The Wedge has already acquired two groups of college followers, the Intelligent Design Undergraduate Research Center (IDURC) and the Intelligent Design and Evolution Awareness (IDEA) Club. The IDURC has become a division of Access Research Network and promotes Wedge books and other products through links to ARN's website and to commercial sites like Amazon.com. It features papers by Wedge members such as Dembski and student ID followers. Also available are "Research Articles" by Wedge members and supporters (including the pseudonymous "Mike Gene"), audio files of conference presentations, and so on. The site posts links to other ID sites and encourages the writing of student papers on ID by paying up to $50 for them, as well as for PowerPoint presentations.[128]

The IDEA Club was begun by students at UC-San Diego, but since the graduation of its student founders, it is now known as the IDEA Center and is affiliated with the Tacoma, Washington, Faith Evangelical Lutheran Seminary, a graduate school that teaches that "God created ex-nihilo and formed the universe in the six literal days as described in Genesis 1." IDEA Club founder Casey Luskin is an IDEA Center administrator. The IDEA Center's advisory board consists of Wedge members Phillip Johnson, William Dembski, Michael Behe, Jonathan Wells, Jay Wesley Richards, Mark Hartwig, and Francis Beckwith; Dennis Wagner, executive director of Access Research Network; and young-earth creationist John Baumgardner, all of whom also serve as the speakers bureau.

The purpose of the center is to spread ID at high schools and universities by starting campus chapters: "The IDEA Center itself grew out of the IDEA Club at UC San Diego and there is now also a new chapter at Vanderbilt University [where all the student officers are science majors and the faculty advisor is a professor of microbiology and immunology]. . . . We aim to help other students to do the same at their own schools—whether public, private, high school, college, secular, or religious." According to the IDEA Center, chapters exist at the University of Texas-Dallas, California State University-Sacramento, Long Beach (California) City College, UC-San Diego (the original chapter), the University of Hawaii-Hilo, Pulaski Academy in Arkansas (pre-kindergarten through grade 12), and South Lehigh High School in Center Valley, Pennsylvania, (which is only a few miles from Bethlehem, where Michael Behe works and resides). Chapters were being planned at the University of San Francisco (Paul

Chien's university) and the University of Wisconsin-Eau Claire. An "off-campus" chapter exists in Baraboo, Wisconsin.[129]

The Wedge has always had as a goal the insertion of ID courses into the university curriculum, and its success in having ID taught by faculty at religious colleges such as Oklahoma Baptist and Biola University is no surprise. However, a startling announcement appears in the 1999 Discovery Institute *Journal*: "Wonderful developments in new scientific critiques of Neo-Darwinism and the materialism it underwrites are being aired in many locations, including, at last, the curricula in some state universities."[130] This is true, up to a point. The Wedge has been getting help from supporters inside two public universities: Jed Macosko, a CRSC fellow with a Ph.D. in chemistry from UC-Berkeley who also did postdoctoral work there, and his father, Chris Macosko, a chemical engineering professor at the University of Minnesota, have taught courses on ID at Berkeley and Minnesota, respectively. The teaching of these courses was announced in the American Scientific Affiliation (ASA) newsletter on two occasions, and both are clearly referred to as courses on ID. The March/April 2000 issue announced that "U. of MN chemical engineering professor, Chris Macosko, has been teaching a class on evidence for design in nature, as an under-division honors colloquium." In the January/February 2001 issue, ASA announced that Chris Macosko was "on sabbatical from the U. of Minnesota and currently co-teaching his son Jed's DeCal course on ID."[131]

The Macoskos managed to insert these courses into programs that teach subjects outside the official catalogue offerings. In fall 1998, while still a graduate student, Jed Macosko taught "ChemE 198," specifically titled "Evidence for Design in Nature?" as one of Berkeley's "DeCal" courses under the sponsorship of Professor Jeffrey Reimer in the Department of Chemical Engineering.[132] The regulations governing the teaching of DeCal courses, classified by UC-Berkeley as "Independent or Group Study" courses and which are *outside* the Academic Senate–approved curriculum, allow academic credit for such courses toward a bachelor's degree under conditions established in the regulations.[133] According to Macosko's syllabus, his course carried two credits, although (according to Prof. Reimer) they were not science credits. (Macosko was also conducting research in a small lab.)[134] The syllabus shows that although an attempt was made to present the appearance of balance, the course was clearly a device for introducing Wedge authors William Dembski, Phillip Johnson, Charles Thaxton, Michael Behe, and J. P. Moreland. (Macosko also included selections by creationist Hugh Ross.) Students were asked to compare explanations in biology textbooks and other scientific sources with the work of Wedge members, to identify the "assumptions" of assigned authors, and to view and criticize the movie *Inherit the Wind*. A follow-up assignment on the latter asked students to decide whether the movie is a true representation of the Scopes trial or merely propaganda. Disparagement of *Inherit the Wind* is a regular Wedge activity. Phillip Johnson does it repeatedly in the effort to discredit what he insists is a false stereotype of creationism and recommends the film's use with students:

> One of the things I recommend that parents or teachers do with high school students in particular is rent a copy of the movie *Inherit the Wind*. . . . It is really very good in terms of Hollywood propaganda. Propaganda against Christianity. . . . But the thing that is really good about it is if you go through it in detail and show how the propaganda is done. . . . And how facts are twisted and how it is presented to be a wonderful propaganda vehicle and a smear on Christians. . . . So a few teaching examples like that and young people not only learn a lot about the evolution controversy but they learn a lot about propaganda and advertising. And then they become protected, inoculated, vaccinated against the effects of this kind of thing. So it is a wonderful kind of education to do that.[135]

Johnson was a guest lecturer in Macosko's class, as were other professors, and students were especially encouraged to invite friends to class when guests were lecturing.[136]

Macosko taught the course again at Berkeley in fall 2000, this time with his father, who was on sabbatical at Berkeley and upon whose Minnesota course Jed's was directly modeled.[137] Posted among Jed's course materials on his ChemE 198 website was a homework assignment apparently used by his father in a 1994 University of Minnesota polymer chemistry class, as indicated in the assignment's heading. This assignment, a problem on the origin of life, asks students to calculate the number of different configurations the enzyme ribonuclease A could make since Earth's beginning, assuming that the planet is five billion years old. The solution is provided, indicating to students that theories suggesting that life's essential chemicals formed randomly are not scientifically defensible. (Please see chapter 5 discussion of the straw-man maneuver, the Basic Argument from Improbability.) The assignment lists three references for students, one of which is a standard polymer chemistry text. The other two are the early ID book by Charles Thaxton, Walter Bradley, and Ray Olsen, *The Mystery of Life's Origins* (1984), and a popular creationist book, *The Biotic Message*, by Walter ReMine (1993). A sure sign of the intent of Jed's course to advance ID was the plan announced by Access Research Network to market the video: "Our plans and goals for 2001 include . . . [v]ideo recording of UC Berkeley course 'Life: By Chance or Design.'"[138] According to the Christian Broadcasting Network, Jed now teaches ID and biology at Seventh-Day Adventist La Sierra University in California.[139]

Chris Macosko had taught his ID class at Minnesota as a freshman honors colloquium, "Origins: By Chance or Design?" in fall 1999. Such colloquia carry credit toward graduation.[140] The course received favorable publicity from Maureen Smith in the *Kiosk*, a university faculty-staff newsletter; Smith points out that honors colloquia "offer faculty members the chance to teach something a little different from what they always teach."[141] According to Smith, although Macosko accepts the old age of the earth, he "has doubts himself about the theory of evolution, especially 'the step from no life to life,' and the origin of the first cell. 'No way are we getting close to getting a theory to the origin of life,' [Macosko] says. . . . Readings for the class include both evolutionists (Charles Darwin himself, Richard Dawkins, Stephen Jay Gould) and critics (Michael Behe, Phillip Johnson). Macosko says he was careful to offer 'authoritative readings on both sides every week.'" He also showed the movie *Inherit the Wind*: "Students say the script was written to show creationists as the bad guys and Darrow as a hero. 'In some sense the movie has become reality in our culture,' Macosko says." Macosko taught the course again in fall 2001.[142]

Although the Macoskos' courses appeared to allot time to evolution, the intent to cast heavy doubt on evolutionary biology was perfectly obvious, as was the religious motivation of the entire undertaking; and Chris Macosko holds the familiar creationist view that evolution is religion: "Like many intelligent-design advocates, Mr. Macosko argues that the belief that life's complexity can be explained through chance and natural selection is in itself a form of faith. 'It's really the religion of naturalism,' he says."[143] Macosko apparently sees his honors colloquium as an antidote to that "religion" and hopes to enable others to counteract it: he conducted a discipline-related workshop on engineering at the God and the Academy conference, Georgia Institute of Technology, in June 2000; several Wedge members participated in this conference. (See chapter 7.) According to the conference website, the purpose was to teach academics how to integrate their religion, including, of course, intelligent design, into the teaching of their respective disciplines:

> Christian scholars are integrating their faith with their research, knowledge and influence with astonishing results. . . . *[N]aturalists have conceded the possibility of intelligent design in the origin of life*. . . . Now is the time to apply the principles of academic integration learned from these successes. Now is the time for Christian faculty to unite within their disciplines and across disciplines to present Christ to the cultures of the new millennium.[144] (emphasis added)

Potential attendees were invited to "discover ways in which to integrate your worldview within appropriate areas of your discipline" and to "join . . . hundreds of professors" who would "map out how they can impact their students, their universities, and their fields of study."

Meanwhile, in lieu of getting ID courses included in standard university science curricula, Wedge members are apparently helping to steer interested students toward ID supporters and fellow Wedge members in universities, enabling students to identify them as ID mentors. Ray Bohlin uses his Probe Ministries website to do this. Probe is oriented toward high school and college students and is apparently a popular site for young evangelicals. For example, the Chapel on the Campus, which serves Louisiana State University students, has posted links to Probe Ministries, where students can easily access ID materials.[145] In November 2000 Bohlin posted a letter from a potential recruit: "I am interested in graduate school . . . if I could find a school and professors that are a little more user friendly. I would like to hear more of what you have to say along the lines of Intelligent Design professors." Bohlin was eager to help, supplying names of Wedge scientists and supporters as possible mentors whom the student could check out:

Mike Behe—Lehigh University
Scott Minnich—University of Idaho
Martin Poenie—University of Texas at Austin
Dean Kenyon—San Francisco State University [Kenyon has since retired.]
Paul Chien—University of San Francisco

Bohlin wrote, "I don't know . . . about these guys [*sic*] need or desire for graduate students but . . . Poenie and Minnich have active research programs utilizing graduate students. Behe has cut back . . . his research to focus on promoting intelligent design, so I'm not sure where he is at in being able to support graduate students." Bohlin informed the potential recruit of the CRSC affiliation of Behe, Minnich, Kenyon, and Chien, supplied relevant Internet addresses to facilitate contact, and recommended that he read *Darwin's Black Box* "ASAP."[146]

The UT-Austin connection, first noted in chapter 7, surfaces again in Bohlin's reply to the student. Probe maintains an on-campus study center, the Probe Center, at Austin, "offering friendship and encouragement to Christians struggling with how to be faithful to Christ in the midst of the often hostile and lonely university environment." This message of "encouragement" typifies the Religious Right's portrayal of secular universities (and secular society in general) as antagonistic to Christians. Probe capitalizes on this anti-secularism (throwing in a bit of religious exclusionism) and specifies itself as a UT evolution/creation resource in its message to students:

Where do I go at UT to . . .

- Get information about other religions and cults on campus
- Discover the problems with the theory of evolution and the evidence for creation . . .
- Come to understand what I believe and why so that I can witness to my atheistic friends and professors[147]

3. *The Wedge wants to cultivate the support of university administrators and financial donors.*

> We believe that, with adequate support, we can accomplish many of the objectives . . . in the next five years (1999–2003). . . . For this reason we seek to cultivate and convince influential individuals . . . college and seminary presidents and faculty . . . and potential academic allies (Wedge Document).

As shown earlier, the Wedge has secured financial support from benefactors such as Howard Ahmanson and the Templeton Foundation (see later discussion). The Wedge's stability will surely depend on continued funding from such donors. On the increase of funding in 1999, Stephen Meyer announced, "We not only have a larger program than before, the existence of 'outyear' funding means greater long term stability."[148]

William Dembski obtained Baylor University President Robert Sloan's support for the establishment of the Polanyi Center and attributed this support, according to *Houston Press*, to Sloan's endorsement of his work: "'[Sloan] liked my stuff. He made it clear that he wanted to get me on the faculty in some way." The article continues, "Three years later, the president offered Dembski not just a position at Baylor but an independent center dedicated to studying the relationship between science and religion, and to furthering Dembski's own work in intelligent design."[149] Sloan's forceful defense of the Polanyi Center during the controversy over its establishment was recounted in the *Houston Chronicle*: "[Sloan] said alumni, students and parents have 'overwhelmingly' supported the goals of the Polanyi Center, but he would still back the center even without such support."[150] Sloan at the time reinforced his moral support of the center with a promise of financial backing for Dembski and Gordon: "Baylor spokesman Larry Brumley said the university will pay for Dembski's salary after the [John Templeton Fund] grant expires next year, and that it is paying Gordon's salary."

As the *Chronicle* also revealed, the John Templeton Fund has been a source of support: "[Dembski's] salary at the Polanyi Center is paid by a $75,000 grant from the John Templeton Fund, distributed through the Discovery Institute."[151] Regarding the controversy over the Polanyi Center's presence at Baylor, the *Chronicle* pointed out that although Sloan refused a request from the Faculty Senate to dissolve the center, he established a "nine-member committee of scholars primarily from outside Baylor to examine whether the Polanyi Center can contribute to constructive dialogue." Sloan's enthusiasm for Dembski cooled following Dembski's intemperate behavior after the External Review Committee issued its report (which was itself remarkably *temperate* under the circumstances). Yet even after Dembski's removal from the directorship of the Polanyi Center and the latter's absorption into the university's Institute for Faith and Learning (October 2000), Dembski continues his employment at Baylor, apparently for at least the remaining years of his 1999–2004 contract.

The steady flow of "soft" money has been critical to Dembski's ability to devote himself to the pursuit of his degrees and ID interests. According to his May 2002 curriculum vitae, he has secured a stream of fellowships and financial awards almost without interruption since 1982, when he graduated from college. He has actually engaged in teaching for only about five of those years, always in short stints of one to two years.[152] Judging from Dembski's c.v.—short on academic employment but long on apologetics and ID activities (lectures, conferences, and publications)—it is clear that he is a full-time Christian apologist and creationist.

4. *The Wedge must acquire physical bases of operation.*

The Polanyi Center was established in October 1999, giving the CRSC its first physical base outside Seattle, or, as William Dembski described the center in a November 17, 2000, Metaviews post, "the first intelligent design think-tank at a research university."[153] Although the Polanyi Center is defunct, and despite his demotion, Dembski still sees the success of ID at least among some faculty and with Baylor's president as proof of its ability to survive in *any* academic setting. He conveyed as much to the *Houston Press*: "Initially Dembski thought that if an intelligent-design center could be successful anywhere, it would be at Baylor. Now, he thinks that if an intelligent-design center could be successful at Baylor, it could succeed anywhere. . . . 'Baylor is—I didn't fully realize this—the bastion for the moderates where anything that smacks of fundamentalism, creationism, just sends people through the roof.'"[154]

In a December 21, 2000, UPI press release, while Baylor spokesman Larry Brumley said that "some bridge-building has to be done within the faculty" for Dembski's work to continue there, Dembski indicated that he would not relinquish his position: "I have offers from some Christian colleges . . . but to go there would mean handing a victory to my opponents here. My contract with Baylor runs for another four years."[155] And after July 2001, he made optimistic public statements on at least three occasions about improvements in his situation there, remarking that he would like to stay despite the opposition, as in an interview with Donald Yerxa: "A lot of good things have happened, and things are looking much better."[156] The CRSC has not established anything like the Polanyi Center in any other university, religious or secular. However, at Baylor and other universities where the CRSC has held conferences, Wedge members have accomplished the next best thing: they have brought the secular universities to *them*, by inviting mainstream scholars and scientists to participate. Whether it has a permanent base on a campus or simply operates as heretofore, through friendly, preexisting religious organizations such as Christian Leadership Ministries, by holding conferences on university campuses such as UT-Austin, Baylor, and Yale, the Wedge raises its profile and projects an *image* of scientific relevance—the falsity of that image doesn't really matter.

The Importance of "Academic" Conferences to the Wedge Strategy

The Wedge's holding highly publicized conferences on university campuses has *not* advanced ID in scientific circles or in the larger academic community; but these conferences have been critical to its strategy. Except for the November 2000 Yale conference, at which no pretense of even-handedness was made, the Wedge has made a point, as we have seen, of inviting selected opponents. Yet none of the pro-evolution invitees has been or is at all likely to accept ID, and none of the pro-ID speakers is likely to join the evolution side as a result of these conferences. Nobody (at least, no one identifiable) has undergone any such conversion. The purpose of the conferences is clearly neither the advancement of scientific knowledge nor the disinterested scrutiny of philosophical and theological issues. (Exactly the same is true of the Intelligent Design Network's Darwin, Design and Democracy conferences in Kansas; these are Wedge conferences as well.)

Although the university is a common setting for scholarly conferences, the Wedge has a goal other than scholarly debate: conferences on university campuses are important because they are accessible to large numbers of students, a key subpopulation for CRSC aims. When asked in an interview what kinds of readers he was anticipating for *Reason in the Balance*, Phillip Johnson said, "One audience is certainly Christian thinking people, including particularly college and graduate school students as well as faculty."[157]

Given the Wedge's close partnership with religious organizations having faculty mentors on hundreds of campuses, university students are a rich source of recruits. After the conference attendees depart, the students remain to be cultivated by sympathetic faculty and resident Wedge members (such as Robert Koons at UT-Austin, and Dembski and Gordon at Baylor). The Wedge Document refers in Phase II to the cultivation of future talent, and Johnson has referred optimistically to the prospect of gaining new members from the student population:

> [I]ncreasing numbers of high-school and college students come to the classroom already knowing that there are reasonable grounds for dissent, advocated by persons [such as Dembski, Behe, et al.] with impressive scientific academic credentials. . . . When the National Academy [of Sciences] dodges all the tough questions with evasive platitudes, *it effectively teaches independent-minded students to regard the pronouncements of science educators with no more trust than they regard political or commercial advertisements*. Eventually the scientific community will pay a high price for this campaign of prevarication. . . .
>
> I measure our success in two ways. First, many thousands of high-school and college students are reading our literature, and are responding very favorably. . . . The most talented of these will be the Wedge members of the future. Second, the Darwinists are completely unable to meet our challenge at the intellectual level, and scarcely try. . . . Once independent-thinking young people have read the dissenting literature, they are not likely to be impressed with the evasive statements of the Darwinist establishment.[158] (emphasis added)

The Wedge's conferences have also enabled the movement to feed off the reputations of famous people who attend, *even if these attendees have done so as hostile witnesses*. The April 2000 Nature of Nature conference featured major academic and scientific figures, and Discovery Institute publicized this in an April 7, 2000, news release: "Among the participants are two Nobel prize winners, Steven Weinberg and Christian de Duve, as well as noted scientists Alan Guth, Simon Conway Morris and others. . . . 'This is going to be the greatest collection of minds on the subject of directionality versus contingency in the natural sciences,' said [William] Dembski."[159] Weinberg's unequivocal rejection of all ID arguments is, of course, well known to professional scientists.[160]

There were other notable attendees; however, the fact that the Wedge invites opponents to participate in their conferences does not mitigate the advantage derived from such an arrangement—indeed, it is an integral part of the strategy. The Wedge's own agenda *requires* that they have a foil: the dynamic of conflict produced by confronting the opposition adds momentum to their movement. As stated in the Wedge Document, the presence of opponents actually serves their purpose: they are thereby able to construct a seemingly respectable platform for the discussion of ID theory, but with the ulterior motive of confrontation rather than collaboration, as stated in Phase III: "[W]e will move toward direct confrontation with the advocates of materialist science through challenge conferences in significant academic settings. . . . *The attention, publicity, and influence of design theory should draw scientific materialists into open debate with design theorists, and we will be ready*" (emphasis added). Clearly, the *substance* of scientific opposition to ID is irrelevant. The "debate" itself is all that matters.

Wedge leaders can subsequently construct websites around these conferences, as they did with the Mere Creation conference, the NTSE conference, and the Nature of Nature conference (until it was removed from Baylor's server after Dembski's demotion). They can market the conferences even after they are over, as they are doing with Design and Its Critics. (Intelligent Design Network markets on its website audio- and videotapes of its Darwin, Design, and Democracy III conference, including a "School Board Set.") They can publish conference papers as books, as they did following the Mere Creation and the Darwinism: Science or Philosophy? conferences (and planned to do after the symposium in China until their scheming was discovered). And they can bring the conferences to virtually anyone through audio files and CD-ROMs of the proceedings, as they have done with the Yale conference.[161] They can be quite certain that either the vocal opposition or—much more commonly—the utter indifference to ID of respectable invited scientists will make only one impression on the audience the CRSC really addresses: the sympathetic, conservative Christians identified in the Wedge Document as the Wedge's "natural constituents."

The impression created will be of the scientific legitimacy of ID as an alternative to "Darwinism."

There is no better statement of what having a platform either created or shared by their opponents means to the Wedge than the one offered by Greg Metzger of InterVarsity Press, which has published many of the Wedge's books. Metzger assessed the benefits to the Wedge of sharing a platform on the 1997 PBS *Firing Line* debate:

> There was no way to lose . . . this debate. . . . [We were] being invited, at the initiative of Barry [Lynn] who contacted Buckley, to air our views on a nationally televised . . . show with three of our leading lights . . . —Behe, Johnson, and Berlinski—accompanied by a veritable cultural icon, William Buckley. This legitimizes . . . I.D. . . . unlike any other . . . event that has yet occurred. Here you have nobody who could in any way be called, nor were they called during the debate, a fundamentalist of the 6-day variety. You had brilliant minds . . . in a civil setting. That, in and of itself, is a major victory that will last in the cultural imagination long after the specific points are forgotten. . . . [I]f a Ken Woodward of Newsweek or a Peter Steinfels from New York Times or anybody in the media of that ilk watches this broadcast it will be emblazoned in their minds that the terms of the debate have changed. . . .
>
> We must rush into this opening and press our points with continued vigor. . . .
>
> So be heartened, David, Phil and Michael. Your labor was . . . not in vain.[162]

5. *The Wedge seeks to exploit its presence in higher education to impress the public.*

Academic credentials are the ticket to success for the Wedge, and members take every opportunity to publicize their own. An example is a short article by Ray Bohlin entitled "Mere Creation: Science, Faith & Intelligent Design." In five pages of commentary, he never mentions a CRSC fellow (and he refers to six of them) *without* noting the academic credentials, as in the following: "So said Dr. Henry F. Schaefer III, professor of chemistry at the University of Georgia, author of over 750 scientific publications, director of over fifty successful doctoral students, and five-time Nobel nominee."[163]

This technique is central to the CRSC's tactics for promoting its brand of creationism. If any creationists are going to merge successfully into the academic mainstream, only the CRSC types, with an array of more or less legitimate academic credentials, stand any chance of accomplishing it. But they must convince uninformed observers who do not understand, or are unaware of, or do not care about their underlying political and religious agenda—or the actual status of their "science"—that they have paid their academic dues. The old-style, young-earth creationism is unlikely to be tolerated on mainstream campuses, even religiously affiliated ones like Baylor. Therefore, the CRSC creationists are the small subset who have taken the time and trouble to acquire legitimate degrees, which provide them with cover after they join university faculties. Johnson explains this in his *Communiqué* interview:

> CJ: Along those lines, what encouragement would you offer to a young student of science—let's say a young lady beginning a Ph.D. program in microbiology at a major university?
>
> Phil: We have a wonderful example here in Michael Behe . . . in what he is able to do while retaining a well funded lab and standing in the scientific world. . . . I think if we're clever enough in quoting the arguments and keeping people in the conversation and so on, and reassuring them that they can doubt Darwinism and still practice science just as well as ever—that it doesn't mean they are going to give up science and, you know, start thumping bibles instead or whatever—I think there'll just be a growing number of people who will get used to that conversation in that element. Behe has so far been able to maintain his standing, and he's getting invitations everywhere. Once you get someone like that breaks the ice, then there are opportunities for more people. So, I don't think you need to be in despair, *but you need to use a lot of tact and judgment and keep your head down while you're getting your Ph.D. in a lot of places—because there is dogmatism, but there are ways to overcome that.*[164](emphasis added)

By keeping their "head[s] down" at universities where they teach and study, intelligent design creationists can blend smoothly into the academic population by compartmentalizing their creationism—separating it from what they do professionally—or by cloaking it in technical, esoteric, and therefore academically palatable, language. Nothing is easier on the contemporary university campus than for scientists to talk about ostensible science to nonscientist colleagues.

There are examples of all styles of fitting in among the CRSC fellows whose work we have discussed in chapters 4 and 5. Academic expertise is sufficiently fragmented now for this tactic to work perfectly. They present little or no risk of embarrassment to their universities and increase their chances of being tolerated, at least by administrators and colleagues who are either directly sympathetic, or unaware of their agenda, or scientifically uninformed. (And because of the explosion of scientific research in all fields relevant to evolution, few faculty members outside the sciences on any campus *are* scientifically informed about evolutionary biology, or geology, or cosmology.) An example of this academic stealth creationism is Robert Koons's hosting of the "Naturalism, Theism and the Scientific Enterprise" conference in 1997 at the University of Texas. Koons acknowledges the advantage of having a sympathetic department head: "I . . . spoke to my department head [Daniel Bonevac] about making the department the official host. My chairman is a good friend of mine (who also happens to be a Christian and is very sympathetic to this sort of thing) and he agreed to attach the department's name to the conference. We didn't get any money from the university, but we did get clerical and administrative support."[165]

William Dembski plays an essential role in advancing the Wedge strategy in academia; the proof of this was his becoming director of the Polanyi Center. He is one of the most "degreed" Wedge members, with two earned doctorates; however, prior to arriving at Baylor, he had never maintained any long-term institutional affiliation except with the CRSC, which remains his *only* long-term affiliation.[166] Dembski's establishing an ID think-tank at Baylor was to be the official beginning of the Wedge's advance into the research university world. A month after being relieved of his duties at Baylor, in a November 17, 2000, Metaviews post, he continued to assert that the Baylor External Review Committee "validated intelligent design as a legitimate form of academic inquiry," while acknowledging that "the committee changed the center's name and took the center's focus off intelligent design" and that "I was finally removed as director."[167] An essential point to understand, however, is that despite Dembski's move to Baylor, he and Phillip Johnson continue to share indistinguishable aims. Each cites the other as a key figure in the ID movement. Johnson refers to Dembski as one of the "key Wedge figures."[168] Dembski cites Johnson as one of the people with whom the movement began and whose *Darwin on Trial* was a "key text" in the movement.[169] Moreover, Johnson has acknowledged as recently as August 1999 his own role as a representative of the movement and its role in carrying the ID debate into higher education, as well as public discussion: "[Evidence for intelligent design] is given in books published by the academic publishers, like Cambridge University Press, and by other scholars, scientists, philosophers in the intelligent design movement, which I represent, and which is carrying this issue into the universities and into the mainstream public discussion."[170]

Full academic legitimacy, not just of ID but of "theistic science," is the Wedge's number one goal.[171] The fact that theistic science will never overthrow legitimate science is irrelevant. Whether Wedge members win or lose on the merits of their arguments in their frequent debates is beside the point. Just getting the subject accepted as academically and culturally *conventional*—even when it is attacked—is the initial step toward victory. As early as 1996, in a review of Del Ratzsch's book, *The Battle of the Beginnings: Why Neither Side Is Winning the Creation/Evolution Debate*, Johnson acknowledged that even carrying the discussion into the *Christian* academic world is a "scandal" but "exciting":

> Our movement is something of a scandal in some sections of the Christian academic world for the same reason that it is exciting: we propose actually to engage in a serious conversation with the mainstream scientific culture on fundamental principles, rather than to submit to the demand that naturalism be conceded as the basis for all scientific discussions. That raises the alarming possibility, as one of Ratzsch's colleagues put it in criticizing me, that "the gulf between the academy and the sanctuary will only grow wider." The bitter feeling that has been spawned in some quarters by that possibility may explain why Ratzsch discusses our group so tentatively, but no matter. What matters for the present is to open up discussion.[172]

He acknowledged this again in 1997 in the *New York Times*: "Mr. [Kenneth] Miller also skewered Mr. Behe's book in a recent review. But that the book was even reviewed is progress in Mr. Johnson's view: 'This issue is getting into the mainstream. People realize they can deal with it the way they deal with other intellectual issues. . . . My goal is not so much to win the argument as to legitimate it as part of the dialogue.'"[173] In spring 1999, Johnson was still describing the ID movement as primarily "destructive" in its function—*admitting that the intelligent design movement so far had produced no "answers" of its own, despite its hope to have some in the future*:

> CJ: So, would it be fair to say that the goal is to undermine or call into question what has generally been accepted in the scientific community rather than purporting your own answers to all of the questions?
>
> Phil: Yes, the starting point is to understand what in the official answers is just dead wrong, because you can't get anywhere until you've made that step. *Now, obviously at some time in the future you hope to get to better answers which are actually true*, and that's a positive program, but you can't begin to work in that direction until you have an acknowledgement that the existing answers are false. You have to get the questions right before you can even determine the falsity of the answers. *So, for the time being, it's primarily a destructive work that's aimed at opening up a closed dogmatic field to new insights.*[174] (emphasis added)

So the Wedge progresses, and getting a foothold in the academic world continues to be crucial to its strategy, as Johnson stressed at D. James Kennedy's February 1999 "Reclaiming America for Christ" conference: "Johnson added that he is happy to be working with university professors, such as Michael Behe of Lehigh University in Pennsylvania. . . . This strategy, he said, 'enables us to get a foothold in the academic world and the academic journals.'"[175] Clearly, Johnson does not expect quantitatively impressive short-term results; rather, he sees intelligent design as a long-term project that will bear fruit after present Wedge members are gone: "I hope we'll be remembered as the pioneers who opened up the criticism and made it possible for the change to occur. It'll take decades . . . and we won't be around to see the final days, but maybe we'll be remembered as among those who started the ball rolling, and that'll be a great satisfaction."[176] In the meantime, the goal is to stay on the offensive and thus wear down the opposition: "Johnson speaks of a Wedge strategy, with himself the leading edge. 'I'm like an offensive lineman in pro football,' he says. . . . 'My idea is to clear a space by legitimating the issue, by exhausting the other side, by using up all their ridicule.'"[177] In Johnson's mid-1999 assessment of the success of the Wedge strategy in *Touchstone*, he remained convinced that Wedge strategists need only be patient, and eventually the academic world will come around: "As the discussion proceeds, the intellectual world will become gradually accustomed to treating materialism and naturalism as subjects to be analyzed and debated, rather than as tacit foundational assumptions that can never be criticized. Eventually the answer to our prime question will become too obvious to be in doubt."[178]

Confident that this strategy will work, Johnson uses Dembski as an example: "I attended a seminar on Dembski's ideas recently at a major university philosophy department where I saw from the reactions how common it is for clever people to deploy their mental agility in the service of obscurity. But Dembski put the concept of intelligent design on their mental maps, and eventually they will get used to it."[179] Clearly, the increasingly vocal and forceful resistance of intellectually qualified people to Dembski's claims does not matter at all to Johnson. Such resistance is almost as good as acquiescence: it makes for "name recognition"! Notably, he does *not* urge that Wedge strategists improve their arguments, or present hard empirical evidence to support their arguments, to secure a place in academic debate.

In the End as in the Beginning: Religion

The Wedge has not once broken its stride over the last ten years, despite its total failure in genuine scientific productivity—and despite the rapidly rising volume of expert criticism of its "science" and accompanying philosophical pretensions. But no matter: in his more candid moments, Johnson admits that this purportedly scientific/academic movement is religious to the core. A movement based on religion does not need the credibility afforded by scientific evidence. At the Reclaiming America for Christ conference, Johnson highlighted again the Wedge's driving religious purpose:

> The objective, he said, is to convince people that Darwinism is inherently atheistic, thus shifting the debate from creationism vs. evolution to the existence of God vs. the non-existence of God. From there people are introduced to "the truth" of the Bible and then "the question of sin" and finally "introduced to Jesus."
>
> "You must unify your own side and divide the other side," Johnson said. He added that he wants to temporarily suspend the debate between the young-Earth creationists, who insist that the planet is only 6,000 years old, and old-Earth creationists, who accept that the Earth is ancient. This debate, he said, can be resumed once Darwinism is overthrown.[180]

Apparently this view is shared by Johnson's Wedge colleagues and is considered its "defining concept":

> My colleagues and I speak of "theistic realism"—or sometimes, "mere creation"—as the defining concept of our movement. That means that we affirm that God is objectively real as Creator, and that the reality of God is tangibly recorded in evidence accessible to science, particularly in biology. We avoid the tangled arguments about how or whether to reconcile the Biblical account with the present

state of scientific knowledge, because we think these issues can be much more constructively engaged when we have a scientific picture that is not distorted by naturalistic prejudice.[181]

As Johnson assessed the state of the Wedge in his keynote speech at IDnet's Darwin, Design and Democracy symposium in Kansas on June 29, 2001, he asserted that "a movement like this doesn't really need to win all its battles. What you find is that after a temporary setback, they're taking two steps forward. They come back strong and more determined to avoid whatever mistakes were made before . . . to have that much more dedication in the future." He also stressed anew the religious goals of the Wedge:

> We founded . . . the Intelligent Design movement . . . to explore and explain the evidence which . . . does point towards the need for a designer, for a Creator. . . .
>
> The second goal was to . . . unify the religious world. By unify, I don't mean . . . everybody signs . . . a statement . . . agreeing on everything. . . . But there should be a central issue that people agree to discuss first . . . the simple question of creation—do you need a Creator to do the creating, or don't you? . . . [T]he immediate response will be that the evidence of science is viewed through the . . . prejudice that natural causes can do and did do the whole job. End of story. And so we thought . . . religious people ought to challenge that. The people of God ought to be unwilling to accept that kind of dogmatic decision. . . .
>
> And the people of God will have had an opportunity to work together for a common result.

And Johnson ended on this decidedly optimistic note:

> So, what is the state of the Wedge? The state of the Wedge is very good. Kansas made a great contribution to that state by raising the issue. . . . It enabled us to organize a wonderful grassroots movement here in Kansas that is going to spread to many other places. . . . The state of the Wedge is fine.[182]

Our hope is that readers will see that Johnson's optimistic assessment of the Wedge's progress and present status is justified, albeit not by the scientific, or philosophical, or legal, or even generally religious merits of his case. In the story of the Wedge to date, we see a demonstration of the power of public relations to shape public opinion and policy on the largest scale—in ways that have nothing to do with the true state of scientific knowledge. And our final hope is that readers will consider seriously the question of what they ought to be doing about it.

Notes

Introduction

1. Donald H. Naftulin, John E. Ware, Jr., and Frank A. Donnelly, "The Doctor Fox Lecture: A Paradigm of Educational Seduction," *Journal of Medical Education* 48 (July 1973), 630–635.

2. Naftulin et al., 631.

3. Naftulin et al., 633.

4. William A. Dembski, "Intelligent Design Coming Clean." Posted on Metaviews in November 2000. Accessed on May 4, 2002, at http://www.arn.org/docs/dembski/wd_id comingclean.htm.

5. Dembski, "Intelligent Design Coming Clean."

6. Phillip E. Johnson, "The Wedge: A Progress Report," Access Research Network. Accessed on April 21, 2001, at http://www.arn.org/docs/johnson/pj_wedge progress041601.htm.

7. See the archive of "Phillip Johnson's Weekly Wedge Update" at http://www.arn.org/wedge.htm.

8. See James Glanz, "Evolutionists Battle New Theory on Creation," *New York Times*, April 8, 2001. Accessed on April 22, 2001, at http://www.nytimes.com/2001/04/08/ science/08DESI.html. See also Teresa Watanabe, "Enlisting Science to Find the Fingerprints of a Creator: Believers in 'Intelligent Design' Try to Redirect Evolution Disputes Along Intellectual Lines," *Los Angeles Times*, March 25, 2001. Accessed on April 22, 2001, at http://www.discovery.org/news/Enlisting Science.html.

9. Robert Wright, "The 'New' Creationism," *Slate*, April 16, 2001. Accessed on April 22, 2001, at http://slate.msn.com/ Earthling/01-04-16/Earthling.asp.

10. "2001 Church/State Legislation," Americans United for the Separation of Church and State, April 18, 2001.

11. Will Sentell, "Baton Rouge Legislator

Calls Theory Racist," *The Advocate*, April 18, 2001. Accessed on April 26, 2001, at http://www.theadvocate.com/news/story.asp?StoryID=20792. At Weston-Broome's April 17, 2001, public meeting, when a questioner asked her what alternatives to teaching evolution she would consider, she mentioned "the design intelligence [*sic*] theory."

12. Intelligent Design Network, "Second Annual IDNet Symposium." Accessed on April 26, 2001, at http://www.intelligentdesignnetwork.org/june_symposim.htm.

13. See references to the two Macoskos' teaching activities in *Newsletter of the American Scientific Affiliation and Canadian Scientific and Christian Affiliation*, January/February 2001. Accessed on April 23, 2003, at http://users.stargate.net/~dfeucht/JAN-FEB01.htm.

14. "2001 Spring Meeting," American Geophysical Union. Accessed on April 26, 2001, at http://www.agu.org/meetings/sm01top.html.

15. National Center for Science Education, "Evolving Banners at the Discovery Institute." Accessed on August 29, 2002, at http://www.ncseweb.org/resources/articles/4116_evolving_banners_at_the_discov_8_29_2002.asp.

1. How the Wedge Began

1. Stephen Goode, "Johnson Challenges Advocates of Evolution," *Insight on the News*, October 25, 1999. Accessed on July 9, 2000, at http://www.arn.org/docs/johnson/insightprofile1099.htm.

2. Nancy Pearcey, "Wedge Issues: An Intelligent Discussion with Intelligent Design's Designer," *World*, July 29, 2000. Accessed on July 27, 2000, at http://www.discovery.org/crsc/CRSCrecentArticles.php3?id=416.

3. Nancy Pearcey, "We're Not in Kansas Anymore: Why Secular Scientists and Media Can't Admit that Darwinism Might Be Wrong," *Christianity Today*, May 22, 2000. Accessed on August 3, 2000, at http://christianityonline.com/ct/2000/006/1.42.html.

4. Lynn Vincent, "Science vs. Science," *World*, February 26, 2000. Accessed on August 30, 2000, at http://www.discovery.org/viewDB/index.php3?program=CRSCstories&command=view&id=148. Wedge member Stephen C. Meyer has recorded his first encounter with Johnson, which occurred during this stay in England, in his article "Darwin in the Dock: A History of Johnson's Wedge," *Touchstone* (July/August 1999). Accessed on March 29, 2002, at http://www.touchstonemag.com/docs/issues/14.3docs/14-3pg57.html.

5. Tim Stafford, "The Making of a Revolution," *Christianity Today* 41 (December 8, 1997). Accessed on February 14, 2000, at http://www.arn.org/johnson/revolution.htm. It is interesting that Johnson's "epiphany"—and his ID crusade—began in 1987, the year the U.S. Supreme Court issued its *Edwards v. Aguillard* ruling outlawing creationism in American public schools. As a law professor, Johnson was certainly aware of this landmark ruling. In fact, in December 1989 he attended a private conference, "Science and Creationism in Public Schools," held to discuss the issue of religion and public schools in the wake of *Edwards*. Other participants were First Amendment scholar Charles Haynes, paleontologist David Raup, and Stephen Jay Gould. The Wedge has crafted its legal arguments regarding the admissibility of ID in public schools with *Edwards v. Aguillard* specifically in mind, in an attempt to evade its constraints against creationism. See Donald A. Yerxa, "Phillip Johnson and the Origins of the Intelligent Design Movement, 1977–1991," *Perspectives on Science and the Christian Faith* 54:1 (March 2002), 47–52. See also CRSC fellows David K. DeWolf and Stephen Meyer's "Intelligent Design in Public School Science Curricula: A Legal Guidebook" at http://law.gonzaga.edu/people/dewolf/fte.htm. See *Edwards v. Aguillard* (1987) at http://caselaw.lp.findlaw.com/scripts/getcase.pl?court=US&vol=482&invol=578.

6. Phillip E. Johnson, "The Wedge: Breaking the Modernist Monopoly on Science,"

Touchstone (July/August 1999). Accessed March 9, 2000, at http://www.touchstonemag.com/docs/issues/12.4docs/12-4pg18.html.

7. Vincent, "Science vs. Science."

8. Tom Woodward, "Meeting Darwin's Wager," Part 2, *Christianity Today*. Accessed on March 27, 2000, at http://www.christianityonline.com/ct/7t5/7t514b.html. William Dembski provides the name of this meeting in his curriculum vitae at http://www.designinference.com/documents/2003.02.CV.htm.

9. "Ad Hoc Origins Committee: Scientists Who Question Darwinism." Accessed on May 24, 2000, at http://www.apologetics.org/news/adhoc.html. Gould's review of *Darwin on Trial*, "Impeaching a Self-Appointed Judge," is online at http://www.freethought-web.org/ctrl/gould_darwin-on-trial.html. Johnson's reply to Gould was printed in *Origins Research* 15:1 (the forerunner of the creationist journal *Origins and Design*) and is online at http://www.arn.org/docs/orpages/or151/151johngould.htm. Both accessed on April 6, 2002.

10. "Ad Hoc Origins Committee." The list of signatories is available at http://www.apologetics.org/news/adhoc.html. See also Pearcey, "We're Not in Kansas Anymore."

11. "Major Grants Help Establish Center for Renewal of Science and Culture," Discovery Institute *Journal* (August 1996). Accessed on July 24, 2000, at http://www.discovery.org/w3/discovery.org/journal/center.html. See published presentations from this conference in *Intercollegiate Review* (spring 1996).

12. Phillip E. Johnson, *Reason in the Balance: The Case Against Naturalism in Science, Law and Education* (Downers Grove, IL: InterVarsity Press, 1995), 208-9.

13. Larry Witham, "Contesting Science's Anti-Religious Bias," *Washington Times*, December 29, 1999. Accessed on June 26, 2002, on the Discovery Institute website at http://www.discovery.org/viewDB/index.php3?program=CRSCstories&command=view&id=65. Even though the Center for the Renewal of Science and Culture—the special, creationist arm of the larger Discovery Institute—is the subject of this study, "CRSC" and "Discovery Institute" (DI) are often used interchangeably, as will occasionally be done here.

14. "Major Grants Help Establish Center for Renewal of Science and Culture," Discovery Institute *Journal* (August 1996).

15. Witham, "Contesting Science's Anti-Religious Bias."

16. This information was at http://www.discovery.org/w3/discovery.org/crsc/crsc96fellows.html but is no longer accessible. Another page from this directory, which appears to have been posted somewhat later than 1996-97, although the year is not specified, lists additional fellows: Walter Bradley, chair, Department of Mechanical Engineering, Texas A & M; John Angus Campbell, professor of speech communications, University of Memphis; William Lane Craig, research professor, Talbot School of Theology; Jack Harris, Ph.D. candidate, University of Washington; Dean H. Kenyon [co-author *Of Pandas and People*], San Francisco State University; Nancy Pearcey, Wilberforce Forum; and Charles Thaxton, Charles University, Prague. Well-known conservative George Gilder is listed as an advisor along with Phillip Johnson.

17. See the "Mere Creation" website for this conference at http://www.origins.org/mc/menus/index.html. See also the article by Jay Grelen, "Witnesses for the Prosecution," *World*. Accessed March 1, 2000, at http://www.worldmag.com/world/issue/11-30-96/national_2.asp. The conference title derives from twentieth-century Christian apologist C. S. Lewis's book *Mere Christianity*. DI has a program entitled "C. S. Lewis and Public Life" to foster understanding of the "connections between faith and public life." See http://www.discovery.org/lewis/purpose.php3. Dembski explains "mere creation": "[F]or C. S. Lewis, 'mere' Christianity signified . . . what minimally one must hold to be a Christian . . . [and] 'mere' creation needs to be interpreted . . . [as] what minimally must be included under a doctrine of creation." The Wedge seeks to emphasize the essentials of a doctrine of creation while de-emphasizing disagreements (e.g., over the age of the earth) in order to unify a "Christian world [which] is badly riven about creation." See

Dembski's "Introduction: Mere Creation" in *Mere Creation: Science, Faith and Intelligent Design* (Downers Grove, IL: InterVarsity Press, 1998), 13.

18. Henry F. Schaefer III, foreword, in *Mere Creation: Science, Faith and Intelligent Design*, ed. William Dembski (Downers Grove, IL: InterVarsity Press, 1998), 9.

19. See http://www.clm.org/ttt/contact.html and http://www.leaderu.com/menus/aboutus.html, respectively. More information on the connection between the Wedge and Christian Leadership Ministries will be given in chapter 9.

20. Neither this conference nor any other CRSC activities have produced any such scientific research, as will be shown.

21. Biola, however, is distinctly *not* a secular university. Its name is derived from the acronym of its former name, the *Bible Institute of Los Angeles*. See http://www.biola.edu/admissions/about/history.cfm. Accessed on April 6, 2002.

22. Scott Swanson, "Debunking Darwin?" *Christianity Today*, January 6, 1997. Accessed on July 25, 2000, at http://www.christianityonline.com/ct/7t1/7t1064.html.

23. The NTSE conference is discussed in the analysis of Wedge activities in chapter 7.

24. Swanson, "Debunking Darwin?"

25. Schaefer, foreword,10–11.

26. Howard Ahmanson's financial support of the CRSC will be discussed in chapters 6 and 9.

27. Phillip E. Johnson, *Defeating Darwinism by Opening Minds* (Downers Grove, IL: InterVarsity Press, 1997), 92.

28. Phillip E. Johnson, *Objections Sustained: Subversive Essays on Evolution, Law and Culture* (Downers Grove, IL: InterVarsity Press), 1998.

29. "Major Grants Help Establish Center for the Renewal of Science and Culture," Discovery Institute *Journal* (August 1996).

2. The Wedge Document: A Design for Design

1. See Chris Mooney, "Survival of the Slickest," *American Prospect* (December 2, 2002), accessed on March 23, 2003, at http://www.prospect.org/print/V13/22/mooney-c.html. See also Barbara Forrest, "The Wedge at Work: How Intelligent Design Creationism Is Wedging Its Way into the Cultural and Academic Mainstream," in *Intelligent Design and Its Critics: Philosophical, Theological, and Scientific Perspectives*, ed. Robert T. Pennock (Cambridge, MA: MIT Press, 2001), 14. For the litany of evolutionary evils, see Christopher P. Toumey, *God's Own Scientists* (New Brunswick, NJ: Rutgers University Press, 1994).

2. Center for the Renewal of Science and Culture, "The Wedge Strategy." Accessed on October 21, 2002, at http://www.public.asu.edu/~jmlynch/idt/wedge.html and http://www.antievolution.org/features/wedge.html.

3. William A. Dembski, ed., *Mere Creation: Science, Faith and Intelligent Design* (Downers Grove, IL: InterVarsity Press, 1998).

4. James Porter Moreland and John Mark Reynolds, *Three Views on Creation and Evolution*, Counterpoint Series (Grand Rapids, MI: Zondervan, 1999). See http://www.amazon.com.

5. "TechnoPolitics: Program No. 734. Airdate: November 15, 1997," and "TechnoPolitics: Program No. 739. Airdate: December 19, 1997," published transcripts. Accessed February 26, 2000, at http://www.arn.org/docs/techno/techno1197.htm and http://www.arn.org/docs/techno/1297.htm.

6. Mindy Cameron, "Theory of 'Intelligent Design' Isn't Ready for Natural Selection," *Seattle Times*, June 3, 2002. Accessed on June 9, 2002, at http://www.seattletimes.nwsource.com/html/editorialsopinion/134466429_mindy03.html.

7. See James Still, "Discovery Institute's 'Wedge Project' Circulates Online," http://www.infidels.org/secular_web/feature/1999/wedge.html. See also "The Wedge: A

Christian Plan to Overthrow Modern Science?" by Keith Lankford at http://www.freethought-web.org/ctrl/archive/thomas_wedge.html. Both accessed June 9, 2002. Although a number of people first received the Wedge Document in early March 1999, it appears to have first been made public on February 5, 1999, when it was posted to the online "Virus" discussion list at http://www.lucifer.com/virus/virus.1Q99/0510.html by Tim Rhodes, a list member. According to Rhodes, the document was "liberated" from the Discovery Institute by someone in Seattle, where the Discovery Institute is located. Accessed on June 9, 2002.

8. James Still to Barbara Forrest, personal communication, February 29, 2000.

9. James Still to Barbara Forrest, personal communication, February 10, 2000.

10. Still, "Discovery Institute's 'Wedge Project' Circulates Online."

11. Still, "Discovery Institute's 'Wedge Project' Circulates Online." Richards also pointed out to Still that "the general concept of the 'Wedge' is described in Phillip Johnson's book *Defeating Darwinism by Opening Minds*."

12. James Still to Molleen Matsumura, personal communication, March 11, 1999.

13. Hal Ostrander, "Intelligent Design Theory: A Powerful Tool in Confronting Darwinism," *SBC Life* (February/March 1998), 10.

14. "Major Grants Help Establish Center for the Renewal of Science and Culture," Discovery Institute *Journal* (August 1996). Accessed April 18, 2000, at http://www.discovery.org/w3/discovery.org/journal/center.html.

15. A website posted by Richard J. Botting, Ph.D., a computer science professor at California State University in San Bernardino, includes the Discovery Institute in a listing of March 1996 new site announcements. Accessed April 19, 2000, at http://www.csci.csusb.edu/doc/1996.03.www.sites.html.

16. "What is The Center for the Renewal of Science & Culture All About?" Accessed March 18, 2000, at http://www.discovery.org/w3/discovery.org/crsc/aboutcrsc.html. This document was no longer available as of May 17, 2000. The directory had been available at http://www.discovery.org/w3/discovery.org/crsc on March 18, 2000, but access was denied as of April 17, 2000. By May 17, 2000, the directory, too, had been removed.

17. "Major Grants Help Establish Center for Renewal of Science and Culture," Discovery Institute *Journal* (August 1996).

18. John G. West, Jr., "The Death of Materialism and the Renewal of Culture," *Intercollegiate Review* (spring 1996), 3. Accessed on August 18, 2002, at http://www.isi.org/publications/ir/spr96/west.pdf. It is consistent with the CRSC's anti-evolution raison d'être that West cites Genesis 1:27 at the beginning of this article: "So God created man in his own image, in the image of God created He him; male and female created He them."

19. "1996–1997 Research Fellows at the Center for the Renewal of Science and Culture." Accessed on March 18, 2000, at http://www.discovery.org/w3/discovery.org/crsc/crsc96fellows.html.

20. Discovery Institute, "Center for Renewal of Science and Culture," Discovery Institute *Journal*. Accessed April 18, 2000, at http://www.discovery.org/w3/discovery.org/journal. There is some discrepancy in the date of this issue. At the URL and on the table of contents page, it is spring 1998. However, on all succeeding pages, it is dated winter 1998.

21. This brochure was in the /w3/ directory, which is no longer available.

22. "Life After Materialism?" Accessed October 16, 2001, at http://www.discovery.org/crsc/materialism.html.

23. See the current list of CRSC fellows at http://www.discovery.org/crsc/fellows/index.html. The list on the CRSC website numbers only forty-three. However, creationist Walter ReMine also holds a CRSC fellowship, as he indicated in his signature on a November 2001 in a letter to Arthur S. Lodge: "Replies to Cosmic Ancestry, 2001," Walter ReMine to Arthur S. Lodge, November 29, 2001. Accessed on September 24, 2002, at http://www.panspermia.org/replies4.htm.

3. Searching for the Science

1. William A. Dembski, "Shamelessly Doubting Darwin," *American Outlook* (November/December 2000). Accessed on December 1, 2000, at http://www.hudson.org/American_Outlook/articles_nov-dec_00/dembski.htm.

2. Ask several professional molecular biologists what this statement means. You will surely find that most or all of them reply, after puzzling over it, "nothing."

3. Center for the Renewal of Science and Culture, "About CRSC." Accessed on September 23, 2000, at http://www.crsc.org/about.html.

4. Center for the Renewal of Science and Culture, "The Wedge Strategy." See the document at http://www.public.asu.edu/~jmlynch/idt/wedge.html and http://www.antievolution.org/features/wedge.html.

5. Center for the Renewal of Science and Culture, "The Research Fellowship Program." Accessed on August 28, 2000, at http://www.crsc.org/fellows/fellowshipInfo.html.

6. "CRSC Innovates in Media and Academia," Discovery Institute *Journal* (spring/winter 1998). Accessed on August 4, 2001, at http://www.discovery.org/w3/discovery.org/journal/spring98.html.

7. These developments are nowhere identified.

8. Center for the Renewal of Science and Culture, "Design Theory: A New Science for a New Century." Accessed on March 29, 2002, at http://www.discovery.org/crsc/index.php3. The word "evidences" (plural) in CRSC's description of its "scientific" research program is an immediate tip-off of the movement's religious motivation. Creationists consistently refer to "evidences" of creation/design rather than merely to "evidence." This oddity is part of evangelical Christian apologetics, exemplified in Probe Ministries' "Christian Apologetics: An Introduction," by Rick Wade: "A true knowledge of God is based upon divine testimony which is accepted by faith, but which is also confirmed for us by *evidences* of various types" (emphasis added). (Both Probe Ministries and "Leadership University," where this document is posted, are close affiliates of the intelligent design movement.) Accessed on August 2, 2001, at http://www.leaderu.com/orgs/probe/docs/apologet.html. Eugenie Scott of the National Center for Science Education points out that "the only people who use 'evidences' (plural) are creationists or people who have spent far too much time reading their literature! 'Evidences' is a term from Christian apologetics." See Douglas Theobald, "29 Evidences for Macroevolution: Scientific Evidences for the Theory of Common Descent with Gradual Modification," The Talk.Origins Archive, at http://www.talkorigins.org/faqs/comdesc/evidences.

9. Phillip E. Johnson, "How to Sink a Battleship: A Call to Separate Materialist Philosophy from Empirical Science," *Real Issue* (November/December 1996). Accessed on January 9, 2001, at http://www.leaderu.com/real/ri9602.html. This article is located on the site of Christian Leadership Ministries' "Leadership University," which describes itself as "Telling the Truth *at the speed of life*."

10. Nancy Pearcey, "Opening the 'Big Tent' in Science: The New Design Movement." Accessed on June 6, 2000, at http://www.arn.org/docs/pearcey/np_bigtent30197.htm. Originally published as "Evolution Backlash" in *World*, March 1, 1997.

11. George W. Gilchrist, "The Elusive Scientific Basis of Intelligent Design," *Reports of the National Center for Science Education* 17:3 (May/June 1997), 14–15. Accessed on August 8, 2001, at http://www.ncseweb.org/resources/rncse_content/vol17/697_the_elusive_scientific_basis_o_12_30_1899.asp.

12. Center for the Renewal of Science and Culture, "As the Millennium Ends: The Promise of Better Science and a Better Culture." Accessed on June 13, 2000, at http://www.discovery.org/w3/discovery.org/journal/1999/crsc.html.

13. Please see the comments of philosopher of science Professor William Wimsatt in chapter 5.

14. See, for example, chapters 10 and 11 of Daniel Dennett's *Darwin's Dangerous Idea* (New York: Simon & Schuster, 1995), or the articles and bitter exchanges following

Maynard Smith's glowing review of Dennett's book in the *New York Review of Books*, November 30, 1995, pp. 46–48.

15. Most of their books have been published by religious presses—Zondervan and InterVarsity Press. However, that is changing. Dembski's *The Design Inference* was published by Cambridge University Press (1998), and Behe's *Darwin's Black Box* was published by The Free Press (1996). As time goes on, the CRSC's goal of wedging into the cultural and academic mainstream is being facilitated by their entry into the publishing mainstream.

16. Center for the Renewal of Science and Culture, "Year End Update" (November/December 1997). Accessed on March 18, 2000, at http://www.discovery.org/w3/discovery.org/crsc/crscnotes2.html. This document is no longer accessible. The "Consultation on Intelligent Design" was also publicized in the Discovery Institute spring 1998 *Journal*: "In November [1997], the Center co-sponsored with Christian Leadership Ministries a 'Consultation on Intelligent Design' in Dallas that brought together more than forty scholars from around the world to share research." Discovery Institute *Journal: Annual Report* (spring 1998), 14. Accessed on June 13, 2000, at http://www.discovery.org/w3/discovery.org/journal/spring98.html.

17. National Institute for Research Advancement to Barbara Forrest, personal communication, January 16, 2001. The NIRA is a "policy research organization established on the initiative of leading figures from the industrial, academic and labor communities . . . [and] is operated through an endowment of capital contributions and donations from both the public and private sectors." It maintains a website on the Japanese government's web domain. The information on DI was accessed on January 4, 2001, at http://www.nira.go.jp/ice/tt-info/nwdtt99/c1233.html.

18. The DI spring 1998 *Journal* was available at http://www.discovery.org/w3/discovery.org/journal/. Accessed on January 16, 2001. Axe's bio was posted on the Center for the Renewal of Science and Culture website, "Senior Fellow Douglas Axe," at http://www.discovery.org/crsc/fellows/DougAxe/ but is no longer available. After he was asked about his bio's remaining on the website as of June 2, 2000, after his listing as a fellow had been removed (Barbara Forrest to Douglas Axe, personal communication, June 2, 2000), the bio, too, was subsequently removed.

19. Douglas Axe to Barbara Forrest, personal communication, June 14, 2000.

20. Douglas Axe to Barbara Forrest, personal communication, June 22, 2000.

21. Center for the Renewal of Science and Culture, "Year End Update," (November/December 1997). Accessed on March 18, 2000, at http://www.discovery.org/w3/discovery.org/crsc/crscnotes2.html. This file is no longer available. The spring 1998 *Journal* containing news about the Consultation on Design (14) was available, without a reference to Axe, at http://www.discovery.org/w3/discovery.org/journal/.

22. William A. Dembski, *Intelligent Design: The Bridge Between Science and Theology* (Downers Grove, IL: InterVarsity Press, 1999), 21.

23. PRG.

24. "Philosophy 333: Evolution and Creation & Liberal Studies 487: Senior Seminar on Evolution and Creation," Instructors James R. Hofmann, CSUF Philosophy Department, and Bruce H. Weber, CSUF Department of Chemistry and Biochemistry. Accessed on February 5, 2001, at http://nsmserver2.fullerton.edu/departments/chemistry/evolution_creation/web/.

25. Message 8881, from Larry Arnhart to Yahoo Evolutionary Psychology discussion group, Monday, December 4, 2000. Accessed on February 5, 2001, at http://groups.yahoo.com/group/evolutionary-psychology/message/8881 (public archive). The abstract to which Dembski refers is available at http://www.sciencedirect.com. Accessed on April 10, 2003.

26. Bryn Nelson, "6 Days of Creation: The Search of Evidence," *Newsday*, March 12, 2002. Accessed on March 14, 2002, at http://www.newsday.com/news/health/ny-dsspdn2621111mar12.story?coll=ny%0D2Dhealth%2Dheadlines.

27. Douglas Axe to Barbara Forrest, personal communication, October 25, 2001.

28. The "Instructions to Authors" on the *JMB* website instruct authors of papers to "supply five keywords after the Summary." Axe's keywords are "exposed residues; protein homologues; molecular evolution; neutral theory; [and] sequence space." The "Aims and Scope" of *JMB* reflect no categories for the publication of ID articles: "Suitable subject areas include: *(a) Genes:* Expression, replication and recombination, sequence organization and structure, genetics of eukaryotes and prokaryotes. *(b) Viruses and Bacteriophages:* Genetics, structure, growth cycle. *(c) Cells and Development:* Developmental biology, organelle structure and function, motility, transport and sorting of macromolecules, energy transfer, growth control. *(d) Proteins, Nucleic Acids and other Biologically important Macromolecules:* Molecular structure, physical chemistry, molecular engineering, macromolecular assembly and enzymology. The *Journal* will not as a rule publish papers which fall outside the areas defined above." *Journal of Molecular Biology,* "Instructions to Authors." Accessed on March 23, 2001, at http://www.academicpress.com/www/journal/mb/mbifa.htm#submission.

29. Gilchrist, "The Elusive Scientific Basis of Intelligent Design."

30. Gilchrist, "The Elusive Scientific Basis of Intelligent Design." Adapted from Table 1. Used with permission. Two indexes were omitted from Gilchrist's table and not used in this survey: Expanded Academic Index was not used since, as Gilchrist says in his article, it covers the humanities and the social sciences; hence, it would not have suited the purposes of this survey. The Life Sciences Collection was not consulted because, according to Gilchrist (telephone interview, July 3, 2000), it has been superceded by other databases.

31. John Lynch to Barbara Forrest, personal communication, July 8, 2000. Dr. Lynch is preparing his survey for publication.

32. The books by Johnson do not include *The Wedge of Truth: Splitting the Foundations of Naturalism* (InterVarsity Press, 2000). Neither does Lynch's survey include Dembski's 2002 Rowman and Littlefield book, *No Free Lunch: Why Specified Complexity Cannot Be Purchased without Intelligence.*

33. John Lynch to Barbara Forrest, personal communication, July 8, 2000.

34. John G. West, Jr., Letter to *Books & Culture* (November/December 1999). Accessed on March 17, 2001, at http://www.christianitytoday.com/bc/9b6/9b6004.html.

35. Berlinski is a mathematician, not a philosopher.

36. For an accessible sample of the hard-hitting analyses of ID arguments by Professor Miller (who is a Christian), see http://biocrs.biomed.brown.edu/Darwin/DI/Design.html. Accessed on April 29, 2002.

37. Laurie Goodstein, "Christians and Scientists: New Light for Creationism," *New York Times,* December 21, 1997. Accessed on August 3, 2001, at http://www.arn.org/docs/fline1297/fl_goodstein.htm.

4. Paleontology Lite and Copernican Discoveries

1. David K. DeWolf, Stephen C. Meyer, and Mark E. DeForrest, "Intelligent Design in Public School Science Curricula: A Legal Guidebook." Accessed on May 22, 2002, at http://www3.baylor.edu/~William_Dembski/docs_resources/guidebook.htm.

2. A long, repetitive argument on how or why the Cambrian fossils support ID has been written by Stephen Meyer, Paul Nelson, and Paul Chien, though not yet (at this writing) published in a scientific journal. "The Cambrian Explosion: Biology's Big Bang," appears intended for publication in a book cited in the DeWolf-Meyer-DeForrest article as upcoming at Michigan State University Press: *Intelligent Design, Darwinism and the Philosophy of Public Education.* (This article was accessed on May 22, 2002, on the Intelligent Design Undergraduate Research Center website at http://www.idurc.org.) The organizer-editor of this volume is a faculty member in rhetoric and communications. None of the authors of the article is a professional paleontologist. Meyer and Nelson are philosophers—

Nelson is, in fact, a young-earth creationist, which lends irony to his co-authoring a paper on fossils the authors acknowledge to be hundreds of millions of years old.

3. An excellent and accessible summary of the evidence and the arguments can be found in the article by Douglas Erwin, James Valentine, and David Jablonski, "The Origin of Animal Body Plans," *American Scientist* (March/April 1997). Accessed on May 22, 2002, at http://www.sigmaxi.org/amsci/articles/97articles/Erwin.html.

4. See Mark Ridley, *Evolution*, 2nd ed. (Cambridge: Blackwell Science, 1996), 537–552. See also Nigel C. Hughes, "Creationism and the Emergence of Animals: The Original Spin," *Reports of the National Center for Science Education*, 20:5 (September/October 2000), for Hughes's explanation of the "spin" CRSC creationists put on the Chengjiang fossils and his correction of their distortion of the fossils' scientific significance. Accessed on September 6, 2000, at http://www.ncseweb.org/resources/rncse_content/vol20/5545_creationism_and_the_emergence__12_30_1899.asp.

5. Graham E. Budd et al., and David J. Siveter et al. "Crustaceans and the 'Cambrian Explosion,'" Technical Comments, *Science* 294 (December 7, 2001), 2047a. Accessed on March 5, 2002, at http://www.sciencemag.org/cgi/content/full/294/5549/2047a.

6. Keith B. Miller, "The Precambrian to Cambrian Fossil Record and Transitional Forms," *Perspectives on Science and Christian Faith* 49 (December 1997). Accessed on March 5, 2002, at http://www.asa3.org/ASA/topics/evolution/PSCF12-97Miller.html.

7. The uniqueness of the Chengjiang fossils, some of which are indeed soft-bodied, lies in the rare circumstances of formation of the rock in which they are entombed.

8. Simon Conway Morris, *The Crucible of Creation: The Burgess Shale and the Rise of Animals* (New York: Oxford University Press, 1998).

9. Stephen Jay Gould, *Wonderful Life: The Burgess Shale and the Nature of History* (London: Hutchinson Radius, 1989).

10. Richard Dawkins, *Unweaving the Rainbow: Science, Delusion, and the Appetite for Wonder* (Boston: Houghton Mifflin Company, 1998), 205.

11. Sean B. Carroll, Jennifer K. Grenier, and Scott D. Weatherbee, *From DNA to Diversity: Molecular Genetics and the Evolution of Animal Design* (Malden, MA: Blackwell Science, 2001); Wallace Arthur, *The Origin of Animal Body Plans* (New York: Cambridge University Press, 1997, 2000).

12. Eleanor Lawrence, "Pushing Back the Origins of Animals," Science Update, *Nature*. Accessed on March 5, 2002, at http://www.nature.com/nsu/990204/990204-4.html.

13. An impartial and knowledgeable account of the technical arguments on these questions of Richard Dawkins and Stephen J. Gould has been written by philosopher Kim Sterelny: *Dawkins vs. Gould: Survival of the Fittest* (Cambridge: Icon Books UK, 2001).

14. Stephen Jay Gould, "Macroevolution," in *The Oxford Encyclopedia of Evolution*, Vol. 1, ed. Mark Pagel (Oxford: Oxford University Press, 2002), E23–28.

15. Cecilia Yau, "The Twilight of Darwinism at the Dawn of a New Millennium: An Interview with Dr. Paul Chien," *Challenger* Magazine (February/March 2000). Accessed on May 22, 2002, at http://www.ccmusa.org/challenger/000203/doc1.html.

16. It is not the "religion" they object to, but to a putative rival to their own religion.

17. Yau, "The Twilight of Darwinism."

18. Yau, "The Twilight of Darwinism." By "the so-called creationists," Chien means *young-earth creationists.* His assertion that his "friends in the network" are not creationists is typical of the CRSC members' effort to disassociate themselves publicly from young-earth creationists, even though within the Wedge strategy itself, Phillip Johnson has urged a partnership with all creationists, despite doctrinal differences, for the purpose of advancing "mere creation." See Nancy Pearcey, "Opening the 'Big Tent' in Science: The New Design Movement," *World*, March 1, 1997. Accessed on May 22, 2002, at http://www.arn.org/docs/pearcey//np_bigtent30197.htm. There are also young-earth creationists within the Wedge itself. See chapter 9.

19. There are many good scientists who are and have been strong critics of important

parts of "standard" Darwinism (e.g., Stephen Jay Gould) *who have not suffered at all*, professionally or personally. This is just more of Chien's conspiracy mongering.

20. Phillip E. Johnson, *Defeating Darwinism by Opening Minds* (Downers Grove, IL: InterVarsity Press, 1997), 61–62.

21. Center for the Renewal of Science and Culture, "The Promise of Better Science and a Better Culture," Discovery Institute *Journal* (spring 1999). Accessed on May 22, 2002, at http://www.discovery.org/w3/discovery.org/journal/1999/crsc.html.

22. For the Chien quotation, see Yau, "The Twilight of Darwinism." According to Dr. David Bottjer, Professor of Earth Sciences (Paleobiology and Evolutionary Paleoecology) at the University of Southern California, "the Chengjiang fossils have been known publicly since at least 1987, with the publication of a paper by Hou Xian-guang. I was shown Chengjiang fossils at the Nanjing Institute of Geology and Palaeontology during a trip there in 1987" (David Bottjer to Barbara Forrest, personal communication, July 5, 2000).

23. "The Explosion of Life," *Real Issue* (March/April 1997). Accessed on May 22, 2002, at http://www.clm.org/real/ri9701/chien.html.

24. Chinese Academy of Sciences, "First Circular" and "Second Circular," International Symposium on the Origins of Animals Body Plans and Their Fossil Records. Accessed May 25, 2000, at http://www.lu.org/symposium/firstcirc.html and http://www.lu.org/symposium, respectively. Chien is also listed as a contact person in a bulletin posted by the Spanish Paleontological Society; his CRSC affiliation is not listed. "International Symposium on the Origins of Animal Body Plans and Their Fossil Records" in "Noticias Paleontológicas," *Boletín de la Sociedad España de Paleontología* (Febrero 1999). Accessed on May 22, 2002, at http://www.uv.es/~pardomv/np/np32/np32_02.html. Nor is his affiliation listed in a notice in the British Micropalaeontological Society's *TMS Newsletter* 59 (Autumn 1998), 11. Accessed on May 23, 2002, at http://www.nhm.ac.uk/hosted_sites/bms/.

25. David Bottjer to Barbara Forrest, telephone interview, May 31, 2000.

26. Evidence of Chien's desire to produce a book from the conference, first suspected by Bottjer while editing the abstract book and even more strongly during the preconference trip to Guizhou, was available months before the conference in a document posted on a German bulletin board hosted by the International Communication Forum in Human Genetics. The document is a letter from "Paul K. Chien, Department of Biology, University of San Francisco," inviting colleagues to "an international symposium on the origin of animal body plans and their fossil records," after which Chien announces, "Contributed papers will be published in a special volume in the year 2000." "International Symposium on the Origins of Animal Body Plans and Their Fossil Records." Accessed on May 22, 2002, at http://www.hum-molgen.de/meetings/meetings/0798.html. A significant feature of this post, added unobtrusively—almost unnoticeably—at the end of the document, is the identity of the person who posted this notice: "Posted by: Jay Richards (for Dr. Paul Chien)." Richards is the program director for the CRSC. However, his posting carries no indication of this affiliation.

27. See Heeren's article in the *Boston Globe*, May 30, 2000, p. E1, entitled "A Little Fish Challenges a Giant of Science," in which he asserts that Chinese scientists have shown that the Chengjiang fossils are "nothing less than a challenge to the theory of evolution" and misquotes scientists David Bottjer and Eric Davidson on the significance of these fossils. Accessed on April 23, 2003, at http://www.calvin.edu/archive/evolution/200005/0275.html. See also "The Cambrian Distortion," (sidebar) *Reports of the National Center for Science Education* 20:5 (September/October 2000), 19, for Bottjer and Davidson's response in their letter to the *Globe*—which the *Globe* did *not* print—protesting Heeren's misquote of them.

28. The placement of Wells's and Nelson's presentations at the end of the conference is in contrast to the placement of their abstracts in the abstract book's table of contents—just after those of the three most prominent scientists at the conference. *International Symposium: The Origins of Animal Body Plans and Their Fossil Records*, ed. Jun-yuan Chen, Paul

K. Chien, David J. Bottjer, Guo-xiang Li, and Feng Gao (Kunming, China: Early Life Research Center, 1999), List of Abstracts.

29. Nigel Hughes to Barbara Forrest, telephone interview, May 30, 2000.

30. David Bottjer to Barbara Forrest, May 31, 2000.

31. David Bottjer to Barbara Forrest, May 31, 2000.

32. Nigel Hughes to Barbara Forrest, May 30, 2000. This is not rare. Serious scientists who are unaware of operations like the Wedge are usually amused at first to learn that creationists *want* to be associated with them in a "professional" way, and then astounded to discover that the creationists are wounded when the professionals refuse.

33. Yau, "The Twilight of Darwinism."

34. Nigel Hughes to Barbara Forrest, personal communication, June 23, 2000.

35. N. C. Hughes, "The Rocky Road to Mendel's Play," *Evolution and Development*, 2 (2000), 63–66.

36. Phillip Johnson, "The Church of Darwin," *Wall Street Journal*, August 16, 1999. Accessed on May 22, 2002, at http://www.arn.org/docs/johnson/chofdarwin.htm.

37. Hughes, "The Rocky Road to Mendel's Play."

38. Nigel Hughes to Barbara Forrest, telephone interview, May 30, 2000.

39. Kevin Padian to Barbara Forrest, telephone interview, May 26, 2000.

40. Nigel Hughes to Barbara Forrest, telephone interview, May 30, 2000, and June 7, 2000.

41. David Bottjer to Barbara Forrest, telephone interview, May 31, 2000.

42. Center for the Renewal of Science and Culture, "Senior Fellow Paul Chien." Accessed on May 22, 2002, at http://www.discovery.org/crsc/fellows/PaulChien/index.html. Chien is no longer the department chairman as his bio states.

43. The Department of Biology, University of San Francisco, "Faculty." Accessed on May 22, 2002, at http://www.usfca.edu/biology/faculty.htm#chien.

44. One of us (PRG) was director and president of the Marine Biological Laboratory, Woods Hole, MA, from 1978 to 1988.

45. "The Explosion of Life," *Real Issue.*

46. Regarding this statement about Chien's fossil collection, Kevin Padian says, "As far as our Chinese colleagues and the Chinese government are concerned, it is illegal to take valuable Chinese fossils out of the country. They belong to the Chinese people. However, it depends on what is classified as valuable and who says so. . . . If these Cambrian fossils are valuable, it's a mystery to me why they are allowed to leave the country. Still, there is no doubt that they are, and similar allowances were made for another recent conference in China" (Kevin Padian to Barbara Forrest, personal communication, July 10, 2000). Allowing that Chien may have *some* Chinese Cambrian fossils, though dismissing the claim that it is the largest collection in North America, Nigel Hughes confirms that the Chinese do issue permits for Western scientists to take materials out of China. Hughes, who visited the fossil site during the symposium in Kunming, China, says that people can get permits, especially if they are associated with a university. At the conference, scientists were issued documents allowing a small amount of material to be taken out of China.

47. Yau, "The Twilight of Darwinism."

48. DeWolf et al., "Intelligent Design in Public School Science Curricula." Accessed on March 5, 2002, at http://arn.org/docs/dewolf/guidebook.htm#notes.

49. One of us (PRG) was among those who, politely, declined the invitation to contribute.

50. ASA *Newsletter* (March/April 2000). Accessed on April 23, 2003, at http://users.stargate.net/~dfeucht/MARAPR00.htm. Pattle Pun, one of the translators, is also a CRSC fellow.

51. Connections are surfacing abroad. William Dembski gave talks sponsored by the Canadian Scientific and Christian Affiliation at the University of Guelph, the University of Toronto, and McMaster University in March 2002. Three fellows of Dembski's new International Society for Complexity, Information and Design (see chapter 7), are David K. Y.

Chiu and Bonnie Mallard at the University of Guelph and David Humphreys at McMaster. Others are from South Korea, England, and New Zealand. Accessed on April 8, 2002, at http://www.csca.ca/dembski.html and http://www.iscid.org/fellows.php. Dembski has made at least two trips to South Korea to the Manmin Research Center, an "international research and educational institute," which "focuses on the numerous miracles performed by Revd. Dr. Lee Jae-Rock" and which, exactly like the former Michael Polanyi Center, "is concerned to see the pursuit of scientific research freed from programmatic external philosophical constraints, particularly those associated with a materialist or naturalist agenda." Manmin representatives attended three Wedge conferences, at least one as presenters: Joon-Ha Hwang, "Supernatural Intervention: An Epistemic Support for Intelligent Design," at the Design and Its Critics conference in June 2000. See http://www.manminresearch.org/html/Mainmcmi.htm and http://www.cuw.edu/Cranach/concurrent_abstracts.htm, respectively. Accessed on April 8, 2002. And most recently, Mark Hartwig reported on the "fledgling intelligent design movement in Brazil," represented by Enezio E. de Almeida Filho, "an active participant in the Wedge down there." See *Weekly Wedge Update*, June 13, 2002. Accessed on June 21, 2002, at http://www.arn.org/docs/wedge/mh_wedge_020613.htm.

52. Michael Behe, *Darwin's Black Box: The Biochemical Challenge to Evolution* (New York: Free Press, 1996), 252.

53. Center for the Renewal of Science and Culture, "Year End Update" (November/December 1997). Accessed on March 18, 2000, at http://www.discovery.org/w3/discovery.org/crsc/crscnotes2.html. This document is no longer available online.

54. Department of Biological Sciences, Lehigh University. Accessed on May 23, 2002, at http://www2.lehigh.edu/page.asp?page=casfaculty#biological.

55. *Ethics & Medics* 23:6 (June 1998).

56. Michael J. Behe, Ph.D., "Randomness or Design in Evolution?" *Ethics & Medics* 23:6 (June 1998), 3–4.

57. See Behe's Lehigh page, accessed on May 23, 2002, at http://www.lehigh.edu/~inbios/behe.html. "Michael J. Behe: Spring 2001 Schedule," Access Research Network. Accessed on May 22, 2002, at http://www.arn.org/behe/mb_schedule.htm.

58. "Michael J. Behe On-line Articles," Access Research Network. Accessed on May 22, 2002, at http://www.arn.org/behe/mb_articles.htm.

59. See *Darwinism: Science or Philosophy? Proceedings of a symposium entitled: "Darwinism: Scientific Inference or Philosophical Preference?" Held on the Southern Methodist University Campus in Dallas, Texas, March 26–28,1992*, ed. Jon Buell and Virginia Hearn (Richardson, TX: Foundation for Thought and Ethics, 1994). Accessed on May 22, 2002, at http://www.leaderu.com/orgs/fte/darwinism/index.html.

60. William J. Bennetta, "Fundamentalists Launch Bogus 'Supplemental Text,'" *Textbook Letter* (March/April 1990). Accessed on May 22, 2002, at http://www.textbook league.org/53panda.htm.

61. See article on the Access Research Network website. Accessed on May 22, 2002, at http://www.arn.org/docs/behe/mb_smu1992.htm.

62. This is probably a reference to the 1994 C. S. Lewis Foundation Summer Institute. Accessed on May 23, 2002, at http://www.cslewis.org:80/conference/history1.html.

63. Michael J. Behe, "Evidence for Intelligent Design from Biochemistry." Accessed on May 22, 2002, at http://www.arn.org/docs/behe/mb_idfrombiochemistry.htm.

64. The review was accessed online on May 22, 2002, at http://www.arn.org/docs/reviews/rev009.pdf.

65. Michael J. Behe, "Histone Deletion Mutants Challenge the Molecular Clock Hypothesis," *Trends in Biochemical Sciences* 15 (October 1990), 374–376.

66. In fact, it is very old as hypotheses go in molecular biology and remains useful for some investigations while requiring major modification for others. It is not "wrong." An authoritative but highly accessible discussion of the "molecular clock" hypothesis and its ap-

plication is available in Richard Dawkins's popular book, *River Out of Eden* (New York: Basic Books, 1995), 42–44. The recently issued, comprehensive *Encyclopedia of Evolution* of Oxford University Press has full, and technically impeccable, material on molecular clocks.

67. Thomas J. Wheeler to Barbara Forrest, personal communication, July 14, 2000.

68. Accessed on May 22, 2002, at http://www.etsu.edu/philos/faculty/niall/complexi.htm.

69. Accessed on May 22, 2002, at http://biomed.brown.edu/Faculty/M/Miller/Behe.html.

70. David W. Ussery, "A Few Links About Michael Behe & *Darwin's Black Box*." Accessed on May 22, 2002, at http://www.cbs.dtu.dk/dave/Behe_links.html. See also David Ussery, "A Biochemist's Response to 'The Biochemical Challenge to Evolution'," *Bios* (July 1998), at http://www.cbs.dtu.dk/dave/Behe_text.html. See also Richard H. Thornhill and David W. Ussery, "A Classification of Possible Routes of Darwinian Evolution," *Journal of Theoretical Biology* 203 (2000), 111–116, at http://www.cbs.dtu.dk/dave/JTB.html. Both accessed on May 22, 2002.

71. John Catalano, "Behe's Empty Box: Introduction." Accessed on May 22, 2002, at http://world-of-dawkins.com/Catalano/box/behe.htm#intro.

72. "Irreducible Complexity and Michael Behe," Talk.Origins Archive. Accessed on May 22, 2002, at http://www.talkorigins.org/faqs/behe.html.

73. Michael J. Behe, "Darwin Under the Microscope," *New York Times*, October 29, 1996. Available at Access Research Network, http://www.arn.org/docs/behe/mb_dm11496.htm. See also Michael J. Behe, "Teach Evolution—and Ask Hard Questions," *New York Times*, August 13, 1999, at http://www.arn.org/docs/behe/mb_ksnytb81399.htm. Both accessed on May 22, 2002.

74. "*Darwin's Black Box*–Michael Behe," interview on "To the Best of Our Knowledge," Hour 2, by Wisconsin Public Radio, January 19, 1997. Schedule accessed on May 23, 2002, at http://wpr.org/book/book97.htm.

75. Michael Behe, "Dogmatic Darwinism," *Crisis* 16:8 (June 1998). Accessed on July 17, 2000, at http://www.catholic.net/rcc/Periodicals/Crisis/1998-06/darwinism.html. Now available at http://www.catholiceducation.org/articles/science/sc0018.html (April 12, 2003). *Crisis* is a conservative Catholic magazine of "politics, culture, and the church," which apparently shares the CRSC's view that American culture is in need of "renewal." In the June 1998 issue, along with Behe's article, is one by Peter Kreeft, a philosopher at Boston College, entitled "How to Win the Culture War." Kreeft asserts that the enemy of American culture is "Not Protestants . . . Not Jews . . . Not Muslims . . . not 'the liberals' . . . not anti-Catholic bigots who want to crucify us . . . not even the media of the culture of death, not even Ted Turner or Larry Flynt or Howard Stern or Disney or Time-Warner . . . not heretics within the Church. . . . Our enemies are demons. Fallen angels. Evil spirits." This is the kind of venue in which Michael Behe defends intelligent design. Needless to say, scientific evidence would not be required here. Kreeft's article was accessed on May 23, 2002, at http://www.ewtn.com/library/ISSUES/CULTURE.HTM.

76. Michael Behe, "Irreducible Complexity: The Biochemical Challenge to Darwinian Theory," approximately 2 hours, 1997. Videocassette. Available at http://www.arn.org/arnproducts/videos/v010.htm. The "live audience at Princeton" in 1997 may have been the "Christian Leadership Ministries/Religious Life Lecture," which Behe delivered at Princeton on November 6, 1997, listed by the Princeton Weekly Bulletin in the November 3–9, 1997, schedule of events at Princeton: "Intelligent Design: Implications for Science and Belief in God." Accessed on May 23, 2002, at http://www.princeton.edu/pr/calendar/97/11-03-97.html.

77. "Molecular Machines: Index of Illustrations, Graphics, and Animations," Access Research Network. Accessed on May 23, 2002, at http://www.arn.org/mm/mm_index.htm.

78. Accessed on May 23, 2002, at http://www.apologetics.org/chronicles/opening

.html. Trinity College maintains a "Center for University Ministries," whose mission is to "expand annual apologetics lectureships to include all 8 of the Ivy League colleges" because "our great concern is that a dear, bold and tactful witness be initiated both locally and worldwide to university students and professors, and to people [and] groups skeptical of the claims of Christianity" (http://www.apologetics.org/societies/trinity.html). The C. S. Lewis Society is "an Evangelical Christian ministry dedicated to (1) evangelizing atheists, university students and professors, and (2) equipping Christians to deal with the contemporary questions and skeptical objections that are posed at the Christian faith" (http://www.apologetics.org/societies/society.html). The marketing of Behe's videotape on their website, obviously not a scientific venue, is clearly considered consonant with their religious aims.

79. Glenn R. Morton, "Waco: The Final Comments." Accessed on April 12, 2003, at http://www.glenn.morton.btinternet.co.uk/wacofinal.htm.

80. Michael J. Behe, "Evidence for Intelligent Design from Biochemistry (From a Speech Delivered at Discovery Institute's God & Culture Conference)," Discovery Institute. Accessed on May 23, 2002, at http://www.discovery.org/viewDB/index.php3?program=CRSC&command=view&id=51.

81. Michael J. Behe, "Molecular Machines: Experimental Support for the Design Inference." Accessed on May 23, 2002, at http://www.arn.org/docs/behe/mb_mm92496.htm.

82. Behe, "Evidence for Intelligent Design from Biochemistry."

83. The reader would need to read those reviews to understand how gross a distortion of their content this is.

84. Kenneth Miller, Brown University, to *Pratt Tribune*, June 28, 2000.

85. See John Catalano, "Some Published Works on Biochemical Evolution," Behe's Empty Box, at http://world-of-dawkins.com/Catalano/box/published.htm. Also available at the Talk.Origins Archive at http://www.talkorigins.org/faqs/behe/publish.html. Both accessed on May 23, 2002.

86. Kenneth Miller to Barbara Forrest, personal communication, July 14, 2000.

87. Muller's early and critical contributions on this subject are discussed in a splendid review of Behe's book by H. Allen Orr, "Darwin v. Intelligent Design (Again)," *Boston Review* (December 1996/January 1997). Accessed on March 5, 2002, at http://bostonreview.mit.edu/BR21.6/orr.html.

88. The articles are on the CRSC website ("Responses to Critics") with other articles published by CRSC fellows, to which there is a link accessed on April 12, 2003, at http://www.crsc.org. Miller has written responses to the first three of Behe's postings, addressing the concerns of each. See Kenneth Miller, "A True Acid Test," at http://BioCrs.biomed.brown.edu/Darwin/DI/AcidTest.html; "Is the Blood Clotting Cascade 'Irreducibly Complex?'" at http://BioCrs.biomed.brown.edu/Darwin/DI/Clotting.html; and "The Mousetrap Analogy or Trapped by Design" at http://biocrs.biomed.brown.edu/Darwin/DI/Mousetrap.html. Accessed on May 23, 2002.

89. Michael Behe, "The Evolution of a Skeptic," interview by Christian Leadership Ministries, *Real Issue* 15:2 (November/December 1996), 8. Accessed on May 23, 2002, at http://www.leaderu.com/real/ri9602/behe.html.

90. "ASBMB Meeting Information." Accessed on March 4, 2002, at http://www.asbmb.org/ASBMB/ASBMBsiteII.nsf/MenuHomePage/Meetings?OpenDocument.

91. Beth McMurtrie, "Darwinism Under Attack," *Chronicle of Higher Education* 48:17 (December 21, 2001), A10. Accessed on March 5, 2002, at http://chronicle.com/free/v48/i17/17a00801.htm.

92. The articles located in the database search are as follows: (1) Behe, Michael J., "An Overabundance of Long Oligopurine Tracts Occurs in the Genome of Simple and Complex Eukaryotes," *Nucleic Acids Research* 23:4 (1995), 689–695; (2) Puhl, Henry L., and Behe, Michael J., "Poly(dA) cntdot poly(dT) Forms Very Stable Nucleosomes at Higher Temperatures," *Journal of Molecular Biology* 245:5 (1995), 559–567; (3) Agarwal,

Seema, and Behe, Michael J., "Non-conservative Mutations Are Well Tolerated in the Globular Region of Yeast Histone H4," *Journal of Molecular Biology* 255:3 (1996), 401–411; (4) Mahloogi, Haleh, and Behe, Michael J., "Oligoadenosine Tracts Favor Nucleosome Formation," *Biochemical and Biophysical Research Communications* 235:3 (1997), 663–668; and (5) Behe, Michael J., "Tracts of Adenosine and Cytidine Residues in the Genomes of Prokaryotes and Eukaryotes," *DNA Sequence* 8:6 (1998), 375–383.

93. For example, Ian Musgrave, "Evolution of the Bacterial Flagella." Accessed on March 6, 2002, at http://minyos.its.rmit.edu.au/~e21092/flagella.htm.

94. Roughly two-thirds of the "People Behind ARN" are CRSC fellows. Accessed on May 23, 2002, at http://www.arn.org/infopage/info.htm.

95. Mike Gene, "Irreducible Complexity and Darwinian Pathways," Access Research Network, Intelligent Design Discussion Forum, 6-13-2000. Accessed on March 5, 2002, at http://www.arn.org/ubb/Forum3/HTML/000019.html. Gene's post is on CRSC's website as "Irreducible Complexity and Darwinian Pathways: Guest Response to Article by R. H. Thornhill and D. W. Ussery." Accessed on May 23, 2002, at http://www.discovery.org/viewDB/index.php3?program=CRSC%20Responses&command=view&id=273. An August 5, 2001, Internet search on "Mike Gene" yielded only his website (www.idthink.net), his ARN posts, and mentions of him on other sites. An August 8, 2001, "whois" search revealed his site's registration to "Mike Gene" at a nonexistent Columbus, Ohio, address and a fictitious phone number (765-4321). Gene removed the site temporarily because "I've made a lot of people mad at me and made many people distrust me" (ARN discussion forum, accessed on April 24, 2002, at http://www.arn.org/cgi-bin/ubb/ultimatebb.cgi?ubb=get_topic;f=13;t=000115). He reposted it later, but alluded to recent discord with the Wedge with a disclaimer: "[T]he hypothesis of Intelligent Design has been intertwined with a socio-political agenda that seeks to insert [ID] into . . . public school curricula. . . . I oppose such actions" (accessed on September 6, 2002, at http://www.idthink.net/links/index.html). He voiced "strong agreement with [Eugenie] Scott and [Glenn] Branch" of the National Center for Science Education in a July 23, 2002, ARN post: "The simple fact that ID has not established itself in the scientific community is all we need to deny ID's entry into a high school curriculum" (accessed on September 6, 2002, at http://www.arn.org/cgi-bin/ubb/ultimatebb.cgi?ubb=get_topic;f=13;t=000215). William Dembski, in an angry post on ARN, responded that he, too, had once believed ID needed more development before being taught in public schools. But having changed his mind after "the dissolution of my ID think tank at Baylor," Dembski announced that "I've come to reject this view entirely." He now believes ID should be taught in public schools in order to break evolution's monopoly, "regardless of what state of formation ID has reached," and criticized Gene's "pseudonymous persona," which shields him from "the brunt of the Darwinian establishment" (accessed on September 7, 2002, at http://www.arn.org/cgi-bin/ubb/ultimatebb.egi?=get_topic;f=13;t=000220). But Gene still appears supportive of ID itself, if not of Wedge politics. See, for example, his "Irreducible Complexity ReVisited." Accessed on September 7, 2002, at http://www.idthink.net/back/ic/index.html. And his pseudonymity has not discouraged ID proponents' support of him. Hill Roberts, a physicist who conducts ID workshops, has links on his "Lord I Believe" site to Gene's ARN posts, which he calls "simply astoundingly wonderful!" Roberts is apparently unconcerned about Gene's identity or his credentials: "These excellent articles fall in the class of 'Boy, I wish I had said that.' I have no clue who Mike Gene is." Accessed on May 11, 2002, at http://lordibelieve.org/page20_00.html.

96. Richard H. Thornhill and David W. Ussery, "A Classification of Possible Routes of Darwinian Evolution," *Journal of Theoretical Biology* 203 (2000), 111–116. Accessed on May 23, 2002, at http://www.cbs.dtu.dk/dave/JTB.html.

97. Gene, "Irreducible Complexity and Darwinian Pathways."

98. Thornhill and Ussery, "A Classification of Possible Routes of Darwinian Evolution."

99. David Ussery to Barbara Forrest, personal communication, July 14, 2000. In the

ARN forum, Gene is not claiming that ID is scientific; indeed, he does not seem concerned with this question. He posted the following comments: "[P]eople want to call ID 'scientific' or 'unscientific' mainly because of cultural reasons, where many view science as some form of authority about the truth of natural history. Since it ain't, it really doesn't matter if ID is 'unscientific'" (online discussion: "Irreducible Complexity and Darwinian Pathways," General ID Discussion, ARN Intelligent Design Discussion Forum, posted 7-5-2000, 8:20 a.m., at http://www.arn.org/ubb/Forum1/HTML/000203.html). He later added, "I am not interested in attaching the label 'science' or 'scientific' to ID. I'm simply interested in the best explanation for natural history, regardless of how it is labeled" (7-5-2000, 1:47 p.m., http://www.arn.org/ubb/Forum3/HTML/000019.html). Ussery, in the same discussion, rejects ID as scientific: "I am not opposed to the concept of intelligent design—but I just don't think that it falls within the realm of science" (7-5-2000, 3:06 p.m.). However, Behe does claim that ID is scientific, making Ussery's responses directly relevant to the way Behe represents ID.

100. David Ussery to Barbara Forrest, personal communication, July 14, 2000.

101. David W. Ussery, "Is Intelligent Design Science?" August 2000. Accessed on May 23, 2002, at http://www.cbs.dtu.dk/dave/Behe_text.html#IsIDscience. Ussery has posted criticisms of Behe's work at this site.

102. See, for recent evidence, Michael Lynch and John S. Conery, "The Evolutionary Fate and Consequences of Duplicate Genes," *Science* 290 (2000), 1151–1156.

103. A clear, simple statement of this situation is in Russell F. Doolittle, "A Delicate Balance," another entry in the effective symposium on *Darwin's Black Box* published in *Boston Review*. Accessed on March 5, 2002, at http://bostonreview.mit.edu/br22.1/doolittle.html.

104. Jerry A. Coyne, "God in the Details: The Biochemical Challenge to Evolution," *Nature* 19 (September 1996). Accessed on March 5, 2002, at http://www.world-of-dawkins.com/Catalano/box/nature.htm.

105. Daniel McShea, "Complexity and Evolution: What Everybody Knows," *Biology and Philosophy* 6 (1991), 303–324.

106. See Scott Gilbert, *Developmental Biology*, 5th ed. (Sunderland, MA: Sinauer Associates, 1997), e.g., pp. 655–726; and Rudolf A. Raff, *The Shape of Life: Genes, Development, and the Evolution of Animal Form* (Chicago: University of Chicago Press, 1996).

107. Coyne, "God in the Details."

108. See Pennock, *Tower of Babel: The Evidence Against the New Creationism* (Cambridge, MA: MIT Press, 1999), 266ff. See also Behe, "Reply to my Critics: A Response to Reviews of *Darwin's Black Box: The Biochemical Challenge to Evolution*," *Biology and Philosophy* 16 (2001), 695.

109. See Richard E. Lenski, Charles Ofria, Robert T. Pennock, and Christoph Adami, "The Evolutionary Origin of Complex Features," *Nature* 423 (2003), 139–145. For an accessible explanation of such work see Carl Zimmer, "Alternative Lifestyles," *Natural History* (May 2001). Accessed on April 22, 2003, on Lenski's website at http://www.msu.edu/~lenski/. See also Volodymyr Dvornyk, Oxana Vinogradova, and Eviatar Nevo, "Origin and Evolution of Circadian Clock Genes in Prokaryotes," *Proceedings of the National Academy of Sciences U.S.A.* 100 (2003), 2495–2500.

5. A Conspiracy Hunter and a Newton

1. William A. Dembski, ed., *Mere Creation: Science, Faith, and Intelligent Design* (Downers Grove, IL: InterVarsity Press, 1998). The title is an echo of C. S. Lewis's *Mere Christianity* (1952), a famous work of passionate Christian apologetics.

2. One of the most interesting such arguments among leading Darwinists is the subject of an accessible but solid little book, Kim Sterelny, *Dawkins vs. Gould: Survival of the Fittest* (Cambridge: Icon Books, 2001).

3. Gregory A. Petsko, "Design by Necessity," *Genome Biology* 2 (2001), comment1010.1—1010.3. Accessed at http://genomebiology.com/2001/2/8/comment/1010.

4. This book, unlike most of Dembski's production, is not overtly a promotion of ID, but rather a logico-mathematical exercise in the detection of design.

5. Robert Koons, a philosopher at the University of Texas-Austin and a frequent collaborator in CRSC ventures, in his dust jacket blurb on Dembski's book, *Intelligent Design*, cited later.

6. Jonathan Wells, "Darwinism: Why I Went for a Second Ph.D.," *Unification Sermons and Talks by Reverend Wells*. Accessed on April 29, 2002, at http://www.tparents.org/library/unification/talks/wells/DARWIN.htm.

7. L. Stavy and Paul R. Gross, "The Protein-synthetic Lesion in Unfertilized Eggs," *Proceedings of the National Academy of Sciences, U.S.A.*, 57 (1967), 735–742; Paul R. Gross, "The Control of Protein Synthesis in Embryonic Development and Differentiation," in *Current Topics in Developmental Biology*, Vol. 2, (New York: Academic Press, 1967), 1–46.

8. For example, Scott F. Gilbert, *Developmental Biology*, 5th ed. (Sunderland, MA: Sinauer Associates, 1997), 480 ff.

9. See, for example, C. Nusslein-Volhard and E. Wieschaus, "Mutations Affecting Segment Number and Polarity in Drosophila," *Nature* 287 (1980), 795–801.

10. W. H. Rodgers and Paul R. Gross, "Inhomogeneous Distribution of Egg RNA Sequences in the Early Embryo," *Cell* 14 (1978), 279–288.

11. Gilbert, *Developmental Biology*, 635–652.

12. Jonathan Wells and Paul Nelson, caption of figure 4 in "Homology: A Concept in Crisis," *Origins and Design* 18:2. Accessed on February 18, 2002, at http://www.arn.org/docs/odesign/od182/hbfig4.htm. Scott Gilbert and Anne M. Raunio, in their preface to *Embryology: Constructing the Organism* (Sunderland, MA: Sinauer Associates, 1997), point out that embryology "is a science of emerging complexity; hence it may be the science of the next century." Wells is apparently astute enough to try to forestall its influence among his followers.

13. Phillip E. Johnson, "Evolution as Dogma: The Establishment of Naturalism," *First Things* (October 1990). Accessed on February 10, 2002, at http://www.arn.org/docs/johnson/pjdogma1.htm.

14. Scott F. Gilbert, John M. Opitz, and Rudolf A. Raff, "Resynthesizing Evolutionary and Developmental Biology," *Developmental Biol*ogy 173 (1996), 357–372.

15. Keith B. Miller, "The Precambrian to Cambrian Fossil Record and Transitional Forms," *Perspectives on Science and Christian Faith* 49 (December 1997). Accessed on February 21, 2002, at http://www.asa3.org/ASA/topics/evolution/PSCF12-97Miller.html.

16. We have discussed this in chapter 4.

17. These arguments are concerned with differential rates of accumulation of mutations and with the computational means for selecting the likeliest patterns of descent from among alternatives. They are *not* about whether genes change in the course of time, providing strong evidence of descent with modification.

18. Eugenie C. Scott, "Fatally Flawed Iconoclasm," *Science* 292 (June 22, 2001), 2257–2258. Accessed on February 15, 2002, at http://www.scienceormyth.org/icons%20of%20evolution.html. Wells's *Icons of Evolution* was published in 2000 by Regnery.

19. A good place to start is David Ussery's exhaustive response at http://www.cbs.dtu.dk/dave/IconsReview.html. Also available is a focused offering by Massimo Pigliucci in *BioScience* 51:5 (May 2001), 411–414. Links to other reviews from experts in one or more of the icons are at http://www.don-lindsay-archive.org/creation/icons_of_evolution.html. Another excellent review is Kevin Padian and Alan Gishlick, "The Talented Mr. Wells," in *The Quarterly Review of Biology* (March 2002). Accessed on April 11, 2002, at http://www.journals.uchicago.edu/QRB/journal/issues/v77n1/770103/770103.web.pdf.

20. The shortest, and pithiest, is Alan Gishlick's "Responses to Jonathan Wells' Ten

Questions to Ask Your Biology Teacher" at http://www.ncseweb.org/resources/articles/7719_responses_to_jonathan_wells3_11_28_2001.asp. A full set of expert responses to the *Icons*, multiply authored and entitled "Icons of *Anti*-Evolution: The Essays," is provided by New Mexicans for Science and Reason at http://www.nmsr.org/text.htm. Both accessed on September 9, 2002.

21. We follow here, approximately, the style of their summaries in the previously cited long review by David Ussery.

22. Wells's obtuseness toward his own discipline, embryology, is highlighted, moreover, by his failure to understand why and how observed similarities and identities—homologies—of the basic processes of morphogenesis, across the phyla of metazoan life, made sense historically and make sense today of the enormous matrix of empirical data in embryology.

23. A clear, concise account of this shift in opinion regarding Earth's early atmosphere is "Reflections from a Warm Little Pond," by David Pacchioli, in *Research/Penn State* 22:1 (January 2001). Accessed on February 22, 2002, at http://www.rps.psu.edu/0101/reflections.html.

24. See a report on this new work, from two different research groups, by Philip Cohen in "Life's Building Blocks Created in Space Simulator," *New Scientist*, March 2, 2002. Accessed on April 11, 2002, at http://www.newscientist.com/news/print.jsp?id=ns99992100.

25. This is a preprint of a paper due to appear in the *Proceedings of the Society of Photo-Optical Instrumentation Engineers*. Accessed on February 22, 2002, at http://www.jmcgowan.com/JigsawPreprint.pdf.

26. Iris Fry, *The Emergence of Life on Earth: A Historical and Scientific Overview* (New Brunswick, NJ: Rutgers University Press, 2000).

27. X. Li, Z.-Y. Zhan, R. Knipe, and D. G. Lynn, "DNA-catalyzed Polymerization," *Journal of the American Chemical Society* 124:5 (February 6, 2002), 746–747.

28. A recent, comprehensive refutation of Wells's *Icons* claims, "Icon of Obfuscation," roughly forty single-spaced pages in small type, is now on the Internet. Written by Nic Tamzek, it consists of eleven chapters, each vetted by an expert on the matter at issue. This site has a rich collection of links to issues related to *Icons*. Accessed on April 9, 2002, at http://www.talkorigins.org/faqs/wells/.

29. The Carnegie Museum of Natural History, which displayed Haeckel's illustrations in a 2001 exhibit, "Art in Nature," pointed out in a press release that "Ernst Haeckel did for Protozoa . . . what Audubon did for birds." The museum displayed on its web page one of Haeckel's beautiful bird lithographs, which is strikingly similar to Audubon's work. Accessed on February 23, 2002, at http://www.carnegiemuseums.org/cmnh/whatsnew/artforms.html.

30. "An obsessive researcher and writer, during the 1860s and 1870s Haeckel produced a collection of enormous books . . . on Darwinism and its ramifications. . . . Often Haeckel would go too far . . . and Darwin's theories were convoluted to merge with Haeckel's often iconoclastic appraisal of human society and the future of mankind. . . . For Darwin, any confusions arising over the meaning of his theories brought him far more anxiety than the trouble-making religious zealots or omission from an Honours List. His own exposition in the *Origin* had been so clear cut and so firmly rooted in detailed experimental data that the adaptation of his purely scientific ideas to sociology or to half-baked pet philosophies upset him. . . . Darwin knew Haeckel was a clever man with a huge surplus of energy, but he found his over-enthusiasm hard to tolerate and often had to put the German scientist right on various points where the theory of evolution had been misinterpreted." Michael White and John Gribbin, *Darwin: A Life in Science* (Dutton, 1995), 232.

31. James Hanken, "Beauty Beyond Belief: Nineteenth Century Scientist and Artist Ernst Haeckel," *Natural History* (December 1998), 56.

32. See, for example, pp. 384ff of Scott F. Gilbert and Anne M. Raunio, *Embryology: Constructing the Organism* (Sunderland, MA: Sinauer Associates, 1997).

33. See Oppenheimer's distinguished essay on von Baer's place in the history of embryology: Jane Oppenheimer, *Essays in the History of Embryology and Biology* (Cambridge, MA: MIT Press, 1967), 136–147.

34. Elizabeth Pennisi, "Haeckel's Embryos: Fraud Rediscovered," *Science* 277 (September 5, 1997), 1435.

35. Michael K. Richardson and Gerhard Keuck, "Haeckel's ABC of Evolution and Development," *Biological Reviews of the Cambridge Philosophical Society* 77 (2002), 495–528.

36. Such misuse of the scientific literature occurred on March 11, 2002, when Stephen C. Meyer and Jonathan Wells, in a "debate" against biologist Kenneth Miller and physicist Lawrence Krauss at an Ohio Board of Education meeting (see chapter 8), presented to the OBOE a bibliography of peer-reviewed publications (with selected quotes and annotations by the Discovery Institute). Although the publications were written by pro-evolution scientists, Wells and Meyer implied that they cast doubt on evolution by prefacing their presentation with the assertion that "the publications represent dissenting viewpoints that challenge one or another aspect of neo-Darwinism (the prevailing theory of evolution taught in biology textbooks), discuss problems that evolutionary theory faces, or suggest important new lines of evidence that biology must consider when explaining origins." This tactic was meant to reinforce Meyer and Wells's plea to allow Ohio schools to "teach the controversy" surrounding evolution. See DI's online version of the bibliography, with a disclaimer *added after the OBOE received it*, at http://www.discovery.org/viewDB/index.php3?program=CRSC&command=view&id=1127. See the National Center for Science Education's analysis of the bibliography *as it was presented to the OBOE* at http://www.ncseweb.org/resources/news/2002/OH/122_intelligent_design_bibliograph_4_5_2002.asp. Both accessed on April 13, 2003.

37. Bruce S. Grant, "Fine Tuning the Peppered Moth Paradigm," *Evolution* 53:3 (1999), 980–984.

38. Bruce S. Grant, "Sour Grapes of Wrath," *Science* 297 (August 9, 2002), 940–941.

39. Bruce S. Grant, "Charges of Fraud Misleading," letter to *Pratt Tribune*, December 13, 2000. Accessed on February 23, 2002, at http://www.pratttribune.com/archives/index.inn?loc=detail&doc=/2000/December/13-653-news92.txt.

40. Frederick C. Crews, "Saving Us from Darwin, Part II," *New York Review of Books*, October 18, 2001. Accessed on October 4, 2001, at http://www.nybooks.com/articles/14622.

41. Intelligent Design Undergraduate Research Center, "Interview with Jonathan Wells, January 2001." Accessed on April 11, 2002, at http://www.idurc.org/wells interview2001.shtml.

42. Extract from testimony given by Michael Behe on June 5, 2001, before the Pennsylvania Senate Education Committee, Harrisburg, PA, 9:00 a.m. to 12:00 p.m. Room 8EA East Wing of the State Capitol. Copy of testimony provided by Prof. Roger D. K. Thomas of Franklin and Marshall College, who also testified at this hearing.

43. Roger D. K. Thomas, "Re Academic Standards for Science and Technology," to Senator James J. Rhoades, Harrisburg, Pennsylvania, June 6, 2001. Accessed on April 11, 2002, at http://www.padnet.org/evol.html#Testimony.

44. Beth McMurtrie, "Darwinism Under Attack," *Chronicle of Higher Education* 48:17 (December 21, 2001), A10. Accessed on April 11, 2002, at http://chronicle.com/free/v48/i17/17a00801.htm.

45. The Wedge proper is best defined as the fellows of the Center for the Renewal of Science and Culture, although there is a widening system of supporters and auxiliaries like the Intelligent Design Network in Kansas. However, not all fellows are equally active in the publishing/public relations campaign. This role is confined to the top echelon of the

CRSC, and of these, almost all publications issue from a relative handful of people. So the rate of publication is astonishing. But it is due partly to the frequent recycling of parts of earlier publications (e.g., see the preface to Dembski's *Intelligent Design: The Bridge Between Science and Theology*), and, in Phillip Johnson's case, the absence of the serious research that mitigates against a book a year.

46. William A. Dembski, "Refuted Yet Again! A Brief Reply to Matt Young," posted on Discovery Institute website on January 25, 2002. Accessed at http://www.discovery.org/viewDB/index.php3?program-CRSC%20Responses&command=view&id=1108.

47. Dembski's involvement is more serious than he reports in the *Research News* interview: "I wear several hats and have done some theological education." One of Dembski's chief areas of interest, arguably the most important since it involves defending the facticity of his religious beliefs, is Christian apologetics, about which he writes frequently. One of his most pressing concerns is the "accommodationism" (to modern ideas, e.g., evolution) in mainline seminaries. (Getting seminaries to "repudiate" Darwinism is a goal in the Wedge Document.) His goal is to steer these seminaries back to orthodox Christianity through evangelical student activism. See his and Jay Wesley Richards's "Introduction: Reclaiming Theological Education," in *Unapologetic Apologetics: Meeting the Challenges of Theological Studies* (InterVarsity Press, 2001).

48. "I am frequently asked what is the latest research that supports intelligent design, and I find myself having to be reticent about who is doing what precisely because of enormous pressure that opponents of design employ to discredit these researchers, undermine their position, and cause them to lose their funding." Accessed on February 2, 2002, at http://www.arn.org/docs/dembski/wd_idcomingclean.htm.

49. The most recent of many summary restatements of Dembski's notion of a design inference found in the Harvard library system in winter 2001 is in *Science and Evidence for Design in the Universe* (2000), published by Ignatius Press for the Wethersfield Institute. His co-authors are familiar: Michael Behe and Stephen C. Meyer. The Institute's Statement of Purpose begins, "The purpose of the Wethersfield Institute is to promote a clear understanding of Catholic teaching and practice and to explore the cultural and intellectual dimensions of the Catholic Faith."

50. William A. Dembski, *The Design Inference: Eliminating Chance Through Small Probabilities* (Cambridge: Cambridge University Press, 1998).

51. Dembski, Behe, et al. deny supporting special creation; but special creation is what "intelligent design" of life amounts to, even if its proponents concede, as some reluctantly do, that "micro-evolution" happens. If life's fundamental differences of body plan are not an automatic result of descent with modification, and if they *are* presaged in a consciously imagined design that was or is imposed upon the history of life on Earth, then that diversity was or is "specially created."

52. An amusing view of Berlinski's qualifications to judge the modesty of Dembski's writing is provided by the book reviews editor of *The Mathematical Intelligencer.* Of Berlinski's popularized calculus book, he says, "Berlinski's greatest friend, but ultimately his worst enemy, is metaphor. The gongorisms that saturate this book actually confound what the author claims as its central mission. . . . The Berlinski rhetoric ultimately becomes suffocating." (Gongorism is a literary style characterized by inflation, affectation, and conceit, or a statement written in such a style.) Jet Wimp, "Mathematics Not," *The Mathematical Intelligencer*, 19:3 (1997), 70–75. See more of Berlinski's scientifically ill-informed rhetoric in the December 1, 2002, February 12, 2003, and April 1, 2003, issues of *Commentary*. One of us (PRG) responded to Berlinski in the March 2003 issue. Excerpts are available on the CRSC's website at http://www.crsc.org/. Accessed on April 23, 2003.

53. Mark Perakh, "A Consistent Inconsistency." Accessed on May 22, 2002, at http://www.talkreason.org/articles/dembski.cfm. This site provides other critiques from Prof. Perakh of the major offerings to date from CRSC writers on intelligent design, including Dembski's *No Free Lunch*. H. Allen Orr, an evolutionary geneticist at the University of Rochester, also highlights Dembski's sense of self-importance: "To appreciate the magni-

tude of Dembski's claims in *No Free Lunch* you need to appreciate the relative modesty of Darwin's claims in the *Origin of Species*. Darwin did not rule out the formal possibility of a designer. Instead, he showed that the (apparent) design residing in organisms could be explained naturally, without recourse to a designer. . . . Dembski's claims are more ambitious. Darwinism, he says, is *formally incapable* of explaining certain features of organisms. . . . Dembski does not mince words: '[I]ntelligent design utterly rejects natural selection as a creative force capable of bringing about the specified complexity we see in organisms.' This is a big claim. And it explains why Dembski gets so much attention." See Orr's review of *No Free Lunch* in the summer 2002 *Boston Review*. Accessed on September 10, 2002, at http://bostonreview.mit.edu/BR27.3/orr.html.

54. Dembski, *Intelligent Design*, 14–15.

55. For example, Ted Honderich, ed., *The Oxford Companion to Philosophy* (Oxford: Oxford University Press, 1995).

56. Elliott Sober, "The Argument from Design," book chapter for *The Blackwell Guide to Philosophy of Religion*. Accessed on February 24, 2002, at http://philosophy.wisc.edu/sober/black-da.pdf.

57. Wesley R. Elsberry, "What Does 'Intelligent Agency by Proxy' Do for the Design Inference?" Review of William A. Dembski's *The Design Inference*. Accessed on October 21, 2002, at http://www.antievolution.org/people/dembski_wa/wre_id_proxy.txt.

58. For an account of Hume and design, in a tough-minded but well-written book that is as opposed to ID as one can get, see pp. 327 ff. in Gary Cziko, *Without Miracles: Universal Selection Theory and the Second Darwinian Revolution* (Cambridge, MA: MIT Press, 1995).

59. There have been, however, modern, purely philosophical objections to Hume's argument; but these are even stronger arguments than Hume's *against* intelligent design. An excellent and diligently stepwise exposition can be found in Elliott Sober, *Philosophy of Biology* (San Francisco: Westview Press, 1993), 27–56.

60. Dembski, *Intelligent Design*, 98–99.

61. William A. Dembski, "The Act of Creation: Bridging Transcendence and Immanence." Accessed on February 24, 2002, at http://www.arn.org/docs/dembski/wd_actof creation.htm. This paper, delivered at the Millstatt Forum in France in 1998, is also chapter 8 of *Intelligent Design*.

62. Gert Korthof, "On the Origin of Information by Means of Intelligent Design." Review of William Dembski's *Intelligent Design*. Accessed on August 14, 2002, at http://home.planet.nl/~gkorthof/kortho44.htm.

63. Steven Weinberg, "A Designer Universe?" Accessed on May 23, 2002, at http://www.physlink.com/Education/essay_weinberg.cfm.

64. James E. Mosimann, *Elementary Probability for the Biological Sciences* (New York: Appleton-Century-Crofts, 1968), §1.3.

65. Henry Morris spends ten pages doing just this in *Scientific Creationism*, Public School Edition (San Diego, CA: Creation-Life Publishers, 1974), 59–69: "Finally, then, the chance that one of these 10^{105} possible combinations [of parts in an organism] will be the correct one is one chance in $10^{158}/10^{105} = 1$ in 10^{53}. This is still an almost infinitesimally small number, actually one chance out of a hundred million billion billion billion billion billion. For all practical purposes, there is no chance at all!"

66. Mark Perakh, "Irreducible Contradiction." Accessed on May 23, 2002, at http://www.talkreason.org/articles/behe2.cfm.

67. We are grateful to Prof. Jeffrey Shallit, who calls attention to this in his collection of useful comments, accessed on May 23, 2002, at http://www.math.uwaterloo.ca/~shallit/quotes.html. Doolittle's remark is made in "Probability and the Origin of Life," in *Scientists Confront Creationism*, ed. Laurie R. Godfrey (New York: W. W. Norton and Company, 1983).

68. William A. Dembski, *No Free Lunch: Why Specified Complexity Cannot Be Purchased Without Intelligence* (New York: Rowman and Littlefield, 2002), 301, 310n.

69. It is impossible for a reader, even a biochemist, to judge the nature and reliability of these "preliminary indications" without understanding the details of the work described in the Axe article in the *Journal of Molecular Biology*, which we cite in chapter 4. Few of Dembski's readers will understand, either.

70. Wesley Elsberry, "Commentary on William A. Dembski's *No Free Lunch: Why Specified Complexity Cannot Be Purchased Without Intelligence*." Accessed on May 23, 2002, at http://www.antievolution.org/people/dembski_wa/rev_nfl_wre_capsule.html.

71. Jason Rosenhouse, "How Anti-evolutionists Abuse Mathematics," *The Mathematical Intelligencer* 23:4 (fall 2001), 3–8. See also Rosenhouse's critique of William Dembski's book *No Free Lunch*: "Probability, Optimization Theory, and Evolution," *Evolution* 56:8 (June 22, 2002), 1721–1722. A more extensive collection of Rosenhouse's essays and book reviews on ID is available at http://www.math.ksu.edu/~jasonr/Evolution.html. See also his weblog, "Science and Politics," at http://scienceandpolitics.blogspot.com. Both accessed on April 15, 2003.

72. "This preservation of favourable individual differences and variations . . . I have called Natural Selection. . . . [I]t implies only the preservation of such variations as arise and are beneficial to the being under its conditions of life. . . . In the literal sense of the word, no doubt, natural selection is a false term. . . . Every one knows what is meant and is implied by such metaphorical expressions; they are almost necessary for brevity." Charles Darwin, *The Origin of Species*, Mentor ed. (New York: Penguin Books, 1958), 89. Accessed on February 25, 2002, at http://www.bbc.co.uk/education/darwin/origin/oos4_1.htm.

73. A very large fraction of the DNA in the genomes of higher organisms does not encode proteins or RNA and does not perform any other currently understood function. But a significant fraction of that DNA consists quite obviously of what were once genes, but are now without function due to accumulation, over vast stretches of time, of errors in replication—"pseudogenes."

74. Korthof, "On the Origin of Information by Means of Intelligent Design."

75. Korthof cites Brian Goodwin's *How the Leopard Changed Its Spots* (Princeton, NJ: Princeton University Press, 2001), a reprint of the original 1994 Phoenix volume. A formal (mathematical) treatment of the broad issue of pattern formation in reaction-diffusion systems, often responsible for striking biological patterns, is by J. C. Eilbeck, "Pattern Formation and Pattern Selection in Reaction-Diffusion Systems," in Brian Goodwin and Peter Saunders (eds.), *Theoretical Biology: Epigenetic and Evoutionary Order from Complex Systems* (Baltimore: Johns Hopkins University Press, 1992). This is a reprint of the original (1989) Edinburgh University Press edition. It is notable that Brian Goodwin is an opponent of standard neo-Darwinism, but also the very opposite of a creationist or a believer in ID.

76. Korthof, "On the Origin of Information by Means of Intelligent Design." Link entitled "Dembski's 'Reply' to My Review," e-mail message from Dembski dated April 16, 2000. Accessed on February 25, 2002, at http://home.wxs.nl/~gkorthof/korthof1.htm#Dembski.

77. Branden Fitelson, Christopher Stephens, and Elliot Sober, "How Not to Detect Design," *Philosophy of Science* 66:3 (1999), 472–488. Also available at http://philosophy.wisc.edu/sober/dembski.pdf. Accessed on February 25, 2002.

78. William A. Dembski, "Another Way to Detect Design?" Other quotations of Dembski's comments *in this section* are from the same source, which makes frequent reference to other Dembski writings. Accessed on February 26, 2002, at http://www.baylor.edu/~William_Dembski/docs_critics/sober.htm.

79. Pennock is the author of the recent, comprehensive analysis of ID creationism, *Tower of Babel: The Evidence Against the New Creationism* (Cambridge, MA: MIT Press, 1999).

80. Robert T. Pennock, "The Wizards of ID: Reply to Dembski." Accessed on February 25, 2002, at http://www.metanexus.net/archives/printerfriendly.asp?archiveid=2645.

81. Pennock, "The Wizards of ID." This exchange is included in Pennock's anthology, *Intelligent Design Creationism and Its Critics: Philosophical, Theological, and Scientific Perspectives* (Cambridge, MA: MIT Press, 2001), 639–667.

82. Perakh, "A Consistent Inconsistency."

83. Dembski, *Intelligent Design*, 160, 167–179. Dembski maintains this position in *No Free Lunch*. His position is, very simply, that "algorithms and natural laws are ideally suited for transmitting already existing CSI. . . . [W]hat they cannot do is explain its origin" (149). This leaves the *supernatural* creation of CSI as the only option.

84. Actually, a similar rule, which Dembski acknowledges as predecessor, was proposed some years ago by Nobel laureate immunologist Peter Medawar, but for different purposes and with different conclusions.

85. Victor J. Stenger, "Messages From Heaven," chapter 4 of *Has Science Found God?* (Amherst, NY: Prometheus Books, 2003).

86. Matt Young, "How to Evolve Specified Complexity by Natural Means." Accessed on May 22, 2002, at http://www.pcts.org/journal/young2002a.html. See also Young's "Note added, March 31, 2002," accessed on May 22, 2002, at http://pcts.org/journal/correspondence/young.02-0331.html..

87. William A. Dembski, "Converting Matter into Mind." Accessed on May 11, 2002, at http://www.arn.org/docs/dembski/wd_convmtr.htm.

88. Richard Dawkins, "The 'Information Challenge': How Evolution Increases Information in the Genome," *Skeptic* 7:2 (1999), 64 ff. Accessed on February 17, 2002, at http://www.skeptic.com/archives41.html.

89. Jeffrey Shallit, "Review of Dembski's *No Free Lunch*," *BioSystems* 66:1–2 (2002), 93–99. Accessed on May 23, 2002, at http://www.math.uwaterloo.ca/~shallit/nflr3.txt.

90. See "William A. Dembski: Curriculum Vitae," at http://www.designinference.com/documents/2003.02.CV.htm. Accessed on April 13, 2003.

91. Posted by Professor William C. Wimsatt to the Yahoo Evolutionary Psychology discussion group at http://groups.yahoo.com/group/evolutionary-psychology/message/18031 on April 4, 2002, and very slightly edited here, according to Wimsatt's suggestion to P. R. Gross posted on April 15, 2002.

92. Van Till's review was accessed on September 9, 2002, at http://www.aaas.org/spp/dser/evolution/perspectives/vantillecoli.pdf. Dembski has responded, not unexpectedly, by accusing Van Till of "invincible ignorance." He also criticizes Van Till for being "steeped in process theology": "Van Till invokes the vocabulary of process theology, which describes God as guiding or persuading creation. But all such talk is empty." In addressing Van Till's theological views, Dembski confirms the criticism that ID is itself a religious movement: if ID were truly a "scientific research program," as Dembski claims, Van Till's theology would be completely irrelevant (as it actually is to his criticisms of *No Free Lunch*). Dembski would have no reason to mention it, or indeed to care about it at all.

93. Richard Wein, "Not a Free Lunch But a Box of Chocolates." Accessed on April 23, 2003, at http://www.talkorigins.org/design/faqs/nfl/.

94. See the www.talkreason.org website listed earlier for Perakh's contributions on intelligent design.

95. Rosenhouse, "Probability, Optimization Theory, and Evolution."

96. Elliott Sober, *Philosophy of Biology* (Boulder, CO: Westview Press, 1993), 54.

6. Everything *Except* Science I

1. Center for the Renewal of Science and Culture, "The Research Fellowship Program." Accessed on April 17, 2002, at http://www.crsc.org/fellows/fellowshipInfo.html.

2. "Major Grants Help Establish Center for Renewal of Science and Culture," Discovery Institute *Journal* (August 1996). Accessed on April 17, 2002, at http://www.discovery.org/w3/discovery.org/journal/center.html.

3. "Center for the Renewal of Science and Culture: Center Innovates in Media and Academia," Discovery Institute *Journal* (spring 1998), 5. Accessed on April 17, 2002, at http://www.discovery.org/w3/discovery.org/journal/. Very little information has been found on the nature of the SG Foundation. The president is Stuart C. Gildred of California, for whom the organization is apparently named. According to Guidestar, the SG Foundation had a total value of $5,738,212 in assets for 1996 (accessed on July 10, 2000). Its 1998 IRS Form 990 shows that at the end of that year the total net assets or fund balances was $8,897,086. No information has been found on how much the SG Foundation gave CRSC.

4. "1998 Letter from the President," Discovery Institute *Journal* (spring 1998), 15.

5. "Discovery Institute, Revenue and Expenses: Fiscal Year Ending Dec. 31, 1997." Accessed on March 24, 2000, at http://www.guidestar.org/. DI's total 1997 revenue was $1,832,398, the difference coming from sources other than contributions.

6. *Journal* (spring 1998), 15.

7. Discovery Institute, Form 990, 1998. Accessed on October 17, 2000 at http://www.guidestar.org.

8. "Major Grants Increase Programs, Nearly Double Discovery Budget," Discovery Institute *Journal* (1999). Accessed on April 17, 2002, at http://www.discovery.org/w3/discovery.org/journal/1999/grants.html. This is apparently the announcement to which Walter Olson refers in the January 1999 *Reason* magazine: "According to the newsletter of the Seattle-based Discovery Institute, California's Ahmanson family, through its Fieldstead & Co. foundation, has donated $1.5 million to the institute's fledgling Center for the Renewal of Science and Culture for a research and publicity program." See "Dark Bedfellows: Postmoderns and Traditionalists Unite Against the Enlightenment" at http://reason.com/9901/co.wo.darkbedfellows.shtml. Accessed on April 17, 2002. See also Steve Benen, "From Genesis to Dominion: Fat-Cat Theocrat Funds Creationist Crusade," *Church and State* (July/August 2000). Accessed on September 15, 2002, at http://www.au.org/churchstate/cs7003.htm.

9. The exact figure, according to information filed by DI with the Washington Secretary of State's office, was $2,366,718 in revenue. Accessed on November 9, 2000, at http://207.153.159.68/sec_state/charities98/detail.tmpl$search?db=charities.wcdb&eqcharity_iddatarq=1292&charity_idsumm=T&searchtitle=.

10. Larry Witham, "Contesting Science's Anti-Religious Bias," *Washington Times*, December 29, 1999. Accessed online on April 17, 2002, at http://www.discovery.org/viewDB/index.php3?program=CRSCstories&command=view&id=65.

11. Financial information for 2000 is available at http://www.guidestar.org/. Accessed on April 14, 2002. The amount of the NCSE's 1998 budget is found in John Cole, "Money Floods Anti-Evolutionists' Coffers," *Reports of the National Center for Science Education* 20:1–2 (January–April 2000), 64–65.

12. Though not as well known as Denton's earlier book, *Evolution: A Theory in Crisis*, *Nature's Destiny* plays a role in the Wedge's agenda. Accessed on April 17, 2002, at https://www.discovery.org/cgi-bin/discoveryWebStore/web_store.cgi?product=(CRSC)&cart_id=. Denton's *Evolution: A Theory in Crisis* and his embrace in *Nature's Destiny* of teleology, the anthropic principle, and anti–natural selection arguments have earned him a firm place in the Wedge. Philip Spieth, a geneticist at the University of California-Berkeley, reviewed *Nature's Destiny*: "[Denton] dismisses out of hand the Darwinian concept of evolution through natural selection on the grounds that organisms are so complex and so 'finely tuned' to their environments with such incredible adaptations that their evolution can only be explained as a directed, purposeful process. Natural selection and 'non-directed models of evolution' are not up to the job in Denton's view. In this respect he echoes the views of 'Intelligent Design' advocates such as Behe. . . . Michael Denton wants to undo the divergence between biology and natural theology, and he wants to impose once again the teleological worldview of natural theology upon the methodology of biological science. . . . [H]e is in step with the group of neocreationists whose tenets are

denoted by the term 'Intelligent Design.' . . . *Destiny* is not a scientific treatise. Rather, it is a book about natural theology." Philip Spieth, "Denton Becomes a (Teleological) Evolutionist," *Reports of the National Center for Science Education* (March/April 1998). Accessed on September 24, 2002, at http://www.ncseweb.org/resources/articles/Spieth-on-Denton.pdf.

13. See William Dembski, "The Intelligent Design Movement," reprinted from *Cosmic Pursuit* (spring 1998), at http://www.arn.org/docs/dembski/wd_idmovement.htm. Accessed on April 17, 2002.

14. Dembski's award was announced by the Templeton Foundation. Accessed on April 17, 2002, at http://www.templeton.org/pcrs_winners.asp. In the 1999 award announcement, the subtitle of the book is *The Science and Metaphysics of Information*. In Dembski's May 2002 c.v., note that the word "science" has been eliminated.

15. On Dembski's "Design Inference Website," *The Design Inference: Making a New Science and Worldview*, has the subtitle *Answering the Toughest Questions About Intelligent Design*, and Dembski has posted the preface and table of contents. In the preface, Dembski tells his readers that "there is good reason to think intelligent design fits the bill as a genuine scientific revolution" and that "it promises to remake science and the world." Accessed on September 16, 2002, at http://www.designinference.com/documents/2002.07.des_rev.htm.

16. See the article at http://www3.baylor.edu/~William_Dembski/docs_resources/guidebook.htm. Accessed on April 15, 2003.

17. See http://www.discovery.org/crsc/fellows/PaulNelson/index.html. Accessed on June 7, 2002. A "monograph," for EM's purposes, may mean anything from a paper to a book-length manuscript. According to information printed on the back of each volume, EM is "an international monograph series for all the evolutionary half of biology . . . in the broadest sense." The back cover also says that "anyone, worldwide" can submit a manuscript and that "there is no payment required from authors, although voluntary payment of page costs is welcome." The EM series is a nonprofit publication with an irregular publication schedule. (Leigh Van Valen to Barbara Forrest, personal communication, December 14, 2000)

18. Both Broom and Hunter are fellows in William Dembski's new organization, the International Society for Information, Complexity, and Design (ISCID). See http://www.iscid.org/fellows.php. Accessed on June 7, 2002. See chapter 7 for information on ISCID.

19. See http://www.arn.org and http://www.discovery.org/cgi-bin/discoveryWeb Store/web_store.cgi.

20. Joseph L. Conn, "God's Air Force: How the National Religious Broadcasters Provide Troops and Ammo for the Religious Right's Christian Nation Crusade," *Church & State* (April 2000). Accessed on April 15, 2002, at http://www.au.org/churchstate/cs4002.htm.

21. A search for the owners of both sites revealed that they are owned by the Discovery Institute. Accessed on June 7, 2002, at http://www.register.com.

22. Hank Hanegraaff, *Bible Answer Man*, interview of Phillip Johnson, December 19, 2000. Accessed on December 29, 2000, at http://www.equip.org/bam/previous.html. No longer available.

23. Center for the Renewal of Science and Culture, "Year End Update" (November–December 1997). Accessed on March 18, 2000, at http://www.discovery.org/w3/discovery.org/crsc/crscnotes2.html. This document is no longer accessible.

24. "Design, Self-Organization, and the Integrity of Creation," Summer Seminar Sponsored by the Fieldstead Institute, Calvin College Seminars in Christian Scholarship. Accessed on April 14, 2002, at http://www.calvin.edu/fss/fieldstd.htm.

25. "Design, Self-Organization, and the Integrity of Creation," Conference Schedule. Accessed on April 14, 2002, at http://www.calvin.edu/fss/dembschd.htm. Graduate Student Scholarship Application, "Design, Self-Organization and the Integrity of Creation" Spring Conference. Accessed on March 12, 2001, at http://www.calvin.edu/fss/

dembgrad.htm. Robert Pennock includes Plantinga and Ratzsch among the important members of the ID movement in *Tower of Babel: Evidence Against the New Creationism* (MIT Press, 1999), 29. Ratzsch participated in the seminal 1996 Mere Creation Conference (http://www.origins.org/mc/menus/speak.html). Plantinga published an article in the Wedge's creationist journal: "Methodological Naturalism?" *Origins and Design* 18:1 (winter 1997); accessed on September 16, 2002, at http://www.arn.org/docs/odesign/od181/methnat181.htm.

26. "Defending the Christian Worldview," 2000 Spring Course Descriptions. Accessed on May 15, 2000, at http://www.tis.edu/MACOURSES/SP2000DESCRIPTION.htm#intelligentDesign. No longer available.

27. William A. Dembski and Jay Wesley Richards, "Introduction: Reclaiming Theological Education," in *Unapologetic Apologetics: Meeting the Challenges of Theological Studies* (InterVarsity Press, 2001), 14. Accessed on April 17, 2002, at http://www.gospelcom.net/ivpress/title/exc/1563-I.pdf.

28. Dembski and Richards, "Introduction," 11 and 15. Dembski and Richards also disavow, without discussion of its full meaning, "fundamentalism, which assumes all conceptual problems facing Christianity are easily resolved." They regard it as an obstacle to transforming "mainline seminaries in particular and the secular academic world in general" (14).

29. Dembski and Richards, "Introduction," 22.

30. Dembski and Richards, "Introduction," 24.

31. See Dembski's biographical sketch on the InterVarsity Press website at www.gospelcom.net/cgi-ivpress/author.pl/author_id=889. Dembski points out Hodge's role in the construction of design arguments in "The Intelligent Design Movement" at http://www.arn.org/docs/dembski/wd_idmovement.htm. According to Jonathan Wells, who wrote his doctoral dissertation about Charles Hodge, the latter repudiated Darwin's theory of evolution on theological grounds. See Jonathan Wells's review of Hodge's book, *What Is Darwinism? And Other Writings on Science and Religion*, at http://www.arn.org/docs/odesign/od172/wells172.htm. All accessed on April 14, 2002.

32. Dembski and Richards, "Introduction," 12.

33. Phillip E. Johnson, foreword, *Unapologetic Apologetics*, 9. Accessed on April 14, 2002, at http://www.gospelcom.net/cgi-ivpress/book.pl/toc/code=1563.

34. Johnson, foreword, 9–10.

35. Rob Boston, "Missionary Man," *Church & State* (April 1999). Accessed on April 14, 2002, at http://www.au.org/churchstate/cs4995.htm.

36. Dr. Kaita is a professor in the Department of Astrophysical Sciences and head of a plasma physics laboratory at Princeton University. See his CRSC bio at http://www.crsc.org/fellows/RobertKaita/index.html. Accessed on September 16, 2002.

37. "Biology Textbook Panda-monium," American Scientific Affiliation *Newsletter* (March/April 1995). Accessed on April 15, 2002, at http://mcgraytx.calvin.edu/ASA/docs/asa_doc22.txt.

38. Accessed on January 19, 2000, at http://www.baylor.edu/~polanyi/purpose.htm. This page is no longer available.

39. Phillip E. Johnson, "How to Sink a Battleship: A Call to Separate Materialist Philosophy from Empirical Science." Accessed on April 17, 2002, at http://www.origins.org/mc/resources/ri9602/johnson.html.

40. Foundation for Thought and Ethics, Richardson, Texas, to "Friend of the Foundation," May 1990, and Foundation for Thought and Ethics, Richardson, Texas, to unspecified recipients, March 8, 1995. For background on the attempted distribution of *Pandas* in Plano, Texas, see Jay Wexler, "Of Pandas, People, and the First Amendment: The Constitutionality of Teaching Intelligent Design in the Public Schools," *Stanford Law Review* 49:2 (January 1997), 439–440.

41. Beth McMurtrie, "Darwinism Under Attack," *Chronicle of Higher Education* 48:17

(December 21, 2001), A8. Accessed on April 15, 2002, at http://chronicle.com/free/v48/i17/17a00801.htm.

42. New Mexicans for Science and Reason, "Special Interest Group Bombards NM Schools with Copies of Behe's 'Darwin's Black Box,'" *Hot News of the Month*, April 5, 2002. Accessed on May 13, 2002, at http://www.nmsr.org/omdahl.htm.

43. Steve Brugge to John Omdahl, personal communication, April 3, 2002.

44. Dave Thomas to Barbara Forrest and Steve Brugge, personal communication, May 7, 2002.

45. New Mexicans for Science and Reason, "Special Interest Group Bombards NM Schools."

46. "Darwin's Black Box Sent to Schools," Creation Science Fellowship of New Mexico *Newsletter*, 13:4 (April 2002). Accessed on May 11, 2002, at http://www.swcp.com/creation/news/0204.htm.

47. Steve Brugge to John Omdahl, personal communication, April 7, 2002.

48. Dave Thomas to Barbara Forrest, personal communication, May 7, 2002.

49. New Mexicans for Science and Reason, "Special Interest Group Bombards NM Schools."

50. The registration information was obtained on March 11, 2002, at www.register.com.

51. See the FTE website at http://www.fteonline.com./index.htm. Accessed on April 15, 2002.

52. The CRSC's science education website was posted at http://www.discovery.org/crsc/scied/. Since approximately summer 2001, it has been inaccessible, as we will discuss shortly.

53. Accessed on November 11, 2000, at http://www.discovery.org/crsc/scied/present/topics/update.htm. At the time of access teachers and students could also order Wedge books from the "Science Education Book Store" at http://www.discovery.org/crsc/scied/bookstore/index.html.

54. Accessed on November 11, 2000, at "Our Mission" at http://www.discovery.org/crsc/scied/what/topics/mission.htm.

55. The objectives and the lesson plan were available at that time at http://www.discovery.org/crsc/scied/evol/cambrian/object/index.html and http://www.discovery.org/crsc/scied/evol/cambrian/resource/lesson/index.html, respectively.

56. Accessed on November 11, 2000, at www.discovery.org/crsc/scied/what/index.html.

57. This bibliography was available prior to its restriction at http://www.discovery.org/crsc/scied/evol/cambrian/resource/biblio/index.htm.

58. This page was at http://www.discovery.org/crsc/scied/evol/cambrian/resource/scientist/index.htm.

59. As of April 15, 2003, this page and others on the Cambrian explosion were still available at http://www.discovery.org/crsc/scied/evol/cambrian/curr/01intro/anarr/04.htm, although there was no link to the pages from CRSC's main page.

60. See, for one example, the sophisticated but highly accessible discussions by Mark Perakh of Dembski's idea of "complexity." Perakh's "A Consistent Inconsistency: How Dr. Dembski Infers Design," is at http://www.nctimes.net/~mark/bibl_science/dembski.htm. Perakh evaluates the work of other Wedge members at http://www.nctimes.net/~mark/bibl_science/behe2.htm and http://www.nctimes.net/~mark/bibl_science/johnson.htm. Other assessments of Dembski's work compiled by Wesley Elsberry may be consulted at http://www.antievolution.org/people/dembksi_wa/. All accessed on October 21, 2002.

61. This link was accessed on March 24, 2001, at http://www.discovery.org/crsc/scied/what/topics/link.htm.

62. "Our Curriculum is Scientifically Accurate." Accessed on March 24, 2001, at http://www.discovery.org/crsc/scied/what/topics/missiontopics/accurate.htm.

63. "Our Curriculum is [*sic*] Legally Permissible in Public Schools." Accessed on

March 24, 2001, at http://www.discovery.org/crsc/scied/what/topics/missiontopics/legal.htm.

64. "Web Lowers Political Hurdles." Accessed on March 24, 2001, at http://www.discovery.org/crsc/scied/present/topics/political.htm. The CRSC Science Education page "Web Curriculum Comes with Interactive Media Power" also posted links under "Other Web Curriculum Examples" at http://www.discovery.org/crsc/scied/present/topics/interact.htm. Accessed on March 24, 2001. One link was entitled "Infectious AIDS: Have We Been Misled?" This link led to the site of Berkeley scientist Peter Duesberg, who has been criticized for his continued, fact-resistant claims to the effect that "there is no virological, nor epidemiological, evidence to back-up the HIV-AIDS hypothesis." Accessed on April 17, 2002, at http://www.duesberg.com/about/index.html.

65. Accessed on November 9, 2002, at http://www.discovery.org/crsc/scied/present/topics/political.htm. The Wedge book *Intelligent Design in Public School Science Curricula: A Legal Guidebook* is available on the web free of charge as a long paper of the same title at http://www.arn.org/docs/dewolf/guidebook.htm. It is also available under a different title, "Teaching the Controversy: Darwinism, Design and the Public School Science Curriculum," at http://law.gonzaga.edu/people/dewolf/fte.htm. In this book/paper David K. DeWolf and Stephen Meyer, with Mark DeForrest, argue that the 1987 U.S. Supreme Court ruling *Edwards v. Aguillard* permits the teaching of ID as a scientific theory, an argument clearly meant to assure teachers that they are on safe legal ground if they incorporate ID into their public school science curricula. In "Evolution of a Controversy" in the November 1999 *American Bar Association Journal*, John Gibeaut reveals that in this paper DeWolf "offers tips for those wanting to get around the Supreme Court. He counsels against the obvious faux pas, such as directly citing creationism or beliefs held by some that the Earth is only 6,000 years old." (Accessed on May 13, 2002, at http://www.abanet.org/journal/nov99/11fevolv.html.) Also available on the web is "A Note to Teachers," in which Stephen Meyer and Mark Hartwig assure teachers that "in its landmark ruling on the Louisiana Balanced Treatment Act, the United States Supreme Court . . . affirmed that teachers already had the flexibility to teach non-evolutionary views and present scientific evidence bearing on the question of origins." Accessed on April 15, 2002, at http://www.arn.org/docs/orpages/or151/151teachnote.htm#anchor448859. This paper is included in *Of Pandas and People* (Foundation for Thought and Ethics, 1993).

66. See Keas's CRSC biographical sketch at http://www.discovery.org/crsc/fellows/MichaelNewtonKeas/index.html. His "Self-Profile" and "Progress Report for 2000–2001" were posted at http://www.okbu.edu/academics/natsci/hp/keas/grow/growagree/profile.htm and http://www.okbu.edu/academics/natsci/hp/keas/grow/growagree/report current.htm. Accessed on April 15, 2002.

67. "ID Colleges," Access Research Network." Accessed on October 1, 2000, at http://www.arn.org/college.htm.

68. Keas posts syllabi and course schedules at http://www.okbu.edu/academics/natsci/us/312/general/syl.htm and http://www.okbu.edu/academics/natsci/us/312/index.htm, respectively. The PowerPoint files were available on the latter page. Accessed on April 15, 2002.

69. See "Videos from ARN," Access Research Network. Accessed on April 12, 2002, at http://www.arn.org/arnproducts/vidinfo.htm. Other ARN products may be accessed from this site. See the text of the Alabama textbook disclaimer and an analysis of it that biologist Kenneth Miller wrote for the National Center for Science Education at http://www.ncseweb.org/resources/articles/1910_dissecting_the_disclaimer_2_7_2001.asp. Accessed on June 27, 2002. In 2001, the Alabama Board of Education voted to replace the original disclaimer with a slightly less blatant but still objectionable one. See "Alabama Keeps Evolution Warning on Books" at http://www.cnn.com/2001/fyi/teachers.ednews/11/09/evolution.ap/. Accessed on June 27, 2002.

70. See Dennis Wagner, "Put Another Candle on the Birthday Cake," *Origins Research* 10:1 (spring/summer 1987). Accessed September 16, 2002, at http://www.arn.

org/docs/orpages/or101/101wagnr.htm. *Origins Research*, SOR's publication, has become the Wedge's journal, *Origins and Design*.

71. See the information about the curriculum at http://www.realscience-4-kids.org/. Accessed on June 8, 2002.

72. "Access Research Network 2000/2001 Annual Report."

73. Letter accompanying 2000/2001 Annual Report, August 2001.

74. See Jeremy Alder, "The Coming Revolution: It's Up to Us," *Student Wedge Update*, May 15, 2001, at http://www.idurc.org/swu/swu-051501.shtml. See also "New Student Division at Access Research Network," *ARN Announce* 15 (May 31, 2001) at http://www.arn.org/announce/announce0501no15.htm. See also letters between Micah Sparacio, the IDURC web master, and Jonathan Wells on Access Research Network at http://www.arn.org/docs/wells/jw_significancepm.htm. All accessed on September 15, 2002.

75. "About the IDEA Club: A Club History, Explanation, and Philosophy." Accessed on March 24, 2001, at http://www-acs.ucsd.edu/%7Eidea/about.shtml#history. A report on Dembski's lecture to the IDEA club was posted at http://www-acs.ucsd.edu/~idea/dembski.htm. Accessed on September 5, 2001.

76. All of these articles are available on the Access Research Network website.

77. See http://www.arn.org/pearcey/nphome.htm. Accessed on April 17, 2002.

78. See this list at http://www.worldmag.com/world/issue/12-04-99/cover_1.asp. Accessed on April 16, 2002. (Readers must register to view *World* articles online.) Gene Veith is the director of the Cranach Institute at Concordia University in Mequon, Wisconsin, which sponsored the June 2000 Design and Its Critics conference and has close ties to the CRSC. See CI's letters of protest regarding Baylor University's demotion of William Dembski at http://institutes.cuw.edu/cranach/dembski_protest.htm. Accessed on April 17, 2003.

79. Skip Porteous, "Bush's Secret Religious Pandering," *Freedom Writer*, Institute for First Amendment Studies (September/October 2000), at http://www.ifas.org/fw/0009/bush.html. Accessed at the Internet Archive on April 20, 2003, at http://www.archive.org/web/web.php. See also Frederick Clarkson, "Christian Reconstructionism," *Public Eye* 8:1 and 2 (March/June 1994), at http://www.publiceye.org/magazine/v08n1/chrisre1.html. See Clarkson's more recent update, "The Culture Wars Are Not Over," *Public Eye* 15:1 (spring 2001), at http://www.publiceye.org/magazine/v15n1/PE_v15n1_TOC.htm#TopOfPage. Both accessed on September 16, 2002.

80. Third-party coverage of ID relating to the Kansas controversy, as well as op-eds by Wedge members, is also posted on the Access Research Network website. Accessed on April 16, 2002, at http://www.arn.org/kansas899.htm.

81. See "Breaking Ranks with Darwinian Orthodoxy," *Origins & Design* 17:2. Accessed on April 16, 2002, at http://www.arn.org/docs/odesign/od172/ranks172.htm.

82. "Dissent on Descent," Ethics and Public Policy *Newsletter* (summer 1996). Accessed on March 6, 2000, at http://www.eppc.org/newsletters/newssu96.html#6.

83. See Ronald Bailey, "Origin of the Specious," *Reason* (July 1997). Accessed on April 16, 2002, at http://www.reason.com/9707/fe.bailey.html. Bailey comments specifically on the alignment of Himmelfarb, Kristol, and Bethell on the side of ID.

84. Hilton Hotel sales staff to Barbara Forrest, personal communication, February 2, 2001. The New York Media Lunch announcement was accessed on the Discovery Institute web page on October 4, 2000, at http://www.discovery.org.

85. Lynn Vincent, "Science vs. Science," *World*, February 26, 2000. Accessed on April 16, 2002, at http://www.discovery.org/viewDB/index.php3?program=CRSCstories&command=view&id=148.

86. See http://www.hoovers.com/co/capsule/8/0,2163,47948,00.html for the announcement of Moon's purchase of UPI. Veteran UPI reporter Helen Thomas resigned when Moon bought the news service. See http://www.amarillonet.com/stories/051700/usn_LA0727.shtml. All accessed on April 16, 2002. See Frederick Clarkson's devastating

analysis of the Unification Church's involvement in American conservative politics in "Americans for Theocratic Action: Rev. Sun Myung Moon, 'Family Values,' and the Christian Right—One Dangerous Theocrat," in *Eternal Hostility: The Struggle Between Theocracy and Democracy* (Monroe, ME: Common Courage Press, 1997).

87. See Access Research Network's page referring to this debate. Accessed on April 17, 2002, at http://www.arn.org/fline1297.htm#anchor22822.

88. Phillip Johnson, "*Monkey Trial* Airs on PBS," *Weekly Wedge Update*, February 25, 2002. Accessed on April 16, 2002, at http://www.arn.org/docs/pjweekly/pj_weekly_020225.htm.

89. The *Talkback Live* transcript is available at http://www.arn.org/docs/kansas/talkback81699.htm. Accessed on April 17, 2002. The *Nightline* transcript was accessed on August 3, 2000, at http://www.abcnews.go.com/onair/nightline/transcripts/nl000727_trans.html. Access has expired.

90. Matt Carter, "Lab Scientists Challenging Darwin," *Tri-Valley Herald*, September 26, 2001. Accessed on April 16, 2002, at http://www.reviewevolution.com/press/fromPress_ScienChalDarwin.php.

91. See "Executive Summary," in *Getting the Facts Straight: A Viewer's Guide to PBS's Evolution* (Seattle: Discovery Institute Press, 2001), 9 and 15. Accessed on September 16, 2002, at http://www.reviewevolution.com/viewersGuide/viewersGuide.pdf.

92. DI's ad is at http://www.reviewevolution.com/press/pressRelease_100Scientists.php. Accessed on June 8, 2002. NCSE's Project Steve links are posted at http://www.ncseweb.org/article.asp?category=18. Dembski's response appeared on DI's website on March 19, 2003, at http://www.discovery.org/viewDB/index.php3?command=view&id=1393&program=CRSC%20Responses.

93. Glenn Branch, "The Abatement," and Skip Evans, "Doubting Darwinism through Creative License," *Reports of the National Center for Science Education* 21:5 and 6 (September–December 2001), 10–14 and 22–23.

94. "Leaked PBS Memo Reveals Improper Political Agenda Behind *Evolution* Series," Discovery Institute press release. Accessed on September 15, 2002, at http://www.reviewevolution.com/press/pressRelease_LeakedPBSMemo.php. The Wedge's entire anti-PBS campaign, including all of the information used in the account, is archived on the website they devote to it at http://www.reviewevolution.com. ARN's "Response to the PBS *Evolution* Project" is at http://www.arn.org/pbs_evolution0901.htm. Accessed on April 16, 2002. The National Center for Science Education provides excellent, extensive coverage of the Wedge's attack on the PBS series in its September-October 2001 *Reports of the National Center for Science Education*.

95. See "Creationism v. Evolution: Will Religion or Science Prevail?" *Justice Talking*, April 10, 2000, at http://www.justicetalking.org/viewprogram.asp?progID=135.

96. See Dobson's website at http://www.family.org/resources/itempg.cfm?itemid=1588. *U.S. New & World Report*'s cover story on Dobson, "A Righteous Indignation," appeared in the May 4, 1998, issue.

97. Lawrence Lerner, *Good Science, Bad Science: Teaching Evolution in the States*, Thomas B. Fordham Foundation, September 2001. Accessed on September 15, 2002, at http://www.edexcellence.net/library/lerner/gsbsteits.html. Neither U.S. Newswire press releases nor DI press releases are archived on their respective websites. All of DI's U.S. Newswire releases are available through Lexis-Nexis at http://web.lexis-nexis.com.

98. "U.S. Newswire Strikes New Content Agreement with America Online." Accessed on December 28, 2000, at http://www.usnewswire.com/about/3quart.htm.

99. Joseph Farah, "WorldNetDaily's Explosive Growth," *WorldNetDaily*, June 1, 2000. Accessed on April 17, 2002, at http://www.worldnetdaily.com/news/article.asp?ARTICLE_ID=14981.

100. Accessed on April 17, 2002, at http://www.discovery.org/viewDB/index.php3?program=CRSC&command=view&id=480.

101. See "Tools for Change." Accessed on April 17, 2002, at http://www.iconsof evolution.com/tools/.

102. Judith Anderson and John Wiester, "Dealing with the Media in the Science Textbook Controversy," Access Research Network. Accessed on April 17, 2002, at http://www.arn.org/docs/pc1110.htm.

103. At this writing, William Dembski has one "mainstream" philosophy article, which, based on the descriptors, appears to be about design: "Randomness by Design," *Nous* 25:1 (1991). Earlier, he published an article in the *Scottish Journal of Theology*, "Schleiermacher's Metaphysical Critique of Miracles," 49:4 (1996). Dembski and Stephen Meyer published "Fruitful Interchange or Polite Chitchat? The Dialogue Between Science and Theology," *Zygon* 33:3 (1998). Dembski also published one article, "Uniform Probability," in the *Journal of Theoretical Probability* (1990) and two, "Converting Matter into Mind: Alchemy and the Philosopher's Stone in Cognitive Science" and "Intelligent Design as a Theory of Information" in *Perspectives on Science and the Christian Faith* (1990 and 1997, respectively). William A. Dembski, "Curriculum Vitae," May 2002. Accessed on September 16, 2002, at http://www.designinference.com/documents/05.02.CV.htm.

104. The spring 1998 DI *Journal* was accessed at http://www.discovery.org/w3/discovery.org/journal. See "Welcome to an Evolving Publication," on Access Research Network's *O&D* web page at http://www.arn.org/docs/odesign/od171/editorial171.htm. Accessed on March 14, 2000. See John Cole, "*Origins & Design*: A Journal, Not Just a Debate Ploy!" *Reports of the National Center for Science Education* (winter 1996), 4–5. For Schaefer's foreword, see *Mere Creation: Science, Faith and Intelligent Design* (InterVarsity Press, 1998), 10–11. For Johnson's announcement of the launching of *O&D*, see "How to Sink a Battleship" at http://www.origins.org/mc/resources/ri9602/johnson.html. For *O&D* staff listings, see "Publication Information" at http://www.arn.org/odesign/odmasthead.htm.

105. See the journal at http://www.iscid.org/pcid.php. Accessed on April 17, 2003.

106. See www.arn.org/authors.htm at Access Research Network.

107. See these "offices," accessed on June 8, 2002, at http://www.facultylinc.com/personal/facoffice.nsf.

7. Everything *Except* Science II

1. George Bishop, "What Americans Really Believe (and Why Faith Isn't as Universal as They Think)," *Free Inquiry* (summer 1999), 38. Accessed on September 24, 2002, at http://www.secularhumanism.org/library/fi/bishop_19_3.html. See also George Bishop, "The Polls—Americans' Belief in God," *Public Opinion Quarterly* 63 (1999), 421–434.

2. Phillip E. Johnson, "The Wedge: Breaking the Modernist Monopoly on Science," *Touchstone* (July/August 1999), 19. Accessed on April 18, 2002, at http://www.touchstone mag.com/docs/issues/12.4docs/12-4pg18.html.

3. George Bishop, "The Religious Worldview and American Beliefs About Human Origins," *Public Perspective* 9:5 (August/September 1998), 39–40. In "Back to the Garden," *Public Perspective* 11:3 (May/June 2000), Bishop offers some analysis of why so many Americans continue to believe the literal biblical account: "[F]or nearly half the adult public in this country *the authority of the Bible* on this question remains an article of faith. And therein lies a clue as to why so many Americans continue to believe in the creationist account of human origins: the authoritarian syndrome—an inclination to be influenced by trusted authorities in our lives" (21). Accessed on September 27, 2002, at http://www.ropercenter.uconn.edu/pubper/pdf/pp113b.pdf. See also remarks by Bishop in Kenneth Chang, "Evolutionary Beliefs," ABCNEWS.com at http://abcnews.go.com/sections/science/DailyNews/evolutionviews990816.html. See also Richard Morin, "Can We Believe in Polls About God?" *Washington Post*, June 1, 1998, at http://washingtonpost.com/wp-srv/politics/polls/wat/archive/wat060198.htm. Both accessed on September 25, 2002.

4. Phillip Johnson, "Re: reality check on Jasanoff v. Levitt," Internet discussion group, May 31, 1999. Accessed on July 17, 2000, at http://vest.gu.se/vest_mail/1194.html. This site is no longer available.

5. "Zogby's Final Report," Discovery Institute press release, September 24, 2001. Accessed on April 18, 2002, at http://www.reviewevolution.com/press/pressRelease_100Scientists.php. A copy of the crosstabs is also available at this site.

6. Glenn Branch, "The Rising Tide," *Reports of the National Center for Science Education* 21:5–6 (September–December 2001), 9.

7. Stephen C. Meyer, "Teach the Controversy," *Cincinnati Enquirer*, March 30, 2002. Accessed on April 19, 2002, at http://www.discovery.org/viewDB/index.php3?command=view&id=1134&program=CRSC. DI commissioned a similar poll in Ohio in May 2002, using questions 1 and 2 from the September 2001 poll. The Ohio poll narrative selectively emphasizes data, in a way similar to the earlier poll. Likewise, the crosstabs show the same across-the-board, ID-favorable high percentages as the 2001 poll, thus reflecting the same flaws. See chapter 8 for information on the Wedge's maneuverings in Ohio, which prompted DI's commissioning of the poll. Chris Mooney's scrutiny of Zogby's polling methods yielded findings mirroring our analysis of DI's poll. Mooney found, for example, that a poll commissioned by an animal rights group produced inflated results that the media reported uncritically: "By hiring the renowned pollster John Zogby, the group had essentially purchased an objective fact." Citing DI's poll, Mooney notes the effect of Zogby polls' wording on results: "[N]umerous Zogby Polls for . . . special interests . . . relied on creative phrasing to give the impression of wide public support for the view that the . . . client is promoting. . . . Zogby is . . . trading on his reputation as a legitimate, media-certified pollster to help groups disseminate inflated claims about public opinion based on inventive wording." See "John Zogby's Creative Polls," *The American Prospect* (February 1, 2003). Accessed on March 29, 2003, at http://www.prospect.org/print/V14/2/mooney-c.html.

8. David Applegate, "Evolution Opponents Hold Congressional Briefing," American Geological Institute, Government Affairs Program SPECIAL UPDATE, May 11, 2000. Accessed on April 19, 2002, at http://www.agiweb.org/gap/legis106/id_update.html.

9. William A. Dembski, "Disbelieving Darwin—and Feeling No Shame!" Accessed on April 19, 2002, at http://www.discovery.org/viewDB/index.php3?program=CRSC&command=view&id=163. Despite his disavowal of any affinity between ID and theistic evolution, Dembski, like Johnson, conveniently includes theistic evolutionists among those whom the polls show to be "committed to some form of intelligent design." For his *denial* of compatibility between ID and theistic evolution, see "What Every Theologian Should Know About Creation, Evolution and Design," at http://www.origins.org/offices/dembski/docs/bd-theologn.html. Such denials reflect the Wedge's true position. But Dembski has not hesitated to make opportunistic assertions about the compatibility between ID and theistic evolution when it was useful to do so. In *Mere Creation: Science, Faith and Intelligent Design*, he says, "Intelligent design is logically compatible with everything from utterly discontinuous creation (e.g., God intervening at every conceivable point to create new species) to the most far-ranging evolution (e.g., God seamlessly melding all organisms together into one great tree of life)" (19).

10. "Darwin, Science, & Going Beyond the Culture Wars," A Presentation by the Fellows of the Discovery Institute's Center for the Renewal of Science and Culture. Accessed June 14, 2000, at http://www.discovery.org/crsc/scied/downloads/.

11. People for the American Way, "Evolution and Creationism in Public Education: An In-depth Reading of Public Opinion," Results of a Comprehensive National Survey, March 2000. The poll, along with other relevant literature, is available at http://www.pfaw.org/pfaw/general/default.aspx?oid=2095. Accessed on September 24, 2002.

12. James Glanz, "Poll Finds that Support Is Strong for Teaching 2 Origin Theories," *New York Times*, March 11, 2000. Accessed on April 19, 2002, at http://www.nytimes.com/library/national/science/031100sci-evolution-poll.html. See Molleen Matsumura's in-

terpretation of the PFAW poll: "Poll Reveals Most Americans Support Evolution Education," *Reports of the National Center for Science Education* (November/December 1999). (*Reports* bears an earlier date than the poll, but because of NCSE's publication schedule, it actually came out after the poll.)

13. PFAW, "Evolution and Creationism in Public Education," 6. The fact that half of the American people are unfamiliar with the *term* "creationism" is itself ambiguous. It surely cannot be said that half of the American people are unfamiliar with the *concept* of divine creation, which is what the term largely represents in the public mind and which, when the term is defined for them, is understood to signify the concept most already possess. PFAW defined the term for respondents: "Creationism is the belief that God created human beings in their current form. Creationism rejects the idea that human beings evolved from less advanced forms of life over millions of years." Of those who *have* heard of the term, 95% understand it as the divine creation of humans, with 59% of these understanding it strictly as creation according to the Bible in the last 10,000 years, and 36% understanding it in a looser sense, as not necessarily happening exactly as the Bible says. The lack of familiarity detected by the poll is very likely the mere lack of familiarity with the term itself rather than with the concept or the belief; thus, it does not signify anything that could be interpreted optimistically by evolution proponents.

14. PFAW, 38.

15. Bishop, "What Americans Really Believe." For this information, Bishop cites the 1993 International Social Survey Program, Inter-University Consortium for Political and Social Research, University of Michigan. For his table of figures, he cites Tom Smith, "Environmental and Scientific Knowledge Around the World," GSS Cross-National Report No. 16, January 1996, National Opinion Research Center, University of Chicago.

16. PFAW, 40–42.

17. Deborah Jordan Brooks, "Substantial Numbers of Americans Continue to Doubt Evolution as Explanation for Origin of Humans," The Gallup Organization, March 5, 2001. Accessed on March 7, 2001, at http://www.gallup.com/Poll/releases/pr010305.asp.

18. Focus on the Family, "Keeping the Darwinists Honest [An Interview] with Phillip Johnson," *Citizen* (April 1999). Accessed on April 19, 2002, at http://www.family.org/cforum/citizenmag/features/a0005469.html.

19. Johnson, *Defeating Darwinism*, 93.

20. Tom Woodward, "Ruse Gives Away the Store: Admits Evolution Is a Philosophy," *Real Issue* (November/December 1994). Accessed on April 19, 2002, at http://www.leaderu.com/real/ri9404/ruse.html. Despite the spin Woodward puts on Ruse's remarks, nobody with knowledge of Michael Ruse's career and work in philosophy of biology has, or has ever had, the slightest reason to think that he has doubts about the scientific integrity of Darwinism. See Ruse's response to the distortion of his remarks in "Booknotes," *Biology and Philosophy* 8 (1993), 353–358. See also Eugenie Scott, "Cold Comfort for Creationists in Ruse Talk," *Reports of the National Center for Science Education* 13:2 (summer 1993), 10–11.

21. Johnson, "The Wedge: Breaking the Modernist Monopoly on Science."

22. Ray Bohlin, "Mere Creation: Science, Faith & Intelligent Design," Probe Ministries. Accessed on April 19, 2002, at http://www.probe.org/docs/mere.html.

23. See Rich McGee, "Mere Creation and the Two Tasks: Exercising the Balance Between Winning the Spirit and the Mind," on the Mere Creation website. Accessed on April 19, 2002, at http://www.origins.org/mc/resources/ri9602/mcgee.html.

24. See "Conference Overview" at http://www.origins.org/mc/menus/overview.html. See also Christian Leadership Ministries' special edition of *The Real Issue* devoted to the Mere Creation conference. Accessed on April 19, 2002, at http://www.leaderu.com/real/ri9602.html. The URL in item 6 is to a now-defunct page on the website of Internet Christian Library, an Internet archive and ministry started by the Institute for Christian Leadership. See http://www.iclnet.org. Accessed on April 19, 2002. ICL "provided the first online 'home' for . . . Christian Leadership Ministries." Stephen Meyer published an

article, "Open Debate on Life's Origin," in the spring 1995 ICL publication *Faculty Dialogue*, which is archived on the ICL site. Accessed on April 19, 2002.

25. Koons ignores the fact that training in biology and the philosophy of science has never been a criterion for joining the ID movement. Phillip Johnson and David DeWolf are lawyers. Other CRSC fellows are theologians. Koons would qualify as a part of the movement for no other reason than his CRSC fellowship, even if he had not engaged in other activities supportive of ID. These fellowships are awarded for the support of *any* CRSC-approved activity that advances ID; this is obvious in light of the absence of any productive scientific research. However, Koons has written a thinly disguised apology for "theistic science," the new paradigm with which the Wedge hopes to replace the current naturalistic one. See Robert Koons, "Making Progress in the Origins Debate," *Real Issue* (March/April 1997), in which he says that "the vast number of new anthropic coincidences discovered in recent years strongly confirms the theistic paradigm and disconfirms its naturalistic competitor." This edition of *The Real Issue* is devoted entirely to ID-related articles. Accessed April 19, 2002, at http://www.leaderu.com/real/ri9701.html. Koons is also a fellow of William Dembski's new ID organization, the International Society for Complexity, Information, and Design (ISCID), discussed later in this chapter. See Koons's paper "Are Probabilities Indispensable to the Design Inference?" in the first issue of ISCID's journal. Accessed on May 14, 2002, at http://www.iscid.org/pcid/2002/1/1/pcid_contents_2002_1_1.php.

26. See "Great Beginnings: UT Origins Conference Opens Doors to Dialogue," *Real Issue* (March/April 1997). Accessed on April 19, 2002, at http://www.leaderu.com/real/ri9701/koons.html.

27. See the Veritas Forum schedule at http://www.utexas.edu/staff/fscf/schedule.html. Accessed on April 20, 2002. Koons put out a call for papers at http://www.dla.utexas.edu/depts/philosophy/faculty/koons/ntse/ntsecall.html. Accessed December 26, 2000. The call for papers is archived on the site of the American Scientific Affiliation. Accessed on April 20, 2002, at http://www.asa3.org/archive/evolution/199608/0235.html.

28. See Scott Swanson, "Debunking Darwin?" *Christianity Today*, January 6, 1997. Accessed on April 20, 2002, at http://www.christianityonline.com/ct/7t1/7t1064.html.

29. The DI summer 1997 *Journal* was accessed at http://www.discovery.org/w3/discovery.org/journal/announcedsum97.html.

30. Robert Koons is listed as a contact person on CRSC fellow William Lane Craig's web page at "Leadership University." Accessed April 20, 2002, at http://www.leaderu.com/offices/billcraig/menus/schedule.html.

31. See Cogdell's and Mills's signatures on Ad Hoc Origins Committee letter at http://www.apologetics.org/news/adhoc.html. See Martin Poenie listed in the Yale ID conference brochure at http://www.rivendellinstitute.org/business/conf.pdf. See Ray Bohlin's affiliation with Probe Ministries at http://www.probe.org/support/speakers.html. See Gordon Mills's ARN biosketch at http://www.arn.org/mills/gmhome.htm. All accessed on May 14, 2002. See list of *Pandas* reviewers in *Of Pandas and People*, 1993, iii.

32. Phillip Johnson, "Interview with Christian Book Distributors, Inc." Week of August 14, 2000. Accessed on April 20, 2002, at http://www.arn.org/docs/johnson/pj_cbinterview800.htm.

33. This announcement was on the MPC's "Events and Programs" website at Baylor; the site was removed after the dissolution of the center in October 2000. Accessed on March 19, 2000, at http://www.baylor.edu/~polanyi/events_progs.htm. This page and other Polanyi Center web pages are available from the Internet Archive at http://www.archive.org/web/web.php.

34. The John Templeton Foundation was another co-sponsor, and the MPC received assistance from Baylor University and Baylor's Institute for Faith and Learning. The co-sponsorship and assistance were acknowledged in an April 10, 2000, news article by DI. Accessed on April 20, 2002, at Lexis-Nexis.

35. Johnson may indeed have helped with the planning of this conference since his

previous involvement in such activity is documented. CRSC fellow Nancy Pearcey wrote in March 1997, after the NTSE conference, that Johnson "worked behind the scenes . . . to organize a series of conferences." Pearcey, "Opening the 'Big Tent' in Science: The New Design Movement," *World*, March 1, 1997. Accessed on March 25, 2001, at http://www.arn.org/docs/pearcey/np_bigtent30197.htm.

36. Nobelist Steven Weinberg recently issued a stinging, and brilliant, dismissal of the main arguments of ID. See Weinberg, "A Designer Universe," on the PhysLink online physics and astronomy education and reference list, April 27, 2002, edition. Accessed on April 25, 2002, at http://www.physlink.com/Education/essay_weinberg.cfm.

37. Nancy Pearcey, "We're Not in Kansas Anymore: Why Secular Scientists and Media Can't Admit that Darwinism Might Be Wrong," *Christianity Today*, May 22, 2000. Accessed on April 20, 2002, at http://www.christianitytoday.com/ct/2000/006/1.42.html. A different view of the conference's success at enhancing the credibility of ID comes from Glenn Morton, who attended and wrote a detailed account for the January–April 2000 *Reports of the National Center for Science Education*. While sympathetic to the establishment of the Michael Polanyi Center and complimentary of the general quality of the conference, Morton cites the poor presentations offered by the ID proponents, with the exception of Paul Nelson in Morton's opinion: "The conference also exposed some of the weaknesses in the ID movement to many of their supporters—something the ID advocates would not want. . . . It was starkly clear to most of the attendees that *the ID movement offered no research program, avoided making empirical predictions, and basically engaged in philosophizing about, rather than explaining, the nature of Nature*" (13) (emphasis added). See also Morton's report on the conference in the July/August 2000 ASA *Newsletter*. Accessed on April 20, 2002, at http://www.toolcity.net/~dfeucht/JULAUG00.htm.

38. Notes compiled on May 10, 2000, by David Applegate, director of the American Geological Institute's Government Affairs Program, who attended the briefing. Cited with permission. Phillip Johnson's public explanation of Dembski's absence from the briefing contradicts other accounts, including Dembski's own in an article by Lauren Kern based on an interview with Dembski: "'One of the things we were very clear about from the beginning,' says [Donald] Schmeltekopf [Baylor provost], 'was that the work of Dembski and Gordon did not have underneath it a political agenda of some kind; that is, to get into textbook wars and creationist politics and that kind of thing.' To that end, Baylor administrators pressured Dembski not to attend a May bipartisan congressional briefing by the Discovery Institute. . . . Dembski was surprised by Baylor's limitation of his 'academic freedom.' He had made no secret of his association with the Discovery Institute, which considers the 'wedge strategy' one of its primary projects. . . . Johnson considers Dembski to be a key wedge figure." Lauren Kern, "In God's Country," *Houston Press*, December 14, 2000. Accessed on April 27, 2002, at http://www.houstonpress.com/issues/2000-12-14/feature2.html.

39. Johnson, "Interview with Christian Book Distributors, Inc."

40. The DAIC conference site was accessed on April 20, 2002, at http://www.cuw.edu/Cranach/design_welcome.htm.

41. "Design and Its Critics," notice and call for papers on the Discovery Institute website. Accessed on February 10, 2000, at http://www.discovery.org/comingEvents/ConcordiaCall/index.html. "Design and Its Critics," notice and call for papers on Access Research Network. Accessed on April 6, 2000, at http://www.arn.org/events.htm.

42. See the Cranach Institute's mission statement and description at http://institutes.cuw.edu/cranach/index.htm and http://institutes.cuw.edu/cranach/about.htm, respectively. See both protest statements at the Cranach Institute website at http://institutes.cuw.edu/cranach/dembski_protest.htm. Accessed on April 18, 2003.

43. "Brief Statement of the Doctrinal Position of The Lutheran Church-Missouri Synod: Of Creation," [Adopted 1932]. Accessed on January 5, 2001, at http://www.lcms.org/belief/doct-03.html.

44. Dr. A. L. Barry, "Should Evolution Alternative Be Taught?/Yes: Make Room for Intelligent Design," Fort Wayne, IN, *Journal-Gazette*, September 3, 2000. See the official statement online at "Office of the President: Statements," The Lutheran Church-Missouri Synod, at http://www.lcms.org/president/statements/intelligentdesign.asp. Dr. Barry is now deceased. See http://www.lcms.org/news/barryd.html. Accessed on April 20, 2002. Dr. Barry's statements about intelligent design were posted on the websites of the Good Shepherd Lutheran Church and Preschool in Roanoak, VA, and Concord, a Lutheran confessional association in Texas, indicating that they were publicized by individual churches and other affiliated groups. Accessed on April 20, 2002, at http://www.luther95.com/GSLCP-RVA/webelieve_wa_CreationAndEvolution.html and http://www.concordtx.org/msnews/creat.htm, respectively.

45. Membership and financial figures were cited at http://www.cuis.edu/ftp/lcmsnews/999402-LCMS_STATISTICS_1999.-001009. Accessed on September 12, 2001. Although the LCMS membership has decreased since 1998, it still represents a major Christian denomination and controls considerable financial resources. Members contributed more than $1 billion to LCMS in 1999.

46. The only non-Wedge member on the schedule was Martin Poenie, a University of Texas professor who had participated at UT-Austin with Robert Koons in Veritas Forum activities. Accessed on December 9, 2002, at http://www.utexas.edu/staff/fscf/schedule.html. See a report on the Yale conference by the "Intelligent Design Undergraduate Research Council," a student group of CRSC followers. Accessed on December 15, 2000, at http://www.idurc.org/news.html. Audio files of conference presentations are available at the CRSC website. Accessed on March 25, 2001, at http://www.discovery.org/crsc/.

47. Phillip E. Johnson, "How to Sink a Battleship: A Call to Separate Materialist Philosophy from Empirical Science," Final Address at the Mere Creation Conference at Biola University, 1996. Accessed on February 6, 2000, at http://www.origins.org/mc/resources/ri9602/johnson.html.

48. The brochure is downloadable from the Rivendell Institute website at http://www.rivendellinstitute.org/business/conf.pdf. For information about the Rivendell Institute, see http://www.rivendellinstitute.org/business/aboutusa.htm. Accessed on April 18, 2003.

49. Alan Gishlick to Barbara Forrest, personal communication, December 18, 2000. Gishlick, who at the time of the conference was completing his doctoral degree in paleontology at Yale, attended both night sessions and daily concurrent sessions of the conference. Gishlick began a postdoctoral position at the National Center for Science Education in 2001.

50. Joseph D'Agostino, "Yale's Roots Show at Intelligent Design Conference," *Human Events Online*, November 17, 2000. Accessed on January 1, 2001, at http://www.humaneventsonline.com/articles/11-17-00/yale.html. See also D'Agostino's "Conservative Spotlight: Center for the Renewal of Science and Culture" in the same issue.

51. It is not unthinkable that even an elite university with strong sciences, such as Yale, offers recruiting opportunities for the Wedge. Three fellows at the Center for the Renewal of Science and Culture are Yale graduates: David K. DeWolf, J. Budziszewski, and Jonathan Wells. See their biographical sketches accessed on September 12, 2001, at http://www.crsc.org/fellows/.

52. "Coming Events," in *Newsletter of the American Scientific Affiliation and Canadian Scientific and Christian Affiliation* 44:5 (September/October 2002), 7. Accessed on September 25, 2002, at http://www.asa3.org/ASA/newsletter/sepoct02.pdf.

53. The conference schedule and attendees were accessed on October 12, 2002, at http://www.iscid.org/rapid/schedule.html and http://www.iscid.org/rapid/attendees.html. See also Dembski's keynote speech, "Becoming a Disciplined Science: Prospects, Pitfalls, and Reality Check for ID," accessed on April 18, 2003, at http://www.designinference.com/documents/2002.10.27.Disciplined_Science.htm.

54. Phillip Johnson, "Conferences at Calvin College, Kansas City, and Elsewhere,"

Weekly Wedge Update, June 4, 2001. Accessed on April 3, 2002, at http://www.arn.org/docs/pjweekly/pj_weekly_010604.htm.

55. This announcement was accessed on April 11, 2002, at http://www.discovery.org/.

56. Intelligent Design Network, "DDD III Program." Accessed on April 20, 2002, at http://www.intelligentdesignnetwork.org/DDDIIprogram.htm.

57. Audiotape of Roger DeHart's presentation, July 15, 2000.

58. The account of the song's performance—complete with lyrics—was part of Phillip Johnson's July 2, 2001, *Weekly Wedge Update* at http://www.arn.org/docs/pjweekly/pj_weekly_010702.htm. The estimated number of attendees was provided by Jack Krebs of Kansas Citizens for Science, who was the only non-ID presenter at the 2001 conference.

59. Pat McLeod, "ASA Lectures Big in Bozeman," ASA *Newsletter* (July/August 1999). Accessed on September 13, 2001, at http://www.toolcity.net/~dfeucht/JULAUG99.htm.

60. Paul Nelson, "Bill and Paul's Excellent Adventure in Boston," ASA *Newsletter* (September/October 1999). Accessed on September 13, 2001, at http://www.toolcity.net/~dfeucht/SEPOCT99.htm.

61. Melody Wang, "Guest Lecture on Existence of God Stimulates Debate," *UCLA Daily Bruin*, February 3, 2000. Accessed on July 6, 2000, at http://www.dailybruin.ucla.edu/db/issues/00/02.03/news.discussion.html.

62. Jay Richards by way of Billy Grassie, "Discovery Institute Presents Dr. William Dembski," Metanews, February 7, 2000. Accessed on September 13, 2001, at http://www.meta-list.org/archives/message_fs.asp?list=news&listtype=Magazine&action=sp_simple_archive_&page=7&ARCHIVEID=2084&searchstring=Richards%20Dembski%20Washington.

63. "Paul Nelson to Speak at University of Colorado," *ARN Announce* 7, September 17, 2000. Accessed on September 13, 2001.

64. "Dembski and Kauffman Square Off in New Mexico," *Weekly Wedge Update*, November 19, 2001. Both this column and the November 27 one were accessed on April 18, 2002, at http://www.arn.org/johnson/wedge.htm. The UNM lecture co-sponsors were listed in a press release at http://panda.unm.edu/CAS/Pressrel.htm. Accessed on April 20, 2002. For a different perspective on the Dembski-Kauffman "square-off," see Dave Thomas, "Dembski Talks on 'Darwin's Unpaid Debt'—But It's Dembski's Note That's in Arrears," adapted from *NMSR Reports* 7:12 (December 2001). Accessed on September 24, 2002, at http://www.nmsr.org/dembski.htm.

65. Creation Science Fellowship of New Mexico, *Newsletter* 13:2 (February 2002). Accessed on April 20, 2002, at http://www.swcp.com/creation/news/0202.htm.

66. "Tentative Schedule for 2001," Access Research Network. Accessed on September 26, 2000, at http://www.arn.org/docs/johnson/pj_sched.html.

67. Johnson's lecture "The Real Evolution/Creation Debate" was sponsored by the UNM Honors Program, the Department of Psychology, and the School of Law. UNM Public Affairs News Release, February 19, 2001. Accessed on September 13, 2001, at http://www.unm.edu/news/Releases/Feb19video.htm. See "UNM Splits Lecture Halls over How Life Came to Be," *Albuquerque Tribune*, February 20, 2001. Accessed on September 13, 2001, at http://www.abqtrib.com/archives/news01/022001_evo.shtml. Johnson's stroke and discontinuance of his column were announced, respectively, in his September 25, 2001, and April 9, 2002, *Weekly Wedge Update*. Accessed on April 20, 2002, at http://www.arn.org/johnson/wedge.htm. Johnson now writes the bimonthly "Leading Edge" column in *Touchstone*. See "Scouts and God" (March 2003) at http://www.touchstonemag.com/docs/issues/16.2docs/16-2pg16.html.

68. John Mark Reynolds, "Que Res Vitas? Phil Johnson Takes His Case to the East," *Origins and Design* 16:1. Accessed on April 20, 2002, at http://www.arn.org/docs/orpages/or161/pjlect.htm.

69. *Indiana Law Annotated*, September 18, 1995. Accessed on April 20, 2002, at http://www.law.indiana.edu/pubs/ila/0904.html.

70. "1998 Annual Christian Legal Society Conference," October 22–25, Lisle, Illinois. Accessed on June 2, 2000, at http://www.lifetimecassettes.com/assoc/order-forms/cls98.html.

71. "The Promise of Better Science and a Better Culture," Discovery Institute *Journal* (1999). Accessed on December 21, 2000, at http://www.discovery.org/w3/discovery.org/journal/1999/crsc.html. The Federalist Society's Religious Liberties Practice Group *Newsletter* published Meyer's testimony (winter 1998). Accessed on April 20, 2002, at http://www.fed-soc.org/Publications/practicegroupnewsletters/religious%20liberties/testimony-relv2i3.htm.

72. Opening statement, testimony of Stephen C. Meyer, U.S. Commission on Civil Rights Hearing, August 21, 1998, Seattle, Washington. Accessed on December 11, 2000, at http://www.arn.org/docs/meyer/sm_uscom.htm. (The entire transcript is available here. The quote's obvious punctuation errors are in the transcript.)

73. "Meyer Testifies Before US Commission on Civil Rights," ASA *Newsletter* (January/February 1999). Accessed on May 14, 2002, at http://www.toolcity.net/~dfeucht/JANFEB99.htm.

74. For background on the Freiler case, see Barbara Forrest, "At the Front in Tangipahoa Parish," *Reports of the National Center for Science Education* 17:5 (September/October 1997), 5–6, and Barbara Forrest and Molleen Matsumura, "Supreme Court Rejects Evolution Disclaimer," *Reports of the National Center for Science Education* 20:1–2 (January/April 2000), 4–5. See the Louisiana ACLU's press release at http://www.laaclu.org/News/1999/Tangipahoa%20August%2016%20Update.htm. See Molleen Matsumura, "Eight Significant Court Decisions" at http://www.ncseweb.org/article.asp. For the Fifth Circuit court rulings, see http://laws.lp.findlaw.com/getcase/5th/case/9730879cv0v2.html. All accessed on April 18, 2003.

75. Phillip E. Johnson, "What (If Anything) Hath God Wrought? Academic Freedom and the Religious Professor," *Academe* (September/October 1995). Accessed on January 21, 2001, at http://www.arn.org/docs/johnson/aaup.htm.

76. See Rep. Mark Souder, "Intelligent Design Is Not a Science," *Congressional Record* (H4480, June 14, 2000). The text of Souder's remarks is available at Access Research Network at http://www.arn.org/docs/idushouse_700.htm. Accessed September 14, 2001. The last name of the second contributor to the letter is left out of Congressman Souder's published version. In response to a request for the complete name, Souder's office staff supplied the name "Robert George" as the "Robert * * *" in Souder's acknowledgment. Phone call to Barbara Forrest from Congressman Mark Souder's staff, August 4, 2000. The only "Robert George" in the Princeton University online directory is Robert P. George, McCormick Professor of Jurisprudence at Princeton University, who served from 1993 to 1998 as a Bush appointee to the U.S. Commission on Civil Rights. Prof. George was one of the Civil Rights Commissioners at the August 1998 hearing in Seattle at which Stephen Meyer testified. His service as a civil rights commissioner is documented in George's "Statement on Free Exercise of Religion in Public Schools," *Religious Liberties News* (spring 1999), the newsletter of the Federalist Society's Religious Liberties Practice Group. Accessed on April 20, 2002, at http://www.fed-soc.org/Publications/practicegroupnewsletters/religious%20liberties/statement-religiousv3i1.htm. Prof. George was appointed in January 2002 to President George W. Bush's Bioethics Council. "President Names Members of Bioethics Council," Statement by the Press Secretary, January 16, 2002. Accessed on May 14, 2002, at http://www.whitehouse.gov/news/releases/2002/01/20020116-9.html.

77. See the description of the book at Access Research Network's purchasing page at http://www.arn.org/arnproducts/books/b031.htm. A long article of the same title is available on William Dembski's Baylor website at http://www3.baylor.edu/~William_Dembski/docs_resources/guidebook.htm. Bradford is quoted in the March/April 2000

ASA *Newsletter* at http://users.stargate.net/~dfeucht/MARAPR00.htm. Accessed on April 18, 2003.

78. David K. DeWolf, Stephen C. Meyer, and Mark Edward DeForrest, "Teaching the Origins Controversy: Science, Or Religion, Or Speech?" *Utah Law Review* 39 (2000), 92–93. Available at http://www.arn.org/dewolf/ddhome.htm. Accessed on April 20, 2002. Wedge member Francis J. Beckwith recently added both a law review article and a book to the Wedge's legal arsenal. See "A Liberty Not Fully Evolved? The Case of Rodney LeVake and the Right of Public School Teachers to Criticize Darwinism," *San Diego Law Review* 39 (2002). See also his *Law, Darwinism and Public Education: The Establishment Clause and the Challenge of Intelligent Design* (New York: Rowman and Littlefield, 2003).

79. David K. DeWolf, "Discussing the Law that Supports Scientifically Accurate Teaching," *Banned in Burlington: The Uncensored Facts*, meeting held at Burlington-Edison (WA) High School, May 16, 2001.

80. Accessed at http://www.discovery.org/crsc/scied/present/topics/political.htm and http://www.discovery.org/crsc/scied/what/topics/missiontopics/legal.htm, respectively. This site is now "currently under development."

81. This PowerPoint slide show was available for downloading at http://www.discovery.org/crsc/scied/downloads/download.htm. Accessed on December 13, 2000.

82. William A. Dembski, "Science and Design," *First Things* (October 1998). Accessed on January 27, 2001, at http://www.firstthings.com/ftissues/ft9810/dembski.html.

83. See Richard Weikart, "The Roots of Hitler's Evil," *Books & Culture* (March/April 2001). Accessed on April 20, 2002, at http://www.christianitytoday.com/bc/2001/002/6.18.html. See other book reviews by Weikart in *Origins & Design*. Accessed on March 17, 2001, at http://www.arn.org/docs/odesign/od181/weikart181.htm. A March 17, 2001, search of *Historical Abstracts* yielded five titles by Weikart, two concerning Darwinism in Germany: (1) "The Origins of Social Darwinism in Germany, 1859–1895," *Journal of the History of Ideas* 54 (1993), 469–488; and (2) "Socialist Darwinism: Evolution in German Thought from Marx to Bernstein," Ph.D. dissertation, University of Iowa, 1994. *Dissertation Abstracts International* 55:8 (1995), 2534-A.

84. Schloss's funding was announced in "CRSC Innovates in Media and Academia" in the spring 1998 Discovery Institute *Journal* at http://www.discovery.org/w3/discovery.org/journal/. See Schloss's biographical sketch at http://www.crsc.org/fellows/JeffreySchloss/index.html. Both accessed on September 25, 2002.

85. Schloss's listing in the "By Chance or Design?" conference was accessed at http://www.atlantaapologist.org/conference.html on April 20, 2002.

86. Schloss is listed in the Yale conference brochure, "Science and Evidence for Design in the Universe."

87. See Dembski's comment in "Intelligent Design Coming Clean," a November 17, 2000, Metaviews posting, accessed on September 14, 2001, at http://www.arn.org/docs/dembski/wd_idcomingclean.htm.

88. Beth McMurtrie, "Baylor Faculty Objects to New Center on Religion and Science," *Chronicle of Higher Education*, May 5, 2000. Accessed on April 20, 2002, at http://chronicle.com/weekly/v46/i35/35a01901.htm.

89. Michael Polanyi (1891–1976), born in Budapest, was a distinguished physical chemist who worked and achieved fame in Germany, but left in 1933 to settle at Manchester University in England. His interests shifted steadily toward social and political science, so that he was for the latter part of his life a professor of social studies. He is an important scientific figure, respected among some libertarian-minded conservatives because of his overriding interest in the concepts and the realities of political and intellectual freedom.

90. See Larry Brumley, "Dembski Relieved of Duties as Polanyi Center Director," Baylor University press release, October 19, 2000, at http://pr.baylor.edu/story.php?id=002390. Relevant posts by Dembski and the Baylor administration, as well as articles in the Baylor *Lariat*, are archived by Wesley Elsberry at http://www.antievolution.org/

people/dembski_wa/mpc.html. For a report on the controversy by a Baylor faculty member, see Richard Duhrkopf, "Baylor University Faculty Object to Polanyi Center," *Reports of the National Center for Science Education* (January–April 2000). See also Eugenie Scott, "Baylor's Polanyi Center in Turmoil," at http://www.ncseweb.org/resources/rncse_content/vol20/3979_baylor39s_polanyi_center_in_12_30_1899.asp. A statement by Robert Baird, former Baylor Faculty Senate president, is at http://www3.baylor.edu/Fac_Senate/5_00_newsletter.htm. Lewis Barker, formerly on the psychology faculty at Baylor (now at Auburn University), posted a December 10, 2000, statement on Dembski's demotion at http://www.egroups.com/message/evolutionary-psychology/9011. All accessed on April 18, 2003. A great deal of coverage of the controversy surrounding the MPC can be found on the web with the aid of a search engine.

91. See William Dembski, "Intelligent Design Coming Clean," Metaviews, November 17, 2000, at http://www.arn.org/docs/dembski/wd_idcomingclean.htm. Accessed on September 14, 2001. The committee's report was much more circumspect. The "External Review Committee Report" actually says, "Within the broad range of issues that bear on the relationship between the sciences and religion, those raised by recent work on the criteria appropriate to claims of intelligent design could well find a place." The committee also stated that "research on the logical structure of mathematical arguments for intelligent design . . . have a legitimate claim to a place in current discussions of the relations of religion and the sciences." This does not "validate intelligent design as a legitimate form of academic inquiry." See "The External Review Committee Report, Baylor University," October 16, 2000. Accessed on September 24, 2002, at http://pr.baylor.edu/pdf/001017polanyi.pdf. Former MPC Associate Director Bruce Gordon, who became "Interim Director of the Baylor Science and Religion Project," wrote a column for the Templeton Foundation's *Research News and Opportunities in Science and Technology* after Dembski's demotion. Ironically, though he had also been a CRSC fellow, Gordon stated that design theory "has been hijacked as part of a larger cultural and political movement" and had "no business making an appearance [in public schools] without broad recognition from the scientific community." He added that ID should be viewed as "at best a supplementary consideration . . . alongside . . . neo-Darwinian biology and self-organizational complexity theory." This apparent recantation can be plausibly interpreted as damage control. See Bruce Gordon, "Intelligent Design Movement Struggles with Identity Crisis," *Research News and Opportunities in Science and Theology* (January 2001), 9. Accessed on September 24, 2002, at http://www.asa3.org/ASA/topics/Apologetics/ResearchNews1-01Gordon.html.

92. These aims were stated on the following Polanyi Center web pages, respectively: "Purpose of Center," accessed on August 27, 2000, at http://www.baylor.edu/~polanyi/purpose.htm; "Events and Programs," accessed on March 19, 2000, at http://www.baylor.edu/~polanyi/events_progs.htm; and "Home Page," accessed on April 17, 2000, at http://www.baylor.edu/~polanyi/. These pages are available at the Internet Archive at http://www.archive.org/web/web.php.

93. If ID were ever to become the "dominant perspective in science," as the Wedge Document projects, it could only be by virtue of its fruitfulness as a "specific research tool," hence the MPC's inclusion of this goal.

94. Accessed on April 17, 2000, at http://www.baylor/edu/~polanyi.

95. See Blair Martin, "Polanyi Center's Future Is Unclear," *Baylor Lariat*, October 24, 2000: "[Dr. Michael] Beaty [director of Baylor's Institute for Faith and Learning] said Dembski will now serve as associate research professor in conceptual foundations of science within the university's Institute of Faith and Learning, where he will devote himself to the research of intelligent design and can serve the remainder of his five-year contract." Accessed on April 18, 2003, at http://www3.baylor.edu/~Lariat/Archives/2000/20001024/art-front02.html.

96. "Purpose of Center," Michael Polanyi Center. Accessed on January 19, 2000, at http://www.baylor.edu/~polanyi/purpose.htm.

97. A picture and discussion of the "suite" were posted on January 2, 2002, on the ARN discussion list at http://www.arn.org/ubb/Forum1/HTML/001378-2.html.

98. ISCID's launch was announced at ARN on December 5, 2001. Accessed on April 21, 2002, at http://www.arn.org/docs2/news/launchofiscid120501.htm. A January 8, 2002, search of its databases by the New Jersey State Treasury Department, Division of Revenue, Business Services Bureau, yielded no registration data for ISCID. The database includes nonprofits and all other organizations registered to do business in New Jersey. Phone interview with Barbara Forrest, January 8, 2002. For ISCID's financial information, see www.guidestar.org. ISCID's announcement of its non-profit status and suggested donation amounts were accessed on March 31, 2003, at http://www.iscid.org/about.php and http://www.iscid.org/donations.php.

99. ISCID home page. All information about ISCID is available at http://www.iscid.org/. Accessed on April 20, 2002.

100. William A. Dembski, "Topic: PIPR: Principle of Inflated Probabilistic Resources," posted on February 25, 2002. Accessed on February 25, 2002, at http://www.iscid.org/ubb/ultimatebb.php?ubb=get_topic&f=6&t=000002. A monthly prize of $100 is awarded to the person who "started the most intensely discussed thread" on the Brainstorms list.

101. Dembski also invited Richard Dawkins to become a fellow. Dawkins declined: "You have invited me to contribute to your journal, and to become a fellow of your society. I decline both invitations, having no desire to be used in your publicity campaign (the Wedge, isn't that what it's called? . . .) I am now sixty years old and increasingly conscious that time is finite. The thing I really resent about you and Johnson and Behe is that you are a gratuitous waste of precious time." Reprinted in *The Link*, Newsletter of Kansas Citizens for Science 1:3 (December 2001), 2. Accessed on April 21, at http://www.kansasscience.com/uncommon/LINK/LinkV1N3print.pdf.

102. The summer 2003 conference announcement was accessed on April 21, 2002, at http://www.iscid.org/conferences.php. The announcement of the essay awards was accessed on June 24, 2002, at http://www.iscid.org/essaycontests.php. Information on ISCID's online conferences and workshops was accessed on March 31, 2003, at http://www.iscid.org/conferences.php and http://www.iscid.org/workshops.php.

103. Macosko's and Sparacio's activities have already been documented. John Bracht's membership in the IDEA Club was listed at http://www-acs.ucsd.edu/~idea/infotheory.htm. James Barham was listed on the 2001 Calvin College Fieldstead seminar program at http://www.calvin.edu/fss/dembschd.htm. The source of the Nelson/Wells paper was at http://www.iscid.org/ubbcgi/ultimatebb.cgi?ubb=get_topic&f=10&t=000009&p=. All accessed on April 21, 2002. The list of editorial board members was accessed on April 1, 2003, at http://www.iscid.org/pcid.php.

8. Wedging into Power Politics

1. Jay Grelen, "Witnesses for the Prosecution," *World*, November 30, 1996. Accessed on March 1, 2000, at http://www.worldmag.com/world/issue/11-30-96/national_2.asp. Johnson repeated this in 1999 in *Insight on the News*: "I get asked all the time, 'How can you do this when you're not a scientist?' The answer is that it is not mainly about science. It is about a certain way of thinking." Accessed on July 9, 2000, at http://www.arn.org/docs/johnson/insightprofile1099.htm.

2. Niles Eldredge notes the correlation between creationism and conservative politics: "In the United States especially, creationism is associated not only most closely with aspects of Christian Fundamentalism, but with conservative (mostly . . . conservative Republican) politics. . . . [P]olitics is the very essence of this conflict. It is the belief that evolution is inherently evil—a belief that stems from religious interpretation, *and therefore poses a threat to the hearts and minds of the populace*, that . . . motivates the vast majority of

the creationists." Eldredge, *The Triumph of Evolution and the Failure of Creationism* (New York: W. H. Freeman, 2000), 11. For a detailed account of the neo-conservative tie to anti-Darwinism, see Ronald Bailey, "Origin of the Specious," *Reason* (July 1997). Accessed on November 3, 2001, at http://www.reason.com/9707/fe.bailey.html.

3. Four of those years have gone by with no sign of the desired recognition from the NSF, or from any other genuine, peer-reviewed, scientific granting agency. The most plausible explanation is that, consistent with the ID record of zero peer-reviewed articles, no grant applications for funding original empirical research have even been submitted.

4. Dembski, introduction to *Mere Creation*, 29. There may be obstacles to the Wedge's getting NSF funding, but Dembski clearly does not see the lack of supporting scientific data as one of them: "In an ideal setting . . . [d]esign would get funding from the NSF to see if there were anything to it. But, instead, the culture war aspect keeps playing in, and it is tough." William Dembski, "Questioning Darwin: William Dembski Discusses Intelligent Design," interview by Donald Yerxa (July 2001), in Templeton Foundation, *Research, News and Opportunities in Science and Theology* 2:3 (November 2001), 13.

5. Bruce Chapman, "From the President, Bruce Chapman: Ideas Whose Time Is Coming," Discovery Institute *Journal* (August 1996). Accessed on July 24, 2000, at http://www.discovery.org/w3/discovery.org/journal/president.html.

6. Phillip Johnson, "The Wedge: Breaking the Modernist Monopoly on Science," *Touchstone* (July/August 1999). Accessed on March 10, 2000, at http://www.touchstonemag.com/docs/issues/12.4docs/12-4pg18.html.

7. See William A. Dembski, "What Every Theologian Should Know About Creation, Evolution and Design," *Princeton Theological Review* 2:3 (1995), 15–21, at http://www.origins.org/offices/dembski/docs/bd-theologn.html. Phillip Johnson said much the same thing in 1993: "[T]he public school system isn't really my venue, it isn't where I want [the issue] argued." Phillip E. Johnson, "NCSE Special: Interview with Phillip E. Johnson," interview by Yves Barbero. Accessed on October 19, 2002, at http://rnaworld.bio.ukans.edu/id-intro/cast/johnson/johnson.htm#Case.

8. The following account was provided by the Burlington-Edison Committee for Science Education, which negotiated with school officials to stop DeHart's use of creationist materials in the public high school (BECSE to Barbara Forrest, personal communication, November 11, 2000). See BECSE's website at http://www.scienceormyth.org for detailed information on the intelligent design episode in Burlington, WA.

9. John Gibeaut, "Evolution of a Controversy," American Bar Association *Journal* (November 1999). Accessed on May 14, 2002, at http://www.abanet.org/journal/nov99/11fevolv.html.

10. See "Inherit the Wind" exam given by Roger DeHart, posted by the Burlington-Edison Committee for Science Education at http://www.scienceormyth.org/test.html. Accessed on May 13, 2002. BECSE also has posted a complete timeline of the DeHart episode at http://www.scienceormyth.org/history.html.

11. The Rutherford Institute was involved in a creationism case prior to assisting DeHart: it provided legal assistance to a Louisiana school board that was sued for adopting an evolution disclaimer. See Barbara Forrest, "At the Front in Tangipahoa Parish," at http://home.austarnet.com.au/stear/barbara_tangipahoa_parish.htm. The Rutherford Institute was founded by John Whitehead, a protegé of R. J. Rushdoony, director of a Christian Reconstructionist organization, the Chalcedon Foundation. For information on the nature of the Rutherford Institute, see Chip Berlet, "Clinton, Conspiracism, and the Continuing Culture War" at http://www.publiceye.org/conspire/clinton/Clinton2_TOC.htm#TopOfPage. See also the profile of the RI by the Institute for First Amendment Studies at the Internet Archive at http://web.archive.org/web/19980204165616/http://ifas.org/fw/9406/rutherford.html. All accessed on April 18, 2003.

12. Gibeaut, "Evolution of a Controversy."

13. "Teacher Gains OK to Use Creationist Text," *Skagit Valley Herald*, June 12, 1999.

Accessed on July 15, 2003, at http://www.skagitvalleyherald.com/articles/1999/06/12/news14820.txt.

14. Beth VanderVeen to Roger DeHart, July 1999. Accessed on May 14, 2002, at http://www.scienceormyth.org/july-1999.html.

15. Burlington-Edison Committee for Science Education, "A Documentation History of Events." Accessed on May 14, 2002, at http://www.scienceormyth.org/history.html.

16. Clifford D. Foster, Jr., [B-E school district attorney], to Richard Berley [ACLU attorney], January 21, 2000. Accessed on May 14, 2002, at http://www.scienceormyth.org/jan-2000.html.

17. "Intelligent Design Backer to Sign Book," *Skagit Valley Herald*, February 8, 2000. Accessed on July 22, 2003, at http://www.skagitvalleyherald.com/articles/2000/02/08/news10766.txt. Dembski participated in this event while employed by Baylor University. Apparently not understanding the political nature of the Wedge and the Michael Polanyi Center's role as an integral part of it when the MPC was established, Baylor Provost Donald Schmeltekopf and the rest of the administration were under the mistaken impression that Dembski was not engaging in political activity such as this book signing for DeHart: " 'One of the things we were very clear about from the beginning,' says Schmeltekopf, 'was that the work of Dembski and Gordon did not have underneath it a political agenda of some kind; that is, to get into textbook wars and creationist politics.' " Lauren Kern, "In God's Country," *Houston Press*, December 14, 2000. Accessed on April 22, 2002, at http://www.houstonpress.com/issues/2000-12-14/feature2.html/1/index.html.

18. DeHart was on the program schedule at http://www.IntelligentDesignnetwork.org/program.htm. Accessed on November 16, 2000. An audiotape of DeHart's presentation, "A Curriculum Module for Teaching the Cambrian Explosion," is now available for sale at http://www.audiomission.com/cgi-bin/CF/viewbuy.cfm?cat=Speaker&cat2=DeHart.

19. Theresa Goffredo, "School Officials Throw Extra Science Materials Out of Class," *Skagit Valley Herald*, May 28, 2000. Accessed on May 29, 2000, at http://www.skagitvalleyherald.com/daily/00/may/28/a1dehart.html.The Wells articles DeHart was denied permission to use were (1) "Haeckel's Embryos and Evolution: Setting the Record Straight," *American Biology Teacher* (May 1999), 345–349, and (2) "Second Thoughts About Peppered Moths," *Scientist*, May 24, 1999, 13. The latter is available at http://www.the-scientist.com/yr1999/may/opin_990524.html. DeHart was also denied permission to use Larry Witham's "Darwin's Icons Disputed: Biologists Discount Moth Study," *Washington Times*, National Weekly Edition, January 25–31, 1999. (Burlington-Edison Committee for Science Education to Barbara Forrest, personal communication, November 25, 2000.)

20. Audiotape of Jonathan Wells at lecture and booksigning for *Icons of Evolution*, Seattle, WA, October 26, 2000.

21. "Banned in Burlington: The Uncensored Facts," Skagit Parents for Truth in Science Education, May 16, 2001. DeWolf's remarks are recorded on the previously cited audiotape.

22. Marina Parr, "Burlington Reassigns Biology Teacher," *Skagit Valley Herald*, July 22, 2001. Accessed on July 25, 2003, at http://www.skagitvalleyherald.com/articles/2001/07/22/news24356.txt.

23. Theresa Goffredo, "Marysville on Edge over New Biology Teacher," *Heraldnet*, August 10, 2001. Accessed on November 4, 2001, at http://www.heraldnet.com/Stories/01/8/10/14189379.cfm.

24. Both letters are available in the *Skagit Valley Herald Archives* at http://www.skagitvalleyherald.com.

25. Burlington-Edison Committee for Science Education to Barbara Forrest, personal communication, November 9 and December 4, 2001, and May 14, 2002. DeHart was listed among Marysville-Pilchuck High School's "Certified Staff" at http://www.msvl.wednet.edu/secondary/MPHS/teachers/al.html. Accessed on December 8, 2001.

26. Brian Kelly, "Controversial Teacher Resigns," *Everett Herald*, May 8, 2002. Accessed on May 14, 2002, at http://www.heraldnet.com/Stories/02/5/8/15465963.cfm.

27. Jade Nirvana Ingmire, "Origins of Life Film to Premiere at SPU," *Falcon*, May 8, 2002. Accessed on July 20, 2003, at http://www.thefalcononline.com/story/2186.

28. Diane Carroll, "Evolution Critic Cheers Board," *Kansas City Star*, August 27, 1999. Accessed in May 2000 at http://www.kcstar.com/item/pages/home.pat,local/3773cd76.830.

29. For information on the Kansas Board of Education's 1999 decision to eliminate most references to evolution from state science standards and to eliminate evolution from the state assessment tests, see "Latest Evolution Battlefield," ABC News, at http://www.abcnews.go.com/sections/science/DailyNews/evolutiondebate990812.html. Accessed on December 11, 2001. For analysis of the Kansas science standards in various stages of revision by creationists and after the February 2001 reinstatement of evolution and Big Bang theory, see "The Kansas Science Standards" at http://www.sunflower.com/~jkrebs/. Accessed on December 8, 2001.

30. *Kansas City Star*, August 27, 1999.

31. Kansas Citizens for Science to Barbara Forrest, personal communication, December 15, 2000. In response to the Kansas BOE's decision, "Kansas Citizens for Science" was formed (http://www.kcfs.org). They have excellent fliers on the Wedge strategy. See also "The Evolution Home Pages" at the University of Kansas at http://rnaworld.bio.ukans.edu/Evolve/Evolve.html. Accessed December 16, 2001.

32. Kate Beem, "Holloway Outpacing Gamble in Race for Campaign Money," *Kansas City Star*, July 25, 2000.

33. For news stories about these elections and other related articles, see http://www.cnn.com/2000/US/08/02/kansas.evolution.01/#1, http://www.cjonline.com/stories/110900/kan_boerace09.shtml, and http://www.cjonline.com/stories/021501/kan_evolution.shtml. Accessed on February 1, 2002.

34. Edward Larson and Larry Witham, "Inherit an Ill Wind," *Nation*, October 4, 1999. Accessed on April 22, 2002, at http://past.thenation.com/issue/991004/1004larson.shtml.

35. Kansas Citizens for Science to Barbara Forrest, personal communication, December 18, 2000. Tom Willis's announcement of the IDnet July 2000 symposium was posted at http://www.csama.org/CSA-GOTO.HTM. Accessed December 19, 2000.

36. See http://www.intelligentdesignnetwork.org. Accessed on November 20, 2000. For a list of IDnet's officers and directors, see http://www.intelligentdesignnetwork.org/people.htm. Another Kansas group, Parents for Objective Science and History (POSH), was formed in January 1999. Originally a young-earth group, it is now an IDnet satellite group. (Kansas Citizens for Science to Barbara Forrest, personal communication, December 15, 2000.)

37. Intelligent Design Network, "IDnet Urges Kansas School Boards to Reject National Science Standards Proposed by Kansas Citizens for Science." Press release, June 8, 2000. Accessed on March 17, 2001, at http://www.intelligentdesignnetwork.org/Press%20Rel%206-8-00.htm.

38. IDnet's letter to the school districts, news release, and "Memorandum and Opinion" are on its website at http://www.intelligentdesignnetwork.org/June%208%20letter%20to%20Boards.htm, http://www.intelligentdesignnetwork.org/Press%20Rel%206-8-00.htm, and http://www.intelligentdesignnetwork.org/legalopinion.htm, respectively. Accessed on December 11, 2001.

39. Floyd Lee, "Evolution Criticized as Lacking Evidence," *Topeka Capital-Journal*, July 16, 2000. Accessed on January 27, 2002, at http://cjonline.com/stories/071700/kan_evolution17.shtml.

40. The Events page is at http://www.intelligentdesignnetwork.org/EVENTS.htm. Accessed on November 20, 2000. Wells's October 2000 lecture was announced here.

41. Jack Krebs, "ID Proponents 'Using' Pratt's Board of Ed," *Pratt Tribune*, November

29, 2000. Accessed on January 27, 2002, at http://www.pratttribune.com/archives/index.inn?loc=detail&doc=/2000/November/29-641-news5.txt. See Calvert's responses to Krebs' letter at http://www.pratttribune.com/archives/index.inn?loc=detail&doc=/2000/December/06-659-news6.txt. Accessed on January 27, 2002. See Krebs's website at http://www.sunflower.com/~jkrebs/.

42. Kansas Citizens for Science to Barbara Forrest, personal communication, December 18, 2000.

43. Elizabeth Pennisi, "Haeckel's Embryos: Fraud Rediscovered," *Science* 277:1435 (September 5, 1997).

44. The standards can be found in Mike Marzolf, "Pratt Board Changes Biology Standards," *Pratt Tribune*, November 28, 2000. Accessed on December 17, 2000, at http://www.pratttribune.com/archives/index.inn?loc=detail&doc=/2000/November/28-589-news1.txt.

45. See Wells's letter at http://www.pratttribune.com/archives/index.inn?loc=detail&doc=/2000/December/06-660-news7.txt. Accessed on January 27, 2002.

46. A fairly complete newspaper account of the Pratt episode can be found at the *Pratt Tribune* archives at http://www.pratttribune.com/.

47. Kenneth Kennedy to Barbara Forrest, personal communication, December 13, 2001.

48. Scott F. Gilbert wrote *Developmental Biology*, 6th ed. (Sinauer Associates, 2000). See the online text at http://www.devbio.com. Kenneth Miller and Joseph Levine have written three high school biology texts for Prentice Hall. See "Textbooks by Ken Miller and Joe Levine" at http://www.millerandlevine.com/km/textbooks/index.html. See also the Access Excellence website at http://www.accessexcellence.org. All accessed on October 19, 2002.

49. "Science–USD #382, Biology: Course Level Outcomes," USD 382 Board of Education, Pratt, Kansas, February 2002. See "USD Revises Curriculum," *Pratt Tribune*, February 26, 2002. Accessed on April 22, 2002, at http://www.pratttribune.com. Some cause for concern remains in the standards. An inspection of referenced URLs revealed two creationist sites. According to Christy Swafford, a Pratt High School science teacher who participated in the 2002 revision, the URLs were recommended by board member Willa Beth Mills and Julie Bohn, a former board member. However, the sites may easily have been mistaken as legitimate by the writing committee, which agreed to include them. One site had posted Jerry Coyne's review of Michael Majerus's *Melanism: Evolution in Action*, but has since replaced it with creationist commentary on Bernard Kettlewell's work. The other URL was to an Access Research Network article first published in *American Biology Teacher* (1993) by creationists Gordon Mills and Walter Bradley. The *ABT* source and unfamiliarity with creationist code talk may have obscured the article's true significance. Various editions of Pratt's Glencoe text have also been poorly rated by the Textbook League at http://www.textbookleague.org/ttlindex.htm, but it is a widely used textbook.

50. Fundraising letter, Free Academic Inquiry and Research Committee, April 9, 2001. FAIR's FEC fundraising reports and other information are online at http://herndon1.sdrdc.com/cgi-bin/fecimg/?C00355701 and http://herndon1.sdrdc.com/cgi-bin/com_detail/C00355701/. Accessed on November 11, 2001.

51. Michael Foust, "Phillip Johnson: Evolution Battles at Baylor, Kan. Could Have Been Won," Southern Baptist Theological Seminary *News and Events* (summer 2001). Accessed on October 19, 2002, at http://www.sbts.edu/news/archives/summer2001/NR12.php. For November 2002 Kansas election results, see Diane Carroll, "Moderates Lose Seats on Kansas Board of Education," *Kansas City Star*, November 6, 2002. Accessed on April 1, 2003, at http://www.kansascity.com/mld/kansascity/news/politics/4453402.htm.

52. See SEAO's web-site at http://www.sciohio.org/start.htm. AFA's web pages are available at the Internet Archive at http://web.archive.org/web/20010928151258/www.afaohio.org. Both accessed on April 18, 2003. SEAO's is not the first effort to intro-

duce ID into Ohio science standards. See the Ohio Academy of Science's March 2001 resolution "Commending the State Board of Education for Supporting the Teaching [of] *Cosmic and Geological Evolution* and Opposing the Forced Teaching of 'Intelligent Design', a Creationist Belief, in Public School Science Education." Accessed on January 15, 2002, at http://www.ohiosci.org/THANKYOURESOLUTION.htm.

53. Phillip Johnson, "Curriculum Controversy in Ohio," *Weekly Wedge Update*, January 17, 2002. Accessed on January 17, 2002, at http://www.arn.org/docs/pjweekly/pj_weekly_020117.htm.

54. "State Board Members on Teaching 'Intelligent Design,'" *Cleveland Plain Dealer*, February 24, 2002. Accessed on April 12, 2002, at http://www.cleveland.com/education/index.ssf?/focuson/more/design.html.

55. "Standards Committee Meeting Minutes," Ohio State Board of Education, January 13, 2002.

56. Francis X. Clines, "In Ohio School Hearing, a New Theory Will Seek a Place Alongside Evolution," *New York Times*, February 11, 2002. Accessed on February 11, 2002, at http://www.nytimes.com/2002/02/11/education/11CREA.html?ex=1014431350&ei=1&en=c1e2fb23fc4a886e.

57. See the Ohio Citizens for Science website at http://ecology.cwru.edu/ohioscience/. Accessed on January 31, 2002.

58. The writing team's first draft was entitled "Ohio's Academic Content Standards for Science: Fall 2001 Review Draft" and was made available for public comment by the Ohio Board of Education. Accessed on April 12, 2002, at http://webapp1.ode.state.oh.us/science_comment/word/Indicator_chart_11-23.doc. SEAO's December 2001 "Suggested Modifications to Draft Indicators" were simultaneously posted on its website. Accessed on January 4, 2002, at http://www.sciohio.org/seaoindi.htm. See also the National Center for Science Education's "An Analysis of Proposed Changes to Ohio Science Standards," at http://www.ncseweb.org/resources/news/2002/OH/884_ncse_analysis_of_ohio_standard_1_2_2002.asp.

59. Phillip Johnson, "More from Ohio," *Weekly Wedge Update*, January 31, 2002. Accessed on February 1, 2002, at http://www.arn.org/docs/pjweekly/pj_weekly_020131.htm.

60. "SEAO Speakers Bureau." Accessed on January 4, 2002, at http://www.sciohio.org/seaospk.htm. Sjogren was assisted by Wedge supporter Robert DiSilvestro of Ohio State University, who headed a list of "Fifty-two Ohio Scientists" who oppose "censorship of scientific views that may challenge current theories of origins." This petition is in a DI March 20, 2002, press release at http://www.discovery.org/news/52ScientistsCallForAcadem.html. Accessed on April 22, 2002. For background on SEAO's formation, see Kate Beem, "Ohio Emerges as Next Battleground for Schools' Evolution Debate," *Kansas City Star*, April 2, 2002. Accessed on April 12, 2002, at http://www.kansascity.com/mld/kansascitystar/2979815.htm.

61. Biosketches of Cochran and Owens-Fink were posted by the Ohio Board of Education at http://www.ode.state.oh.us/board/members/. Accessed on January 31, 2002.

62. David Lore, "Ohio Closer to Standards for Teaching of Evolution," *Columbus Dispatch*, January 31, 2002.

63. "Mission Statement," Ohio Eagle Forum. Accessed on April 12, 2002, at http://www.ohioeagleforum.org/About%20main.htm.

64. Intelligent Design Network, "DDD III Program." Accessed on April 12, 2002, at http://www.intelligentdesignnetwork.org/DDDIIprogram.htm.

65. John Mangels and Scott Stephens, "Evolution Targeted in Curriculum Study," *Cleveland Plain Dealer*, January 15, 2002. Accessed on April 12, 2002, at http://www.cleveland.com/debate/.

66. "Remarks of John H. Calvert, J. D. to the Standards Committee of the Ohio State Board of Education, January 13, 2001." Accessed on January 24, 2002, at http://www.intelligentdesignnetwork.org/ohioboardtalk.htm. Three days after Calvert's January 13 address to the standards committee, the Ohio ACLU issued a press release stating its

concern about "efforts to include 'intelligent design theory' in the proposed science curriculum." "ACLU Expresses Concern and Caution over Intelligent Design," ACLU of Ohio Press Release, January 16, 2002. Accessed on January 25, 2002, at http://www.acluohio.org/press_releases/2002_press_releases/jan_16.htm.

67. Intelligent Design Network, "Form for Comments on Modifications to Draft of the Science Academic Content Standards for Ohio." Accessed on February 2, 2002, at http://intelligentdesignnetwork.org/ohiocomments.asp.

68. "State Board of Education Standards Committee to Hold Meeting March 11 Featuring Science Education Panel Discussion," press release by Ohio Board of Education Standards Committee, February 20, 2002. Accessed on April 13, 2002, at http://www.ode.state.oh.us/news/releases/february02/2-20-02.1.asp.

69. Krauss's comments were recorded by Catherine Candisky in "Design vs. Evolution Discussion Monday," *Columbus Dispatch*, April 13, 2002. Accessed on March 8, 2002, at http://www.dispatch.com/news-story.php?story=dispatch/news/news02/mar02/1131253.html. Durbin's remarks are cited with permission from notes he made at the meeting on March 11, 2002.

70. Stephen C. Meyer, "Teach the Controversy," *Cincinnati Enquirer*, March 30, 2002. Accessed on August 15, 2002, at http://www.discovery.org/viewDB/index.php3?program=CRSC&command=view&id=1134. Meyer's use of the word "Darwinian" is outdated and inaccurate. Evolutionary theory is now far beyond its original Darwinian form. But ID proponents use the term as a catchword because of its negative associations for their intended audience.

71. Liz Sidoti, "Backers of Intelligent Design Say They're Offering Compromise," *Cleveland Plain Dealer*, March 11, 2002. Accessed on March 12, 2002, at http://www.cleveland.com/newsflash/news/index.ssf?/cgi-free/getstory_ssf.cgi?o0330_BC_OH—EvolutionDebate&&news&newsflash-ohio.

72. Science Excellence for All Ohioans, "Suggested Modifications to Draft Indicators," April 1, 2002. Accessed on April 12, 2002, at http://www.sciohio.org/seaoindi.htm. We discuss later in this chapter the reference to the "Santorum language" in #5.

73. "Conservative Lawmakers Backing Bills to Teach Other Theories with Evolution," *Cleveland Plain Dealer*, January 25, 2002. Accessed on February 3, 2002, at http://www.cleveland.com/news/plaindealer/index.ssf?/xml/story.ssf/html_standard.xsl?/base/news/1011954659121881 76.xml.

74. HB 481 and HB 484 were accessed at http://www.legislature.state.oh.us/bills.cfm?ID=124_HB_481 and http://www.legislature.state.oh.us/bills.cfm?ID=124_HB_484, respectively, on February 2, 2002.

75. Stephen Ohlemacher, "Legislators Keep Intelligent Design at Arm's Length," *Cleveland Plain Dealer*, March 13, 2002. Accessed on April 13, 2002 at http://www.cleveland.com/news/plaindealer/index.ssf?/xml/story.ssf/html_standard.xsl?/base/news/1016015493598 7166.xml. See also Eric Meikle, "Ohio Reflections," *Reports of the National Center for Science Education* 22:6 (November–December 2002), 4–5.

76. Johnson is quoted in Mark Athitakis, "Looking for God at Berkeley," *SF Weekly*, June 20, 2001. Accessed on April 22, 2002, at http://www.sfweekly.com/issues/2001-06-20/feature.html/1/index.html. Calvert's assistance in writing the legislation is documented in Beem, *Kansas City Star*, April 2, 2002.

77. Liz Sidoti, "Committee Members Propose Final Changes to Science Standards," *Akron Beacon Journal*, October 10, 2002. Accessed on October 11, 2002, at http://www.ohio.com/mld/beaconjournal/news/state/4256850.htm. See also Susan R. Schell, "State Refuses to Advance Intelligent Design Theory," *Canton Repository*, October 13, 2002. Accessed on October 13, 2002, at http://www.cantonrep.com/cantonrep01/menus.php?ID=66862&r=3&Category=11. See also "Academic Content Standards: Science," Draft, October 2002, Center for Curriculum and Assessment, Office of Curriculum and Instruction, Ohio Department of Education. Accessed on October 17, 2002, at http://www.ode.state.oh.us/academic_content_standards/sciencecontentstd/PDF/9-23-ScienceDraft_Oct.pdf.

78. Lynn Elfner, CEO, Ohio Academy of Science, personal communication to Barbara Forrest, October 18, 2002. See the OAS definition of science in "A Resolution by the Ohio Academy of Science: Advocacy for Teaching *Cosmic, Geological and Biological Evolution* and Opposition to Forced Teaching of Creationist Beliefs in Public School Science Education," at http://www.ohiosci.org/EVOLRESOLUTIONFinalApprovedFeb282000.htm. Accessed on October 18, 2002.

79. See Liz Sidoti, "Evolution to Be Part of State Science Curriculum," Associated Press, October 15, 2002. Accessed on October 18, 2002, at http://web.lexis-nexis.com. See also Scott Stephens, "Panel Approves Science Guidelines," *Cleveland Plain Dealer*, October 15, 2002. Accessed on October 17, 2002, at http://www.cleveland.com/news/plaindealer/index.ssf?/xml/story.ssf/html_standard.xsl?/base/news/103467429114850.xml.

80. Liz Sidoti, "Ohio Students Should Be Taught Evolution," Associated Press, October 14, 2002. Accessed on October 18, 2002, at http://web.lexis-nexis.com. The uncertainty over what individual Ohio school districts might do prompted the American Association for the Advancement of Science to issue a strongly worded resolution against teaching ID. See "AAAS Board Resolution on Intelligent Design Theory" at http://www.aaas.org/news/releases/2002/1106id2.shtml.

81. Discovery Institute, "Ohio Board Backs Academic Freedom and Encourages Critical Analysis of Evolution," *Discovery Institute News*, October 15, 2002. Accessed on October 18, 2002, at http://www.discovery.org/news/ohioBoardBacks.html. DI's press release was accessed on October 17, 2002, at http://www.usnewswire.com/topnews/prime/1015-142.html. These releases are archived at Lexis-Nexis. SEAO's statement, "Latest News on the Standards: Resolution of Intent to Adopt Science Standards," was posted at http://www.sciohio.org/sbe1015.htm. Accessed on October 20, 2002.

82. See "IDnet Congratulates Ohio State Board on Improvements to Ohio Science Standards," News Release, October 17, 2002, at http://www.intelligentdesignnetwork.org/PressRelease101702.htm. See also Larry Witham, "Ohio Schools to Teach Evolution 'Controversy'," *Washington Times*, October 17, 2002, at http://www.washtimes.com/national/20021017-92467660.htm. See also Pamela R. Winnick, "Inherited Debate," *National Review Online*, October 18, 2002, at http://www.nationalreview.com/comment/comment-winnick101802.asp. All accessed on October 18, 2002.

83. Robert Lattimer, speaking at IDNet's Darwin, Design, and Democracy symposium, July 2002. Audio file provided by Jack Krebs, vice-president of Kansas Citizens for Science, August 2002. For information on the board's final vote, see Scott Stephens, "Policy Resolved on Origin of Life," *Cleveland Plain Dealer*, December 11, 2002, at http://www.cleveland.com/news/plaindealer/index.ssf?/base/news/1039602647180400.xml. See Ohio's final science standards at http://www.ode.state.oh.us/academic_content_standards/acsscience.asp#Science_Academic_Content_Standards. See SEAO's call for applicants at http://www.sciohio.org/curmod.htm. All accessed on April 26, 2003.

84. See "IDnet Announces Establishment of IDnet of New Mexico," News Release, Intelligent Design Network, July 23, 2002, at http://www.intelligentdesignnetwork.org/PressReleaseNewMexico.htm. Regarding Cobb County, see Mia Taylor and Mary MacDonald, "Cobb Unanimously Approves Discussion of Other Theories," *Atlanta Journal-Constitution*, September 26, 2002, at http://www.accessatlanta.com/ajc/metro/cobb/0902/27evolution.html. See the new Cobb Co. policy at http://www.accessatlanta.com/ajc/metro/cobb/0902/27text.html. All accessed on October 18, 2002. See also Mary MacDonald, "28 Scholars Back Cobb on Evolution," *Atlanta Journal-Constitution*, September 21, 2002, and Mary MacDonald, "Noted Critics of Evolution Urge Balance," *Atlanta Journal-Constitution*, September 16, 2002. See Discovery Institute's press release on Cobb Co. at http://www.discovery.org/viewDB/index.php3?command=view&id=1267&program=CRSC. See also "Ohio to Allow 'Intelligent Design' Option," *Marietta Daily Journal*, October 16, 2002, at http://apt.mywebpal.com/news_tool_v2.cfm?pnpID=7&CategoryID=89&StoryID=10081830&show=localnews. Both accessed on October 18, 2002. See also Mary McDonald, "Cobb Issues Evolution Guidelines to Teachers," *Atlanta Journal-Constitution*,

January 9, 2003, and "ACLU Sues Over Evolution Disclaimers in Textbooks," accessed on April 3, 2003, at http://www.foxnews.com/story/0,2933,61065,00.html.

85. Trish Gura, "Evolution Critics Seek Role for Unseen Hand in Education," *Nature* 416 (March 21, 2002), 250.

86. This announcement was made in a May 8, 2000, U.S. Newswire press release. Accessed on April 4, 2002, at http://web.lexis-nexis.com.

87. Brownback's participation is cited with permission from notes by David Applegate, director of the American Geological Institute's Government Affairs Program, May 10, 2000. Brownback endorsed creationist Linda Holloway, a Kansas BOE member, in her re-election campaign against fellow Republican Sue Gamble in the 2000 Republican primary. See "Board Member Wins Support," *Topeka Capital-Journal*, April 15, 2000, at http://cjonline.com/stories/041500/kan_holloway.shtml. Accessed on December 27, 2000.

88. Daniel Walker Guido, "Creation Continues to Overcome [D]arwinists' Attacks, Johnson Says," *Baptists & News*, March 28, 2000. Accessed on December 26, 2001, at http://www.baptist.org/Baptists&News/Creationism_012.htm.

89. "Petri Ideas Attracting Respect and Attention," Discovery Institute *Journal* (August 1996), was accessed at http://www.discovery.org/w3/discovery.org/journal/petri.html. See also http://www.discovery.org/fellows/boardOfDirectors.html. Accessed on November 10, 2001. Applegate notes, May 10, 2000.

90. A copy of Souder's address and the Baylor scientists' letter is available at Access Research Network at http://www.arn.org/docs/idushouse_700.htm. Accessed on April 22, 2002.

91. Despite his seniority, Petri did not become chair, but vice chair of the committee. Congressman John A. Boehner (R-OH) became chair. Committee member list, House Education and Workforce Committee, http://edworkforce.house.gov/members/mem-fc.htm. Accessed on March 17, 2001.

92. See "Special Update: Evolution Opponents Hold Congressional Briefing" at http://www.agiweb.org/gap/legis106/id_update.html. See also Applegate's July 2000 *Geotimes* article, "Creationists Open a New Front," at http://www.geotimes.org/july00/scene.html. Accessed on May 13 and July 31, 2000, respectively.

93. David Applegate, "Special Update: Evolution Opponents on the Offensive in Senate, House," American Geological Institute, June 19, 2001. Accessed on September 3, 2001, at http://www.agiweb.org/gap/legis107/evolution_update0601.html.

94. "Better Education for Students and Teachers Act," U.S. Senate, June 13, 2001. Available at http://thomas.loc.gov/cgi-bin/query/R?r107:FLD001:S06148, p. S6147. Accessed on November 16, 2001.

95. "Better Education for Students and Teachers Act," U.S. Senate, June 13, 2001. Available at http://thomas.loc.gov/cgi-bin/query/R?r107:FLD001:S06148, p. 6150. Accessed on November 5, 2001.

96. "Better Education for Students and Teachers Act," U.S. Senate, June 13, 2001. Available at http://thomas.loc.gov/cgi-bin/query/R?r107:FLD001:S06148, p. S6152. Accessed on November 17, 2001.

97. "Senate Roll Call Votes, 107th Congress—1st Session (2001)." Accessed on April 20, 2003, at http://www.senate.gov/legislative/LIS/roll_call_lists/roll_call_vote_cfm.cfm?congress=107&session=1&vote=00182. A breakdown of the vote by yeas and nays is available here.

98. Quoted in David Applegate, "Update on the Challenges to the Teaching of Evolution," December 5, 2001. Accessed on December 12, 2001, at http://www.agiweb.org/gap/legis107/evolution.html.

99. Larry Witham, "Senate Bill Tackles Evolution Debate," *Washington Times*, June 18, 2001. Accessed on November 13, 2001, at http://www.washtimes.com/national/20010618-5554308.htm.

100. See Johnson's *Weekly Wedge Update* for June 11 and June 18, 2001, at http://www.arn.org/johnson/wedge.htm. Accessed on December 15, 2001.

101. Accessed on December 12, 2001, at http://www.answersingenesis.org/docs2001/0623news.asp.

102. "Senate Makes Statement About Evolution-Only Curricula," *Culture Facts*, June 22, 2001. Accessed on November 17, 2001, at http://www.frc.org/get/cu01f4.cfm?CFID=177740&CFTOKEN=47322741.

103. "Darwinian Teaching on Trial: Senators for Sound Science," *World*, June 30, 2001. Accessed on November 5, 2001, at http://worldmag.com/world/issue/06-30-01/opening_2.asp.

104. Paul E. Gammill, Albuquerque, New Mexico, to New Mexico State Board of Education, August 24, 2001, and Paul Gammill to Flora Sanchez, President, New Mexico State Board of Education, January 14, 2002. NM BOE member Marshall Berman, a physicist elected to the board on an anti-creationist platform, succeeded in 1999 in getting the board to reverse its 1996 decision to delete evolution from its science standards. See "New Mexico SBE: Evolution—13!; Creationism—1," by the Coalition for Excellence in Science and Math Education at http://www.cesame-nm.org/announcement/sbe.html. Accessed on November 17, 2001.

105. See "Joint Explanatory Statement of the Committee of Conference," Title I, Part A, where the Santorum amendment is inserted under item 78, dealing with standards and assessment. Accessed on February 3, 2002, at http://edworkforce.house.gov/issues/107th/education/nclb/conference/stateofman/index.htm. The conference committee report is available at http://frwebgate.access.gpo.gov/cgi-bin/getdoc.cgi?dbname=107_cong_reports&docid=f:hr334.107.pdf. Accessed on February 3, 2002.

106. Bruce Chapman and David DeWolf, "Why the Santorum Language Should Guide State Science Education Standards." Accessed on April 18, 2003, at http://www.discovery.org/articleFiles/PDFs/santorumLanguageShouldGuide.pdf.

107. See David Applegate, "Federal Challenges to the Teaching of Evolution," January 2, 2002. Accessed on May 24, 2002, at http://www.agiweb.org/gap/legis107/evolution_congress.html. See "August 2001 Joint Letter from Scientific and Educational Leaders on Evolution in H.R. 1," at http://www.agiweb.org/gap/legis107/evolutionletter.html. Accessed on November 5, 2001. See also David Applegate, "Special Update: Evolution Opponents on the Offensive in Senate, House," at http://www.agiweb.org/gap/legis107/evolution_update0601.html. Accessed on December 15, 2001.

108. Congressional Research Service, Library of Congress, personal communication to Barbara Forrest, January 22, 2002.

109. Prof. Dennis D. Hirsch, "Science vs. Intelligent Design: The Law." Accessed on April 13, 2002, on the Ohio Citizens for Science site at http://ecology.cwru.edu/ohioscience/legal-hirsch.asp. See the National Center for Science Education's release, "Santorum Amendment Stripped from Education Bill," at http://www.ncseweb.org/pressroom.asp?year=2001. Both accessed on April 13, 2002.

110. Catherine Candisky, "Governor Sidesteps Evolution Argument," *Columbus Dispatch*, February 10, 2002. Accessed on March 9, 2002, at http://www.dispatch.com/news-story.php?story=dispatch/news/news02/feb02/1080444.html.

111. Discovery Institute, "Biologist Ken Miller Flunks Political Science on Santorum." Accessed on April 22, 2002, at http://www.discovery.org/viewDB/index.php3?command=view&id=1149&program=CRSC. See Miller's "The Truth About the 'Santorum Amendment' Language on Evolution," at http://www.millerandlevine.com/km/evol/santorum.html.

112. See "Why the Santorum Language Should Guide State Science Education Standards." Accessed on April 28, 2002, at http://www.discovery.org/articleFiles/PDFs/santorumLanguageShouldGuide.pdf.

113. David Lore, "Expert Review Likely to Heighten Debate," *Columbus Dispatch*, June 27, 2002. Accessed on October 19, 2002, at http://libpub.dispatch.com/cgi-bin/documentv1?DBLIST=cd02&DOCNUM=27710&TERMV=198:7:205:9:108229:7:10823

7:8:. While a case directly initiated by ID proponents might not be a very strong one, it would inevitably consume even more time and money, as all court cases do.

114. Congress, Senate, Senator Santorum of Pennsylvania, 107th Cong., 1st sess., *Congressional Record* 147:176 (December 18, 2001), S13377–13378.

115. This release was accessed at U.S. Newswire on April 22, 2002, at http://www.usnewswire.com/topnews/temp/1221-109.html. The same version was accessed on April 22, 2002, at Lexis-Nexis.

116. "Congress Gives Victory to Scientific Critics of Darwinism," *Discovery Institute News*, December 21, 2001. Accessed on December 26, 2001, at http://www.discovery.org/news/congressGivesVictory.html.This release could not be found at Lexis-Nexis, although it was posted on the DI website as "Discovery Institute News."

117. The schedule of writing team committee meetings was posted at http://www.ode.state.oh.us/academic_content_standards/acsscience.asp?. The Calvert-Harris paper is at http://www.intelligentdesignnetwork.org/endingwar.htm. SEAO's website registration was documented on March 11, 2002, at http://www.register.com. The press release by Ohio Rep. John Boehner announcing the president's signing of H.R. 1 was accessed on April 13, 2002, at http://edworkforce.house.gov/press/press107/hr1signing10802.htm.

118. "Federal Santorum Amendment," Science Excellence for All Ohioans. Accessed on January 16, 2002, at http://www.sciohio.org/seaoleg.htm.

119. John Calvert, "Remarks of John Calvert, J. D. to the Standards Committee of the Ohio Board of Education," January 13, 2002. Accessed on February 1, 2002, at http://www.intelligentdesignnetwork.org/ohioboardtalk.htm.

120. Robert Lattimer and John Calvert, "Intelligent Design Is a Matter of Academic Freedom," *Cleveland Plain Dealer*, January 18, 2002. Accessed on January 18, 2002, at http://www.cleveland.com/news/plaindealer/othercolumns/index.ssf?/xml/story.ssf/html_standard.xsl?/base/opinion/1011349850270331 33.xml.

121. Catherine Candisky, "Lawmakers Enter Debate on Science Curriculum," *Columbus Dispatch*, January 24, 2002. Accessed on January 24, 2002, at http://www.dispatch.com/news-story.php?story=dispatch/news/news02/jan02/1047586.html.

122. "Creationism Advocates Tout Dead Santorum Amendment," Ohio Citizens for Science. Accessed on February 3, 2002, at http://ecology.cwru.edu/ohioscience/santorum.asp.

123. Science Excellence for All Ohioans, "Federal Santorum Amendment." Accessed on April 13, 2002, at http://www.sciohio.org/seaoleg.htm.

124. Deborah Owens-Fink, "Is It Science? Faith? Are They Equal Theories of the Origin of Life?" *Akron Beacon Journal*, January 30, 2002. Accessed on January 30, 2002, at http://www.ohio.com/bj/editorial_pages/docs_editorial/011260.htm.

125. Discovery Institute, "Santorum Language on Evolution: Revised Amendment, Congressional Statements," January 31, 2002. Accessed on February 2, 2002, at http://www.discovery.org/viewDB/index.php3?command=view&id=1109&program=CRSC.

126. Rep. John A. Boehner and Rep. Steve Chabot, Washington, D.C., to Jennifer L. Sheets and Cyrus B. Richardson, Jr., Columbus, Ohio, March 15, 2002. The Discovery Institute posted a pdf version of the original letter. Accessed on April 13, 2002, at http://www.discovery.org/news/BoehnerChabotLetterToOhio.html.

127. Prof. Dennis Hirsch to Barbara Forrest, personal communication, May 23, 2002.

128. Rick Santorum, "Illiberal Education in Ohio Schools," *Washington Times*, March 14, 2002, and Edward M. Kennedy, "Evolution Is Designed for Science Classes," *Washington Times*, March 21, 2002. Accessed on April 13, 2002, at http://www.washingtontimes.com/commentary/20020314-50858765.htm and http://www.washingtontimes.com/op-ed/20020321-76780268.htm, respectively. Just a few days before his letter appeared, *U.S. News and World Report* featured Kennedy in the March 11, 2002, issue ("A Liberal in Winter"): "These days, after the battles on civil rights, education, and healthcare, Kennedy says his next frontier is science: 'There are incredible opportunities right now in the life sci-

ences: the human genome, genetic research. In the next century, the breakthroughs are going to be monumental.'"

129. William Dembski, "Edward Kennedy—Expert on Science?" Press Release, March 21, 2002. Accessed on April 13, 2002, at http://www.arn.org/docs2/news/kennedy expertonscience032102.htm.

130. "Minutes of the [Ohio] House Education Committee, Tuesday, March 5, 2002, Representative Jamie Callender, Chairman." Reidelbach also provided to the committee (1) DI's "Santorum Language on Evolution: Revised Amendment, Congressional Statements," (2) DI's Zogby International poll, and (3) DI's "A Scientific Dissent from Darwinism," signed by the "100 scientists." LA HCR 50 was accessed on April 4, 2003, at http://www.legis.state.la.us/leg_docs/03RS/CVT1/OUT/0000K4Q7.PDF.

131. Phillip Johnson, "More from Ohio," *Weekly Wedge Update*. Accessed on February 1, 2002, at http://www.arn.org/docs/pjweekly/pj_weekly_020131.htm.

132. An interesting, possibly coincidental fact is that Michael Behe is one of Santorum's constituents. Behe earlier supported changes in Pennsylvania science standards that would have allowed ID in public school science classes. See his letter in Johnson's June 11, 2001, *Weekly Wedge Update*, "The Pennsylvania Controversy," at http://www.arn.org/docs/pjweekly/pj_weekly_010611.htm. See also Pamela Winnick, "Proposed Rules Boost Teaching of Creationism," *Pittsburg Post-Gazette*, November 29, 2000. Accessed on November 5, 2001, at http://www.post-gazette.com/regionstate/20001129creationism1.asp.

133. Audiotape of Jonathan Wells's lecture and book signing, Seattle, Washington, October 26, 2000.

134. Jonathan Wells, *Icons of Evolution: Science or Myth? Why Much of What We Teach about Evolution Is Wrong* (Washington, DC: Regnery Publishing, 2000), 243–244.

135. "George W. Bush: Running on His Faith," *U.S. News & World Report*, December 6, 1999. Accessed on December 16, 2001, at http://www.usnews.com/usnews/news/991206/bushint.htm. See also CNN's August 27, 1999, story, "Presidential Candidates Weigh in on Evolution Debate," at http://www.cnn.com/ALLPOLITICS/stories/1999/08/27/president.2000/evolution.create/. Accessed on December 16, 2001.

136. Phillip E. Johnson, *The Wedge of Truth: Splitting the Foundations of Naturalism* (Downers Grove, IL: InterVarsity Press, 2000), 78–79.

137. See Stephen Meyer's "What's the Difference?" in *World*, October 21, 2000. Accessed on December 16, 2001 at http://www.worldmag.com/world/issue/10-21-00/closing_1.asp. Marvin Olasky's "Bush Can Still Win" is at http://www.worldmag.com/world/issue/09-23-00/closing_2.asp. Accessed on December 16, 2001.

138. See William A. Dembski and Mark Hartwig, "Fishing for Votes," Discovery Institute website. Accessed on December 16, 2001, at http://www.discovery.org/viewDB/index.php3?command=view&id=536&program=CRSC.

139. Bruce Chapman and Stephen Meyer, "A Plan for Recovery of the Iffy Economy," *Seattle Times*, December 28, 2000. Accessed on December 16, 2001, at http://archives.seattletimes.nwsource.com/cgi-bin/texis/web/vortex/display?slug=bruce28&date=20001228&query=Chapman+Meyer.

140. Hanna Rosin, "Gore Avoids Stance on Creationism," *Washington Post*, August 27, 1999, Page A8. Accessed on December 16, 2001, at http://www.washingtonpost.com/wp-srv/politics/campaigns/wh2000/stories/creation082799.htm.

9. Religion First—and Last

1. Phillip Johnson, "Is God Unconstitutional?" Part I, *Real Issue*, (September/October 1994). Accessed April 26, 2002, at http://www.leaderu.com/real/ri9403/johnson.html.

2. Johnson, "Is God Constitutional?"

3. Phillip E. Johnson, *The Wedge of Truth*, see chapter 1, "Philip Wentworth Goes to

Harvard." For Phillip Wentworth's essay, see http://www.theatlantic.com/issues/95nov/warring/whatcoll.htm. Accessed on March 17, 2001.

4. Johnson, *Wedge of Truth*, 38.

5. Johnson, *Wedge of Truth*, 20.

6. Tim Stafford, "The Making of a Revolution," Access Research Network. Accessed on April 26, 2002, at http://www.arn.org/johnson/revolution.htm. Originally published in *Christianity Today* 41, December 8, 1997.

7. "Keeping the Darwinists Honest," interview with Phillip Johnson, *Citizen* (April 1999). Accessed on January 27, 2001, at http://www.arn.org/docs/johnson/citmag99.htm.

8. "Interview," interview with Phillip Johnson about *The Wedge of Truth* by Christian Book Distributors (via e-mail, August 14, 2000). Accessed on October 14, 2002, at http://www.christianbook.com/Christian/Books/dpep/interview.pl/16559901?sku=22674. "In the beginning was the Word" is the first line in "The Gospel According to St. John." Johnson cites this passage in support of "mere" creation to avoid the disputes that arise when young-earth creationists cite Genesis.

9. Bruce Chapman, "From the President, Bruce Chapman: Ideas Whose Time Is Coming," Discovery Institute *Journal* (August 1996). Accessed on April 26, 2002, at http://www.discovery.org/w3/discovery.org/journal/president.html.

10. Center for the Renewal of Science and Culture, "Major Grants Help Establish Center for the Renewal of Science and Culture," Discovery Institute *Journal* (August 1996). Accessed on April 26, 2002, at http://www.discovery.org/w3/discovery.org/journal/center.html.

11. The "Life After Materialism" page was accessed on April 17, 2000, at http://www.discovery.org/w3/discovery.org/crsc/crsc.html.

12. Laurissa MacFarquhar, "The Gilder Effect," *New Yorker*, May 29, 2000, 110. Included in *The New Gilded Age: The New Yorker Looks at the Culture of Affluence* (e-book), ed. David Remnick (Random House, 2001). Accessed on April 5, 2002, at http://www.randomhouse.com/boldtype/1200/macfarquhar/essay.html.

13. George Gilder, "Freedom from Welfare Dependency," interview by the Acton Institute for the Study of Religion and Liberty, *Religion and Liberty* 4:2 (March/April 1994). Accessed on April 26, 2002, at http://acton.org/publicat/randl/94mar_apr/gilder.html.

14. Johnson believes his work is a calling from God: "I enjoy the struggle immensely, and thank God that I was given this calling." See "Interview with Phillip Johnson About *The Wedge of Truth*."

15. "By Design: A Whitworth Professor Takes a Controversial Stand to Show That Life Was No Accident," *Whitworth Today*, Whitworth College (winter 1995). Accessed on March 24, 2002, at http://www.arn.org/docs/meyer/sm_bydesign.htm.

16. See Lynn Vincent, "Science vs. Science," *World*, February 26, 2000. Accessed on April 29, 2000, at http://www.discovery.org/crsc/CRSCdbEngine.php3?id=148. See also Dembski's "Design as a Research Program: Fourteen Questions to Ask About Design," at http://www.discovery.org/viewDB/index.php3?program=CRSC%20Responses&command=view&id=259.

17. William A. Dembski, "What Can We Reasonably Hope For?" A Millennium Symposium, *First Things* (January 2000). Accessed on April 26, 2002, at http://www.firstthings.com/ftissues/ft0001/articles/dembski.html.

18. This study does not criticize Dembski's right to his personal religious views. Rather, it shows that, Dembski's protestations to the contrary, what he claims to present as science is, after all, religion.

19. William A. Dembksi, "Signs of Intelligence: A Primer on the Discernment of Intelligent Design," *Touchstone* (July/August 1999), 84.

20. Fred Heeren, "The Lynching of Bill Dembski," *American Spectator* (November 2000). Accessed on October 14, 2002, at http://www.spectator.org/amspec/classics/Nov00/heeren0011.htm. Dembski asserted as recently as July 2002 that "the designer need not be a deity. It could be an extraterrestrial or a telic process inherent in the uni-

verse" ("Commentary on Eugenie Scott and Glenn Branch's 'Guest Viewpoint: Intelligent Design Not Accepted by Most Scientists,' 7/2/02," at http://www.designinference.com/documents/2002.07.Scott_and_Branch.htm; accessed on October 14, 2002). But once again, Dembski contradicts himself in another work. In "The Incompleteness of Scientific Naturalism," written for the 1992 "Darwinism: Science or Philosophy?" symposium (the first Wedge conference), Dembski uses his "Incredible Talking Pulsar" to dispel any idea that he is talking about "an extraterrestrial" (which would be a natural entity). (See this paper at http://www.leaderu.com/orgs/fte/darwinism/chapter7.html; accessed on September 17, 2002.) The pulsar is a Dembskian "thought experiment": a "rotating neutron star" emitting a "pattern of pulses in Morse code." It communicates meaningful messages—in English—and identifies itself as "the mouthpiece of Yahweh, the God of both the Old and the New Testaments." Dembski asserts that a naturalistic explanation would not be able to account for the pulsar. But here is how Dembski accounts for it: "It is inescapable that . . . we are dealing with . . . a super-intelligence . . . [not] an intelligence that at this time surpasses human capability, but which in time humans can hope to attain . . . [and not] a super-human intelligence that might nevertheless be realized in some finite rational material agent embedded in the world (say an extraterrestrial or a conscious supercomputer). . . . I mean a supernatural intelligence . . . surpassing anything . . . physical processes are capable of offering . . . [and] anything . . . humans or finite rational agents in the universe are capable of even in principle. . . . What lesson can we learn from the pulsar? I claim we should infer that a designer in the full sense of the word is communicating through the pulsar, i.e., a designer who is both intelligent and transcendent." I.e., God.

21. Jay Wesley Richards, "Naturalism in Theology and Biblical Studies," in *Unapologetic Apologetics: Meeting the Challenges of Theological Studies* (Downers Grove, IL: InterVarsity Press, 2001), 95–96.

22. William A. Dembski and Jay Wesley Richards, "Introduction: Reclaiming Theological Education," in *Unapologetic Apologetics,* 17–18. Accessed on April 23, 2002, at http://www.gospelcom.net/ivpress/title/exc/1563-I.pdf.

23. J. Gresham Machen, *What Is Christianity?* (Grand Rapids, MI: Eerdmans, 1951), 162, quoted in Dembski and Richards, "Introduction: Reclaiming Theological Education," 19.

24. Dembski and Richards, "Introduction: Reclaiming Theological Education," 17.

25. Dembski and Richards, "Introduction: Reclaiming Theological Education," 21.

26. Dembski and Richards, "Introduction: Reclaiming Theological Education," 15.

27. William A. Dembski, "The Task of Apologetics," in *Unapologetic Apologetics,* 43. Accessed on April 23, 2002, at http://www.gospelcom.net/ivpress/title/exc/1563-1.pdf.

28. Jonathan Wells, "Why I Went for a Second Ph.D.," in *Unification Sermons and Talks by Reverend Wells* at http://www.tparents.org/Library/Unification/Talks/Wells/DARWIN.htm. Accessed on March 26, 2001.The "True Parents" are Rev. and Mrs. Moon, hence Moon's being called "Father." Frederick Clarkson, in *Eternal Hostility: The Struggle Between Theocracy and Democracy* (Monroe, ME: Common Courage Press, 1997), says, "The theocratic Unification Church of Sun Myung Moon is also propelled by the politics of demons. In its indoctrination practices, new members are taught that Satan may be working through their own biological parents" (135).

29. Raymond G. Bohlin, "Why Does the University Fear Phillip Johnson?" Probe Ministries. Accessed on April 26, 2002, at http://www.probe.org/docs/philjohn.html.

30. Chapman, "From the President, Bruce Chapman: Ideas Whose Time Is Coming."

31. Witham, "Contesting Science's Anti-Religious Bias."

32. Weyerhauser is listed on FTE letterhead in a 1995 letter from FTE to supporters requesting prayers for its marketing of *Pandas*. See Weyerhauser's obituary in the American Scientific Affiliation's July/August 1999 newsletter at http://www.asa3.org/ASA/newsletter/JULAUG99.htm. See a photo of Weyerhauser with Paul Nelson and ARN

chairman Dennis Wagner at http://www.arn.org/infopage/wagner.htm. Both accessed on April 24, 2002.

33. Stewardship Foundation, "Guidelines," and "History of Foundation." Accessed on April 24, 2002, at http://www.stewardshipfdn.org/page6.html and http://www.stewardshipfdn.org/page3.html, respectively.

34. Maclellan Foundation, "Mission" and "The Maclellan Foundation, Inc. Giving Guidelines." Accessed on April 24, 2002, at http://www.maclellanfdn.org/Pages/2About/abtmsn_b.htm and http://www.maclellanfdn.org/Pages/2About/abgvgd_b.htm, respectively.

35. Teresa Watanabe, "Enlisting Science to Find the Fingerprints of a Creator," *Los Angeles Times*, March 25, 2001. Accessed on November 6, 2001, at http://www.discovery.org/news/EnlistingScience.html.

36. Grantees were listed at http://www.maclellanfdn.org/Pages/3OurGrnt/ourgrusa.htm. Accessed on April 24, 2002.

37. "Buying a Movement: Right Wing Foundations in American Politics," *A Report by People for the American Way*. Accessed on April 24, 2002, at http://www.globalpolicy.org/finance/docs/buymvmnt.htm. See also John Gizzi, "After Henry Salvatori: California's 'Most Generous' Conservative Philanthropists," Foundation Watch, Capital Research Center (October 1998). Accessed on April 24, 2002, at http://rnaworld.bio.ukans.edu/ID-intro/di/fund/CA_Cons_Phil—fw-1098.html. Ahmanson's Fieldstead and Company is not to be confused with the Ahmanson Foundation, a philanthropic foundation started by his father, Robert H. Ahmanson. The elder Ahmanson uses his wealth to fund worthwhile causes such as the arts, health care, and education, and especially UCLA, his alma mater. He received the UCLA Center on Aging's Icon Award on June 8, 2002. Accessed on June 22, 2002, at http://www.aging.ucla.edu/iconawards2002.html.

38. See "Who to Challenge on the Theocratic Right" at http://www.publiceye.org/research/directories/theo_top.html and the National Education Association's "Selected Biographies" at http://www.nea.org/nr/old/nr981001.html, respectively. See also Steve Benen, "From Genesis to Dominion," *Church & State* (July/August 2000) at http://www.au.org/churchstate/cs7003.htm. All accessed on April 18, 2003.

39. See http://www.businesswire.com/webbox/bw.100997/499089.htm. Accessed on April 24, 2002. Business Wire says that Home Savings of America is "one of the nation's largest full-service consumer and small-business banks."

40. See Dembski's course description in the Internet Archive at http://web.archive.org/web/20010210184421/http://www.calvin.edu/fss/field00.htm. Accessed on April 19, 2003. Ahmanson's funding the seminars at the Calvin location is explained by the fact that his wife, Roberta Green Ahmanson, is a Calvin alumna. See http://www.calvin.edu/news/releases/1999_00/alumni.htm. Accessed on April 19, 2003.

41. Joseph Conn to Barbara Forrest, personal communication, June 23, 2000.

42. Gizzi, "After Henry Salvatori."

43. Clarkson, "Laying Down the (Biblical) Law," in *Eternal Hostility*, 77–96. Clarkson's journalism background makes him an expert on radical religious movements: "He was the first to expose the Christian Coalition's plans to take over the Republican Party. . . . He is the founding editor of *Front Lines Research*, a bi-monthly journal on the Radical Right. . . . He has served as a consultant to documentaries and news features . . . for . . . CBS News, PBS, and German and Japanese television. . . . Clarkson is frequently . . . quoted by . . . *The New York Times, The Washington Post, USA Today, The Boston Globe,* and *The San Francisco Examiner*. He has also appeared on numerous . . . news programs, notably NPR's Morning Edition." See "About the Author," in *Eternal Hostility*.

44. Clarkson, "Theocrats in Action," in *Eternal Hostility*, 100, 111. Clarkson notes that Ahmanson and his Capitol Commonwealth Group, which funded conservative California candidates and causes, had by the early 1990s "helped elect one fourth of the 120 members of the [California] Senate and Assembly" (112).

45. Clarkson, "Theocrats in Action," 111.

46. Chapman, "From the President: Ideas Whose Time Is Coming."

47. "William A. Dembski: Curriculum Vitae, September 1998," "William A. Dembski: Curriculum Vitae, November 2001," and "William A. Dembski: Curriculum Vitae, February 2003." Accessed on April 19, 2003, at http://www.leaderu.com/offices/dembski/menus/cv.html, http://www3.baylor.edu/~William_Dembski/cv.htm, and http://www.designinference.com/documents/2003.02.CV.htm, respectively.

48. To see the "virtual offices" of Wedge members at Leadership University, see http://www.leaderu.com/offices. On the SMU conference, see "Table of Contents," *Proceedings of a Symposium Entitled Darwinism: Scientific Inference or Philosophical Preference?* at http://www.leaderu.com/orgs/fte/darwinism/. On the Mere Creation conference, see the comments of Rich McGee, director of International Expansion for CLM, at http://www.origins.org/mc/resources/ri9602/mcgee.html. For *The Real Issue*, "The Mere Beginning: A Special Report," available in pdf, see http://www.leaderu.com/real/ri9602.html. All accessed on April 24, 2002.

49. See "Affiliate Program: Why," at the Christian Leadership Ministries website at http://www.clm.org/menus/affprowhy.html. Accessed on April 5, 2002.

50. Clarkson, *Eternal Hostility*, 107–108.

51. See the Veritas Forum website at http://www.veritas.org. Accessed on April 19, 2003. See also chapter 12, "Ministering with Others on Campus," of *Ministering in the Secular University* at http://www.clm.org/msu/chapter12.html. Accessed on March 26, 2001.

52. For the UCSB Veritas Forum TV schedule, see http://www.id.ucsb.edu/veritas/tv/. See the University of Florida Veritas page at http://nersp.nerdc.ufl.edu/~cff/veritas.html. Both accessed on April 19, 2003.

53. Veritas Forum fundraising letter, July 7, 2000. Accessed on May 16, 2002, at http://www.veritas.org/generic.htm.

54. See the *Wedge of Truth* page at http://www.gospelcom.net/ivpress/wedgeoftruth/ and the *Unapologetic Apologetics* page at http://www.gospelcom.net/cgi-ivpress/book.pl/code=1563. All accessed on May 16, 2002.

55. See IDEA Club founder Casey Luskin's report on the Trinity conference at http://groups.google.com/groups?hl=en&selm=907vob%24hj%241%40news.koti.tpo.fi&rnum=1. For the board of advisors of the Center for University Ministries, see http://www.apologetics.org/societies/trinity.html. For Woodward's winter 2000 *Inklings* newsletter item about the Chinese scientist, see http://www.apologetics.org/news/ink1.html. All accessed on April 24, 2002.

56. William A. Dembski and Mark Hartwig, "God's Fingerprints on the Universe," Focus on the Family interview by James Dobson (October 4–5, 2001). The audiotape of the interview is available at http://www.family.org/resources/itempg.cfm?itemid=2658.

57. Dobson, "God's Fingerprints on the Universe."

58. Rob Boston, "Family Feud," *Church & State* 51:5 (May 1998), 12.

59. Boston, "Family Feud," 14. Alexander-Moegerle worked for Dobson for ten years, helping him set up his television/radio operation. His judgment of Dobson is based on long-term acquaintance and on sources such as "personal recollections from ten years of direct executive suite conversations with Dobson; talks in his office, car, and home . . . [and] hundreds of pages of memoranda from and to Dobson." Gil Alexander-Moegerle, *James Dobson's War on America* (Amherst, NY: Prometheus Books, 1997), 15. For Dobson's influence on Republican Party policy, see James Carney, "The G.O.P. Mantra: Keep Dobson Happy," *Time*, May 11, 1998. For Dobson's statements reflecting his theocratic tendencies, see his contribution to "The End of Democracy? A Discussion Continued," *First Things* 69 (January 1997), at http://www.firstthings.com/ftissues/ft9701/articles/theend.html#dobson. Both accessed on October 14, 2002. In *First Things*, Dobson says that "rulers ultimately derive their governing authority from God. . . . I stand in a long tradition of Christians who believe that rulers may forfeit their divine mandate when they systematically contravene the divine law." This is the language of theocracy.

60. Rob Boston, "Missionary Man," *Church & State* (April 1999). Accessed on March 24, 2002, at http://www.au.org/churchstate/cs4995.htm.

61. "Signs of Intelligence" was available in the Truths that Transform archive at http://www.crministries.org/TTTarchive.htm. Accessed on April 24, 2002.

62. Rob Boston, "D. James Kennedy: Who Is He and What Does He Want?" *Church & State* (April 1999). Accessed on March 24, 2002, at http://www.au.org/churchstate/cs4994.htm.

63. Rob Boston, "Left Behind," *Church & State* 55:2 (February 2002), 9. Accessed on October 14, 2002, at http://www.au.org/churchstate/cs2022.htm. Boston reports that LaHaye "calls church-state separation 'the big lie' and insists, '[W]ith a false interpretation of separation, the humanists have rendered our government almost as secular as Communist China and the former Soviet Union.'"

64. Catherina Hurlburt, "Design, Darwinism and the Freedom to Learn," *Family Voice* (September/October 2001). Accessed on March 22, 2002 at http://www.cwfa.org/library/_familyvoice/2001-09/06-15.shtml.

65. Clarkson, *Eternal Hostility*, 35.

66. See Dembski's November 17, 2000, Metaviews post at http://www.metalibrary.net/id-wd/index-body.html. For physicist Victor Stenger's critique of the Wedge's creationism, see "Intelligent Design: The New Stealth Creationism" at http://www.infidels.org/library/modern/vic_stenger/stealth.pdf. Both accessed on April 24, 2002.

67. Dembski, *Intelligent Design*, 212, 224.

68. "God and Evolution in Kansas Classrooms," *Nightline*, July 27, 2000. Unedited transcript accessed on August 3, 2000, at http://www.abcnews.go.com/onair/nightline/transcripts/n1000727_trans.html. This transcript is not archived. For similar remarks Johnson made at the Capitol Hill Baptist Church in Washington, D.C., in September 1999, see "Johnson Challenges Advocates of Evolution," *Insight on the News*, October 25, 1999, at http://www.arn.org/docs/johnson/insightprofile1099.htm. Accessed on April 26, 2002.

69. See William Dembski, "Intelligent Design Coming Clean," Metaviews, November, 2000. Accessed on April 24, 2002, at http://www.arn.org/docs/dembski/wd_idcoming clean.htm.

70. Robert Pennock, *Tower of Babel: The Evidence Against the New Creationism* (Cambridge, MA: MIT Press, 1999), 227.

71. Ray Bohlin, "Probe Answers Our E-Mail: 'Why Won't You Take a Stand on the Age of the Earth?'" Accessed on October 14, 2002, at http://www.probe.org/docs/e-age.html. On Bohlin's Probe bio page, he again walks the age-of-the-earth tightrope: "The plain language of Genesis 1 seems to teach a recent literal six-day creation. There is much data from science, however, that indicates the universe and earth are billions of years old. . . . This issue should not be the focus of the creation/evolution debate at this time." Probe's "Doctrinal Statement," however, strongly points to the literal interpretation of Genesis: "The final authority of our beliefs is the Bible, God's infallible written Word, the sixty-six books of the Old and New Testaments. We believe that the Bible was uniquely, verbally, and fully inspired by the Holy Spirit, and that it was without error in the original manuscripts." Accessed on October 14, 2002, at http://www.probe.org/docs/bohlin.html and http://www.probe.org/support/doctrinal.html, respectively.

72. Henry M. Morris, "A Young-Earth Creationist Bibliography." Accessed on May 11, 2002, at http://www.trueorigin.org/imp-269a.asp.

73. Gerhard Schönknecht and Siegfried Scherer, "Too Much Coal for a Young Earth?" *Creation Ex Nihilo Technical Journal* 11:3 (1997), 278–282. Accessed on May 11, 2002, at http://www.answersingenesis.org/docs/1233.asp.

74. David K. DeWolf, Stephen C. Meyer, and Mark Edward DeForrest, "Teaching the Origins Controversy: Science, Or Religion, Or Speech?" *Utah Law Review* 39:1 (2000), 93. Accessed on April 24, 2002, at http://www.arn.org/dewolf/ddhome.htm.

75. William A. Dembski, "Signs of Intelligence," Truths that Transform interview by D. James Kennedy (February 25, 2002).

76. Walter L. Bradley and Roger Olsen, "The Trustworthiness of Scripture in Areas Relating to Natural Science." Accessed on March 24, 2002, at http://www.origins.org/offices/bradley/docs/trustworthiness.html.

77. William A. Dembski, "What Every Theologian Should Know About Creation, Evolution and Design." Accessed on January 11, 2002, at http://www.origins.org/offices/dembski/docs/bd-theologn.html. This paper was reprinted from *Princeton Theological Review*, March 1996.

78. Pennock, *Tower of Babel*, 26–27.

79. Dembski, *Intelligent Design*, 66.

80. Bradley and Olsen, "The Trustworthiness of Scripture in Areas Relating to Natural Science." Every assertion made in this statement on statistics, information theory, and simple genetics is, despite the tone of keen authority, either nonsensical or plain wrong.

81. William A. Dembski, *No Free Lunch: Why Specified Complexity Cannot Be Purchased Without Intelligence* (New York: Rowman and Littlefield, 2002), 149.

82. Michael J. Behe, *Darwin's Black Box: The Biochemical Challenge to Evolution* (New York: Simon and Schuster, 1996), 145–151.

83. Michael J. Behe, "The Sterility of Darwinism." Accessed on March 25 2002, at http://www.arn.org/docs/behe/mb_brresp.htm.

84. Henry M. Morris, *Scientific Creationism*, Public School Edition (San Diego, CA: Creation-Life Publishers, 1974), 43–45.

85. Phillip E. Johnson, "Evolution as Dogma: The Establishment of Naturalism," *First Things* (October 1990). Accessed on April 24, 2002, at http://www.arn.org/docs/johnson/pjdogma1.htm.

86. Phillip E. Johnson, *Darwin on Trial* (Washington, DC: Regnery Gateway, 1991), 14.

87. Speech delivered at a symposium at Hillsdale College in November 1992. Later published with other symposium papers in *Man and Creation: Perspectives on Science and Theology*, ed. Michael Bauman (Hillsdale, MI: Hillsdale College Press, 1993). Accessed on April 24, 2002, at http://www.arn.org/docs/johnson/wid.htm.

88. Jonathan Wells, "Issues in the Creation-Evolution Controversies," *The World and I* (January 1996). Available at http://www.worldandi.com/subscribers/1996/January/index.shtml.

89. We are indebted to Cheryl Capra, a science teacher with Education Queensland (Australia) for her advice in clarifying this definition.

90. Phillip E. Johnson, "Designer Genes: Phillip E. Johnson Talks to Peter Hastie," interview by Peter Hastie, *Australian Presbyterian* 531 (October 2001), 4–8. Accessed on October 14, 2002, at http://www.pcvic.org.au/njc/pdf/AP10.01.pdf. See also Phillip E. Johnson, interview by Hank Hanegraaff, *Bible Answer Man* radio program, February 20, 2003.

91. Robert T. Pennock, "The Wizards of ID," in *Intelligent Design Creationism and Its Critics: Philosophical, Theological, and Scientific Perspectives*, ed. Robert T. Pennock (Cambridge, MA: MIT Press, 2001). Pennock's original Metanexus post of this article is available at http://www.metanexus.net/archives/printerfriendly.asp?archiveid=2645. Accessed on October 15, 2002.

92. Morris, *Scientific Creationism*, 11.

93. Johnson, *Darwin on Trial*, 4n.

94. Duane Gish, "Dr. Duane Gish: Crusader," interview by Todd Wood, *Creation Matters* 1:1 (January/February 1996). Accessed on March 26, 2002, at http://www.creationresearch.org/creation_matters/96/cm9601.html.

95. "Law Professor Speaks at Lawrence Church," transcript of speech by Phillip Johnson, Grace Evangelical Presbyterian Church, Lawrence Kansas, April 8, 2000. Accessed on March 26, 2002, at http://www.aletheiaforum.org/4.8.2000johnson.html.

96. William A. Dembski, "The Intelligent Design Movement." Accessed on April 24, 2002, at http://www.origins.org/offices/dembski/docs/bd-idesign.html. Reprinted from *Cosmic Pursuit*, 1998.

97. Dembski, "What Every Theologian Should Know." Accessed on April 26, 2002, at http://www.origins.org/offices/dembski/docs/bd-theologn.html. Robert Pennock also points out ID's rejection of theistic evolution in *Tower of Babel*, 30–31.

98. William A. Dembski, "The Act of Creation: Bridging Transcendence and Immanence," presented at Millstatt Forum, Strasbourg, France, August 10, 1998. Accessed on April 26, 2002, at http://www.arn.org/docs/dembski/wd_actofcreation.htm. For Johnson's 1996 invocation of John 1:1, see "How to Sink a Battleship: A Call to Separate Materialist Philosophy from Empirical Science," at http://www.origins.org/mc/resources/ri9602/johnson.html. For his more recent invocation of it, see *The Wedge of Truth*, 151ff: "The place to begin is . . . the Gospel of John." For his denial on national TV of his invocation of scripture, see the 1997 *Firing Line* debate. When Barry Lynn asked, "Do you believe there were dinosaurs on Noah's ark?" Johnson replied, "I don't make any reference to the Bible.." Johnson issued another denial on CNN's *TalkBack Live* on August 16, 1999: [Bobbie Battista] "Would God like this intelligent designer theory, or would he prefer that it's more literal?" [Johnson] "Well, that's quite another question, and it doesn't go into my position. . . . I don't think there is any need to bring the Bible into this." See transcript at http://www.arn.org/docs/kansas/talkback81699.htm.

99. Back issues of *Creation Matters* are archived at http://www.creationresearch.org/matters.html. See Creation Research Society's sale of Wedge books at http://www.creationresearch.org/Merchant2/merchant.mv. Both accessed on April 19, 2003.

100. Bob Harsh, "The Wedge of Truth," *Origins Insights* (March 2001). Accessed on May 17, 2002, at http://www.csfpittsburgh.org/newsletters.htm.

101. Henry Morris, "Neocreationism," *Impact* 296 (February 1998). Accessed on March 26, 2002, at http://www.icr.org/pubs/imp/imp-296.htm.

102. Ashby Camp, "The Intelligent Design Movement: An Ally?" *Creation Matters* 4:6 (December 1999). Accessed on March 26, 2002, at http://www.creationresearch.org/creation_matters/99/cm9911.html#An%20Ally?

103. Gregory J. Brewer, "The Imminent Death of Darwinism and the Rise of Intelligent Design," *Impact* 341 (November 2001). Accessed on March 26, 2002, at http://www.icr.org/newsletters/impact/impactnov01.html.

104. John D. Morris, "Cracks Are Widening in Evolution's Dam!" *Acts and Facts* 31:5 (May 2002). Accessed in May 2002 at http://www.icr.org/pubs/af/pdf/af0205.pdf.

105. See John Stear, "The Carl Baugh Page" at http://home.austarnet.com.au/stear/carl_baugh_page.htm. Regarding Baugh's questionable scientific degrees, see Glen J. Kuban, "A Matter of Degree: Carl Baugh's Alleged Credentials," at http://www.talkorigins.org/faqs/paluxy/degrees.html. For an example of how Baugh has misrepresented fossil evidence, see Ronnie J. Hastings, "A Tale of Two Teeth," at http://www.talkorigins.org/faqs/paluxy/tooth.html. See Kent Hovind's website at http://www.drdino.com. Regarding Hovind's lack of scientific credentials, see Barbara Forrest, "Unmasking the False Prophet of Creationism: Kent Hovind," *Reports of the National Center for Science Education* 19:5 (September/October 1999), at http://www.ncseweb.org/newsletter.asp?curiss=14. All accessed on April 19, 2003.

106. *Truth and Science on Intelligent Design*, Truth and Science Seminar on Intelligent Design, April 12–13, 2002, videocassette. Dembski's remark about *UFO* magazine apparently refers to Jan Hester's article in the April–May 2001 issue, "Smashing Icons: The Rise of 'Intelligent Design.'" Hester announces that "a crack in the Darwinian edifice has been . . . widened to the point of empirical emergency."

107. The Truth and Science Seminar and speaker schedule were announced at http://www.truthandscience.com. The ID material was on Baugh's website at http://www.creationevidence.org. See links to information about Baugh at http://home.austarnet.com.au/stear/carl_baugh_page.htm. All accessed on April 25, 2002.

108. Carl Wieland, "AIG's Views on the Intelligent Design Movement," Answers in Genesis, August 30, 2002. Accessed on September 2, 2002, at http://www.answersin genesis.org/docs2002/0830_IDM.asp?vPrint=1.

109. AIG sells *Darwin's Black Box* at http://shop.gospelcom.net/cgi-bin/AIGUS. storefront/en/product/10-3-081. AIG's defense of DeHart, "Anti-creationists Threaten Another Teacher's Liberty," is at http://www.answersingenesis.org/docs2/4350news 7-26-2000.asp. An article on Wedge activities in Ohio, "Vote on Evolution Next Monday" (September 7, 2002), is at http://www.answersingenesis.org/docs2002/0907ohio_ vote.asp. For Truman's reviews of Wedge books, see http://www.answersingenesis. org/home/area/magazines/tj/docs/tj_v14n1_designer_science.asp, http://www.answersingenesis.org/home/area/magazines/tj/docs/TJn13v2_design.asp, and http://www.answersingenesis.org/home/area/magazines/tj/docs/v15n1_wedge.asp. "The Berkeley Boat-Rocker" is at http://www.answersingenesis.org/docs/3132.asp. All accessed on October 15, 2002.

110. Lynn Vincent, "Science v. Science," *World*, February 26, 2000. Accessed on April 25, 2002, at http://www.discovery.org/viewDB/index.php3?program=CRSCstories& command=view&id=148. There is occasional grousing by YECs that IDers are not "biblical" enough. And IDers have on at least one occasion referred to Answers in Genesis YEC Ken Ham and his associates as "guitar-strumming hillbillies." See Matt Carter, "Lab Scientists Challenging Darwin: 'Intelligent Design' Theory Supports a Thoughtful Creator," *Tri-Valley Herald*, September 26, 2001. Accessed on November 19, 2001, at http://www. reviewevolution.org/press/fromPress_ScienChalDarwin.php.

111. Johnson, *The Wedge of Truth*, 152–162.

112. Interview of Phillip Johnson by Hank Hanegraaff on *Bible Answer Man* radio program, December 19, 2000. Accessed on December 29, 2000, at http://www.equip.org/bam/previous.html.

113. Nancy Pearcey, "Opening the 'Big Tent' in Science: The New Design Movement," Access Research Network. Accessed on April 26, 2002, at http://www.arn.org/docs/ pearcey/np_bigtent30197.htm. Originally published as "The Evolution Backlash," *World*, March 1, 1997.

114. "The Wedge Strategy: Five Year Strategy Plan Summary—Phase I."

115. Walter Bradley has been hired as a distinguished professor in Baylor's Department of Engineering. See http://ecswww.baylor.edu/ecs/engineering/bradley.htm. Accessed on April 20, 2003. Bruce Gordon stated in his c.v. (as of August 2000) that he had "adjunct faculty" status at Baylor (1999–2000) and was associate director of the Polanyi Center. His website lists two courses he taught in fall 1999 and spring 2000, along with a colloquium for faculty and graduate students he led with William Dembski. As of April 2003, Gordon's c.v. states that he is "Acting Director, The Baylor Center for Science, Philosophy and Religion" and "Assistant Research Professor, Institute for Faith and Learning," the latter title indicating this is not a teaching position. See http://www3.baylor.edu/~Bruce_Gordon/.

116. "Darwin, Science, & Going Beyond the Culture Wars," A Presentation by the Fellows of the Discovery Institute's Center for the Renewal of Science and Culture. Accessed at http://www.discovery.org/crsc/scied/downloads/ on June 14, 2000.

117. Mary Beth Marklein, "Evolution's Next Step in Kansas," *USA Today*, July 19, 2000. Accessed on April 25, 2002, at the Discovery Institute at http://www.discovery.org/ viewDB/index.php3?program=CRSCstories&command=view&id=393.

118. See "ID Colleges," Access Research Network, at http://www.arn.org/college.htm. See also http://www.arn.org/colleges/obu.htm. Both accessed on March 26, 2001.

119. Johnson, "The Wedge: Breaking the Modernist Monopoly on Science."

120. The Waltzer papers were "Positive Evidence for Design in the Natural World" (2000) and "The Olfactory System: A Model for Intelligent Design" (2001). (Waltzer delivered a paper with the latter title at the Calvin Conference.) Abstracts can be found at http://www.msstate.edu/org/MAS/mas00.pdf and http://www.msstate.edu/org/MAS/ mas01.pdf. The abstract of the February 2002 anti-ID paper by John D. Davis, "The De-

signers of 'Intelligent Design,'" is at http://www.msstate.edu/org/MAS/mas02.pdf. All accessed on October 12, 2002.

121. The March 2002 meeting is listed in *Proceedings and Addresses of the American Philosophical Association Pacific Division Seventy-Sixth Annual Meeting Held in Seattle, WA, 27–31 March 2002*, ed. Elizabeth Radcliffe (University of Delaware: American Philosophical Association), 57. Accessed on March 26, 2002, at http://www.apa.udel.edu/apa/publications/proceedings/v75n3/grpprogram.asp#Wednes. According to Anita Silvers, secretary-treasurer of the Pacific Division, the APA does not exercise responsibility or authority over such meetings. Ms. Silvers pointed out the history of philosophical interest in the "argument from design," citing eighteenth-century watchmaker arguments as among the most famous versions. (Anita Silvers to Barbara Forrest, personal communication, September 30, 2002, and October 2, 2002.) The Evangelical Philosophical Society's call for 2003 papers was issued in September 2002, "Evangelical Philosophical Society: Call for Papers."

122. Students of propaganda and disinformation technology will recognize this, not as insight, but as proper appropriation of the most effective methods for convincing people beyond all doubt that the propagandist's case must be true, whether or not it actually is.

123. "Communiqué Interview: Phillip E. Johnson," spring 1999. Accessed on April 25, 2002, at http://www.arn.org/docs/johnson/commsp99.htm.

124. Johnson, "The Wedge: Breaking the Modernist Monopoly on Science."

125. Hank Hanegraaff, *Bible Answer Man* radio interview of Phillip Johnson, December 19, 2000. Accessed on December 29, 2000, at http://www.equip.org/bam/previous.html.

126. William Dembski and Mark Hartwig, "God's Fingerprints on the Universe," interview by James Dobson, October 2001. Audiotape. Dobson's remark about homeschooling should be taken seriously. In his review of Christopher P. Toumey's *God's Own Scientists*, J. W. Haas, Jr., says, "Creationists have great influence in Christian academies and among home schoolers." See "God's Scientists Under the Microscope," *Perspectives on Science and the Christian Faith* 46 (December 1994), 266–268. Accessed on May 15, 2002, at http://www.asa3.org/ASA/PSCF/1994/PSCF12-94Haas.html. Though we do not address the issue here, exposure of homeschoolers to creationism is a concern, given that roughly one million children are now homeschooled. They are likely to be from religiously conservative families, ID's primary constituency, and these families are more likely to be politically active than nonhomeschool ones. See Patricia Lines, "Homeschooling," ERIC *Digest* 151 (September 2001), at http://eric.uoregon.edu/publications/digests/digest151.html. Homeschool websites advocating ID are readily found on the Internet. Considering children in conservative Christian academies where creationism is commonly taught, the number of potential Wedge supporters could be even higher. As of the 1999–2000 school year, there were 773,237 children in conservative Christian schools. See Stephen P. Broughman and Lenore A. Coliacello, "Private School Universe Survey: 1999–2000," (National Center for Education Statistics, U.S. Dept. of Education, August 2001) at http://nces.ed.gov/pubs2001/2001330.pdf. Both accessed on May 17, 2002.

127. Lauren Kern, "In God's Country," *Houston Press*, December 14, 2000. Accessed on April 25, 2002, at http://www.houstonpress.com/issues/2000-12-14/feature2.html/1/index.html.

128. See the site at http://www.idurc.org/index.shtml. Accessed on November 20, 2001.

129. See the Faith Seminary website at http://www.faithseminary.edu/mission.html. See the Idea Center's website at http://www.ideacenter.org/infomenu.htm and its list of advisory board members at http://www.ideacenter.org/advisoryboard.htm. See the page with instructions for setting up student chapters at http://www.ideacenter.org/chapterlocations.htm. Accessed on April 5, 2003.

130. "The Promise of Better Science and a Better Culture," Discovery Institute *Jour-*

nal (1999). Accessed on April 26, 2002, at http://www.discovery.org/w3/discovery.org/journal/1999/crsc.html.

131. These newsletters may be found at http://users.stargate.net/~dfeucht/MARAPR00.htm and http://users.stargate.net/~dfeucht/JANFEB01.htm, respectively. Accessed on April 19, 2003.

132. The DeCal courses, "Special Studies" courses as defined by Berkeley's Academic Senate Regulations, are taught by students who must secure faculty sponsors. See the description of DeCal courses at http://www.OCF.Berkeley.EDU/~decal/about.htm. See the requirements for setting up the courses at the DeCal website, http://www.OCF.Berkeley.EDU/~decal/start.htm. Both accessed on October 15, 2002. Having such courses taught by students has caused some problems. One DeCal course on human sexuality received negative publicity when local news media disclosed that extracurricular sexual activity had occurred at the home of student instructors. Because of concerns over the quality of the course, the faculty sponsor announced a review of its content. See Brittany Adams, "Continuation of Sexuality De-Cal [*sic*] Classes Uncertain," *Daily Californian*, February 15, 2002. Accessed on March 30, 2002, at http://www.dailycal.org/article.asp?id=7723&ref=search.

133. Regulations governing credit for Macosko's and similar courses are found in "Title II. Independent or Group Study: A230. Special Studies." Academic Senate Regulations, University of California-Berkeley.

134. Jed Macosko's syllabus was accessed on December 4, 2000, through his online curriculum vitae at http://www.cchem.berkeley.edu/~yksgrp/jed.html by following the link "Design in Nature" under "Teaching Experience." Macosko also provided a links page at http://sunsite.Berkeley.EDU/cgi-bin/eres/viewcourse.pl?chmz@Zeng198_REIMER, from which students could access mostly pro-ID websites. The sites have now become accessible only with a password. *The Chronicle of Higher Education* covered the Macoskos' teaching activities in Beth McMurtrie, "Darwinism Under Attack," December 21, 2001. Accessed on March 26, 2002, at http://chronicle.com/free/v48/i17/17a00801.htm. *San Francisco Weekly* featured Macosko in "Looking for God at Berkeley," June 20, 2001, at http://www.sfweekly.com/issues/2001-06-20/feature.html/1/index.html.

135. "Law Professor Speaks at Lawrence Church," transcript of speech by Phillip Johnson at Grace Evangelical Prebyterian Church, April 8, 2000. Accessed on March 26, 2002, at http://www.aletheiaforum.org/4.8.2000johnson.html.

136. Jed Macosko, even before becoming a CRSC fellow, was an ID proponent. He indicated his ID sympathies when, as a doctoral student, he wrote a report for the Wedge's journal *Origins & Design* (winter 1999) on the June 1998 Science and the Spiritual Quest conference (a legitimate event), bewailing the absence of ID proponents: "If these opening speakers had taken questions, mine would have been: 'Where are the scientists who question the naturalistic assumptions and conclusions of Darwinism? Why aren't they a part of this discussion?'" See Macosko, "Science and the Spiritual Quest Conference: A Report," *Origins & Design* (19:2). Accessed on January 5, 2001, at http://www.arn.org/docs/odesign/od192/ssq192.htm. Macosko was also a presenter at William Dembski's "Design, Self-Organization, and the Integrity of Creation" conference at Calvin College in May 2001.

137. Jeffrey Reimer to Barbara Forrest, personal communication, December 28, 2000. Chris Macosko was listed as a course coordinator, using his son's e-mail address, in the "Fall 2000 Schedule of Courses" on the DeCal website. Accessed on January 4, 2001, at http://www.ocf.berkeley.edu/~decal/fall2000.htm. Chris Macosko's Berkeley sabbatical and co-teaching of the Berkeley ID course with Jed are announced in the January/February 2001 American Scientific Affiliation *Newsletter* at http://users.stargate.net/~dfeucht/JANFEB01.htm.

138. Access Research Network 2000/2001 Annual Report.

139. See Gailon Totheroh, "The War Between Evolution and God's Intelligent De-

sign," Christian Broadcasting Network. Accessed on October 15, 2002, at http://www.cbn.com/CBNNews/News/020902a.asp.

140. CLA Honors Division to Barbara Forrest, personal communication, January 8, 2001.

141. Maureen Smith, "Origins: By Chance or Design?" *Kiosk*, January 2000. Accessed on January 5, 2001, at http://www1.umn.edu/urelate/kiosk/1.00text/origins.html. See also "Macosko and Monsma Teach the Controversy" in the January/February 2001 ASA *Newsletter* at http://users.stargate.net/~dfeucht/JANFEB01.htm.

142. Smith, "Origins: By Chance or Design?" The fall 2001 course listing was accessed on March 26, 2002, 2002, at http://www.itdean.umn.edu////students/academics/freshsem.html.

143. McMurtrie, "Darwinism Under Attack."

144. See the God and the Academy conference overview at http://leadernet.clm.org/clm/charting.nsf. Accessed on October 16, 2002.

145. Chapel on the Campus, Baton Rouge, LA. Accessed on April 19, 2003, at http://www.thechapelnet.com/college/toughquestions.asp.

146. Ray Bohlin, "Probe Answers Our E-Mail: I'm Interested in Grad School in Intelligent Design." Accessed on May 10, 2002, at http://www.probe.org/docs/e-gradschool.html.

147. See Probe's "Mission" at http://www.probe.org/menus/whatisprobe.html. Information on the Probe Center pages was accessed at http://www.probe.org/probe center/index.html and http://www.probe.org/probecenter/our_emphases.html on May 14, 2002.

148. "Major Grants Increase Programs, Nearly Double Discovery Budget," Discovery Institute *Journal* (1999). Accessed on April 26, 2002, at http://www.discovery.org/w3/discovery.org/journal/1999/grants.html.

149. Kern, "In God's Country."

150. Ron Nissimov, "Baylor Professors Concerned Center is Front for Promoting Creationism," *Houston Chronicle*, July 1, 2000.

151. In the same article Jay Richards, CRSC program director, denies that DI was directing Dembski's work at the Polanyi Center. Richards said, however, that DI hoped intelligent design would be taught along with evolution.

152. "William A. Dembski, Curriculum Vitae, May 2002." Accessed on October 15, 2002, at http://www.designinference.com/documents/05.02.CV.htm.

153. William A. Dembski, "Intelligent Design Coming Clean," posted to Metaviews on November 17, 2000. Accessed on April 26, 2002, at http://www.arn.org/docs/dembski/wd_idcomingclean.htm.

154. Kern, "In God's Country."

155. Uwe Siemon-Netto, "It's Perilous to Ponder the Design of the Universe," UPI press release, December 21, 2000. Accessed on April 26, 2002, at http://www.discovery.org/crsc/CRSCrecentArticles.php3?id=575.

156. "Questioning Darwin: William Dembski Discusses Intelligent Design," interview by Donald Yerxa (Templeton Oxford Summer Seminar, July 2001), *Research, News, and Opportunities in Science and Theology* 2:3 (November 2001), 13. Dembski made similar statements in his October 2001 interview with James Dobson and in the *Durango Herald*, "Researcher Challenges Darwin's Theory of Evolution," October 17, 2001. Accessed on March 26, 2002, at http://www.durangoherald.com/asp-bin/article_generation.asp?article_type=news&article_path=/news/news011017_2.htm.

157. "Reason in the Balance: An Interview with Phil Johnson About His Forthcoming Book," *Real Issue* (March/April 1995). Accessed on April 26, 2002, at http://www.leaderu.com/real/ri9501/reason.html.

158. Johnson, "The Wedge: Breaking the Modernist Monopoly on Science."

159. "International Scientists and Scholars Convene to Discuss the Nature of the

Universe," *Discovery Institute News*, April 10, 2000. Accessed April 26, 2002, at Lexis-Nexis.

160. See Steven Weinberg, "A Designer Universe?" at http://www.physlink.com/education/essay_weinberg.cfm.

161. See Henry Schaefer's statement in the foreword of *Mere Creation* of the Wedge's plan to make use of the World Wide Web for "preparing information usable in the campus environment of a modern university" (11). See the Mere Creation website at http://www.origins.org/mc/menus/index.html. See Robert Koons's posting of the NTSE papers and other information at http://www.utexas.edu/cola/depts/philosophy/faculty/koons/ntse/main.html. See Access Research Network's marketing of the DAIC audiotape at http://www.arn.org/arnproducts/audinfo.htm. See also the "Discovery Institute Audio Center" at http://www.lifeaudio.com/ministries/discoveryinstitute/main/seriesInfo.jhtml?id=209. All accessed on April 26, 2002. The Intelligent Design Network markets its tapes at http://www.intelligentdesignnetwork.org/D3Tapes.htm. Accessed on October 15, 2002.

162. "Report by Greg Metzger," ARK website. Accessed on March 26, 2002, at http://www.arky.org/current/cvedofl.htm.

163. Ray Bohlin, "Mere Creation: Science, Faith & Intelligent Design." Accessed on April 26, 2002, at http://www.probe.org/docs/mere.html.

164. "Communiqué Interview: Phillip E. Johnson."

165. Robert Koons, "Great Beginnings: UT Origins Conference Opens Doors to Dialogue," *Real Issue* (March/April 1997). Accessed on April 24, 2002, at http://www.leaderu.com/real/ri9701/koons.html.

166. See Dembski's curriculum vitae at http://www.designinference.com/documents/2003.02.CV.htm. He has had a series of research fellowships and short-term teaching jobs in numerous places.

167. Dembski, "Intelligent Design Coming Clean."

168. Johnson, "The Wedge: Breaking the Modernist Monopoly on Science."

169. Dembski, "The Intelligent Design Movement."

170. "Kansas Deletes Evolution from State Science Test," *Talkback Live*. Aired August 16, 1999, 3:00 p.m. EST. Accessed on April 26, 2002, at http://www.arn.org/docs/kansas/talkback81699.htm.

171. In a July 2001 interview at Oxford University with William Dembski by Donald Yerxa, when Yerxa asked Dembski about the "notion of 'theistic science,'" Dembski denied any sympathy for it: "I am not at all sympathetic to that because intelligent design is not dealing with a doctrine of creation." "Questioning Darwin: William Dembski Discusses Intelligent Design, " 13. However, even a brief perusal of Dembski's book *Intelligent Design: The Bridge Between Science and Theology* clearly belies this denial. Certainly, Dembski's fellow Wedge member Robert Koons understands ID as theistic science, since he uses the terms interchangeably: "If theistic science or intelligent design theory is to become a progressive research program, it must do more than poke holes in the evidence for Darwinism." See Robert Koons, "Making Progress in the Origins Debate: A Summary of NTSE," at http://www.origins.org/real/ri9701/koons2.html. Accessed on November 21, 2001.

172. Phillip E. Johnson, "Starting a Conversation About Evolution," review of *The Battle of the Beginnings: Why Neither Side Is Winning the Creation-Evolution Debate*, by Del Ratzsch. Accessed on March 26, 2001, at http://www.arn.org/docs/johnson/ratzsch.htm.

173. Laurie Goodstein, "Christians and Scientists: New Light for Creationism," *New York Times*, December 21, 1997. Accessed on April 26, 2002, at http://www.arn.org/docs/fline1297/fl_goodstein.htm.

174. Lawrence, "Communiqué Interview." Accessed on April 26, 2002, at http://www.arn.org/docs/johnson/commsp99.htm.

175. Boston, "Missionary Man."

176. Lawrence, "Communiqué Interview." Evidence of the long-term nature of the

strategy lies in the fact that, except for Johnson, the most important Wedge members, such as Dembski and Meyer, are relatively young.

177. Stafford, "The Making of a Revolution."

178. Johnson, "The Wedge: Breaking the Modernist Monopoly on Science." Johnson's "prime question" is this: "What should we do if empirical evidence and materialist philosophy are going in different directions? Suppose, for example, that the evidence suggests that intelligent causes were involved in biological creation. Should we follow the evidence or the philosophy?" For Johnson, this question is tantamount to asking the academic establishment, "If the evidence suggests intelligent design, should we be genuinely scientific and admit this, or should we be unscientific and refuse to admit it?"

179. Johnson, "The Wedge: Breaking the Modernist Monopoly on Science."

180. Boston, "Missionary Man."

181. Johnson, "Starting a Conversation About Evolution."

182. Phillip Johnson, "The State of the Wedge," delivered at the Darwin, Design, and Democracy II symposium, June 29, 2001. Audiocassette.

Index